Understanding and Managing Power Quality Issues

Other related titles:

You may also like

- PBPO092 | Zobaa and Aleem | Power Quality in Future Electrical Power Systems | 2017
- PBPO036 | Blume and Stephanblome | Voltage Quality in Electrical Power Systems | 2001
- PBPO183 | Lourenço and London Junior | Power Distribution System State Estimation | 2022

We also publish a wide range of books on the following topics:
Computing and Networks
Control, Robotics and Sensors
Electrical Regulations
Electromagnetics and Radar
Energy Engineering
Healthcare Technologies
History and Management of Technology
IET Codes and Guidance
Materials, Circuits and Devices
Model Forms
Nanomaterials and Nanotechnologies
Optics, Photonics and Lasers
Production, Design and Manufacturing
Security
Telecommunications
Transportation

All books are available in print via https://shop.theiet.org or as eBooks via our Digital Library https://digital-library.theiet.org.

IET ENERGY ENGINEERING 189

Understanding and Managing Power Quality Issues

Sincy George, George Sebastian and Vivek Agarwal

The Institution of Engineering and Technology

About the IET

This book is published by the Institution of Engineering and Technology (The IET).

We inspire, inform and influence the global engineering community to engineer a better world. As a diverse home across engineering and technology, we share knowledge that helps make better sense of the world, to accelerate innovation and solve the global challenges that matter.

The IET is a not-for-profit organisation. The surplus we make from our books is used to support activities and products for the engineering community and promote the positive role of science, engineering and technology in the world. This includes education resources and outreach, scholarships and awards, events and courses, publications, professional development and mentoring, and advocacy to governments.

To discover more about the IET, please visit https://www.theiet.org/.

About IET books

The IET publishes books across many engineering and technology disciplines. Our authors and editors offer fresh perspectives from universities and industry. Within our subject areas, we have several book series steered by editorial boards made up of leading subject experts.

We peer review each book at the proposal stage to ensure the quality and relevance of our publications.

Get involved

If you are interested in becoming an author, editor, series advisor, or peer reviewer please visit https://www.theiet.org/publishing/publishing-with-iet-books/ or contact author_support@theiet.org.

Discovering our electronic content

All of our books are available online via the IET's Digital Library. Our Digital Library is the home of technical documents, eBooks, conference publications, real-life case studies and journal articles. To find out more, please visit https://digital-library.theiet.org.

In collaboration with the United Nations and the International Publishers Association, the IET is a Signatory member of the SDG Publishers Compact. The Compact aims to accelerate progress to achieve the Sustainable Development Goals (SDGs) by 2030. Signatories aspire to develop sustainable practices and act as champions of the SDGs during the Decade of Action (2020–30), publishing books and journals that will help inform, develop, and inspire action in that direction.

In line with our sustainable goals, our UK printing partner has FSC accreditation, which is reducing our environmental impact to the planet. We use a print-on-demand model to further reduce our carbon footprint.

Published by The Institution of Engineering and Technology, London, United Kingdom

The Institution of Engineering and Technology (the "**Publisher**") is registered as a Charity in England & Wales (no. 211014) and Scotland (no. SC038698).

First published: 2025

The Institution of Engineering and Technology
Futures Place,
Kings Way,
Stevenage,
Herts, SG1 2UA,
United Kingdom

www.theiet.org

British Library Cataloguing in Publication Data
A catalogue record for this product is available from the British Library

ISBN 978-1-83953-337-2 (hardback)
ISBN 978-1-83953-338-9 (PDF)
ISBN 978-1-80705-107-5 (ePub3)

Typeset in India by MPS Limited

Cover image credit: Jose A. Bernat Bacete/Moment via Getty Images

Contents

About the authors

Sincy George is the former professor and principal at St. Francis Institute of Technology, University of Mumbai, India. Her research interests include power system and power quality. She received a PhD in Power Electronics and Power System from the Indian Institute of Technology, Bombay, India. She handled research projects and published several papers. She is a senior member of the IEEE and fellow of IE (India). She has over 35 years' experience in academia.

George Sebastian is the former associate director of Nuclear Power Corporation of India Limited, India. With over 35 years' experience, he specialized in the design of electrical systems, testing and qualification of equipment viz. EHV transmission, turbo generators and excitation, UPS, converters, SCADA, and protection systems for nuclear power plants. He presented a number of papers at various national and international conferences. He is a Fellow of IE (India).

Vivek Agarwal is a professor at the Indian Institute of Technology Mumbai, India. He is a fellow of the IEEE as well as various key Indian societies. His research interests include power converters, power quality, renewable energy conversion and electric vehicles. He headed several large projects on these topics and holds many patents. He earned his PhD at the University of Victoria, British Columbia, Canada.

Chapter 1

Generation, transmission, distribution and utilization of electric power: an overview of past, present, and future

"We have many a monument of past ages; we have the palaces and pyramids, the temples of the Greek and the cathedrals of Christendom. In them is exemplified the power of men, the greatness of nations, the love of art and religious devotion. But the monument at Niagara has something of its own, more in accord with our present thoughts and tendencies. It is a monument worthy of our scientific age, a true monument of enlightenment and of peace. It signifies the subjugation of natural forces to the service of man, the discontinuance of barbarous methods, the relieving of millions from want and suffering"

— Nikola Tesla at the opening ceremony of the first hydro-electric power station at Niagara on January 12, 1897.[1]

Goals of the chapter

After going through this lesson, the reader shall get a broad idea and should be able to understand the basic principles of the following:

1. History and the development of the electrical grid and various components of it: Generation, transmission, distribution and utilization of electrical energy
2. Isolated and interconnected electrical grid, recent developments – introduction to renewable energy (low inertia) generation and distributed generation
3. Power electronics in the electrical power grid
4. Power system protection, intelligent electronic devices, communication, monitoring, control and SCADA systems
5. Power quality and cybersecurity issues

1.1 Introduction

Today, the electrical power system is considered one of the most crucial and essential infrastructures. It is an "enabler" that provides electric power to

[1] www.teslasociety.com/exhibition.htm

households, educational and research and development (R&D) institutions, hospitals, transportation, industry, business, etc., and is integral part of the life and growth of mankind. So critical is electricity for our lives that it is nearly impossible to imagine even a few hours without electricity. A few minutes of absence or blackout is headline news. It is always there at the flick of a switch at lightning speed or even produced at the instant of its requirement, even at a faraway place. The power flow is instantaneous and finite. Commercially viable storage options for electric energy do not currently exist, and hence, it is becoming increasingly important as an efficient and flexible form of energy in state-of-the-art energy systems. The recent developments in power electronics have helped in the enhancement of efficiency and the formation of a large, interconnected network of power generation to utilization equipment. However, it also affected the power quality, mainly by injecting harmonics into the system. The system today is composed of many agencies for the generation, transmission and distribution of both public and private nature. It is an emerging topic of international cooperation and tie-up for cross-border transmission of power, innovation, research and development for energy security, global sustainability and harmony. The growing relevance of electricity in the world also poses new challenges and threats. It is, therefore, appropriate to give tribute to the early innovators who contributed enormously to the development of electricity networks (power systems) and made our lives simpler.

1.2 History of electric power

The announcement by **Alessandro Volta** (1745–1827), an Italian physicist and chemist, in March 1800 to the Royal Society of London about his invention of the pile or electromotor device, opened a new era of life for mankind [1]. This was followed by **Michael Faraday** (1791–1867), one of the world's greatest experimental physicists and chemists, known as the father of the electric motor, generator, transformer and electrolysis. He wrote the "Law of Induction" [2]. Two measuring units are named in his honour, the Farad (for capacitance) and the Faraday (also called the Faraday constant, the amount of electric charge that liberates one-gram equivalent of any ion from an electrolytic solution). He proved that the electricity induced from a magnet, a battery and static electricity were all the same.

Thomas Alva Edison (1847–1931), an American inventor and businessman, is credited with the development of a practical incandescent electric lamp. The first public demonstration of his lighting system was in December 1879, when the Menlo Park laboratory complex was electrically illuminated. In September 1882, the first commercial power station, at Pearl Street in Lower Manhattan, went into operation, providing light and power to customers in a one-quarter-square-mile area. To supply electric power to the lights within the area, Edison developed a 27-tonne, 100 kW, 110 V Direct Current (DC) dynamo driven by reciprocating steam engines powered by coal-fired boilers [3].

Nikola Tesla (1856–1943), a Serbian immigrant to the United States and an amazing inventor and pioneer of the Alternating Current (AC) system, started his career in Edison's enterprise. Later, he obtained a lot of patents on polyphase AC systems. His partnership with **George Westinghouse** (1846–1914), an industrialist, helped to develop the first hydro-electric power station in Niagara Falls (Adams Plant) that went into operation on 26 August 1895 [4]. The power is transmitted at 11 kV to the City of Buffalo, 32 km away, through overhead lines. The clear advantage of high-voltage AC transmission for long distances influenced the future of the electrical industry worldwide. The development of power systems after 1800 AD is briefly summarized in Table 1.1.

Few images related to the power system development are given in Figures 1.1 and 1.2.

Table 1.1 Historical development of power system

Place	Inventor/Installation	Year
Italy	Alessandro Volta (1745–1827) invented "Pile"/battery	1800
Denmark	Hans Christian Oersted (1777–1865) discovered the magnetic effect of electric current [2]	1820
England	Michael Faraday (1791–1865) produced electricity from magnetism and invented the dynamo, the electric motor (1821) and electromagnetic induction	1831
Milan, Ohio	Thomas Alva Edison (1847–1931) invented the incandescent bulb (filament – carbonized sewing thread that burnt for 13 and 1/2 hrs) and illuminated the Menlo Park laboratory complex	1879
Canada	Canada's first Hydro-electric generating station at Chaudière Falls	1881
Manhattan	Thomas Alva Edison (1847–1931): First commercial power station, 120 V DC, for light and power to a one-square-mile area	1882
Holborn, London	First coal-fired, thermal power station (93 kW, 110 V DC – Edison Electric Lighting Station) for public use opened	1882
Serbia-America	Nikola Tesla (1856–1943) discovered the rotating magnetic field and introduced AC motors and transformers	1882
Ottawa	The Parliament buildings were lighted with electricity on 17 January 1884	1884
Ottawa	First city in the world to light its streets with electricity	1885
Great Barrington	500 V, 133–1/3 Hz, single-phase, 25 HP steam engine. Transmission at 3000 V and step-down to 100 V for lighting. Transmission length: 4000 ft	1886
Ottawa	Electric streetcars introduced	1891
Columbia	George Westinghouse (1846–1914) purchased the manufacturing rights of the AC system from Nikola Tesla (1888), and later, Tesla worked as an employee. Developed a lighting system using a polyphase system at 60 Hz for the World's Columbian Exposition	1893
Niagara	International Niagara Commission decided to implement Tesla's polyphase system at 60 HZ with a 5000 kW, 11000 V, 3-phase, 75 rpm generator	1893

(Continues)

Table 1.1 (*Continued*)

Place	Inventor/Installation	Year
Niagara	Niagara Hydro station #1 was in operation in August 1895, and the first electrical power transmission to Buffalo, 32 km away through 11 kV line	1896
Darjeeling, India	First hydro station in India, Sidrapong Hydel Power Station, 2 × 65 kW, 2300 V, 1 phase and 83.3 Hz. It was reconditioned to 3-phase 50 Hz in 1931	1897
St. Catharine, Ontario	The DeCew Falls Hydro-Electric station (2 × 1500 HP) transmitted at 22,500 volts, 66 2/3 Hz, two-phase, 56 km to Hamilton, Ontario. In 1930 frequency changed to 60 Hz	1898
UK	The world's first integrated 132 kV, 50 Hz grid with Manchester, Leeds, Newcastle, Birmingham, Bristol, London and Glasgow	1935
Sweden	World's first commercial high-voltage direct current (HVDC) link (20 MW, 100 kV utilizing mercury arc valves) between Västervik, on the east coast of Sweden, and Ygne, an island of Gotland in the Baltic Sea – 98 km submarine cable	1954
USA	World's first voltage source converter (VSC) HVDC link, ±200 kV, 400 MW, between Pittsburgh and Potrero Hill, San Francisco	2010
India	800 kV, 6000 MW, 1765 km HVDC Chhattisgarh-Tamil Nadu	2020

(a) (b)

Figure 1.1 *Discovery of electromagnetism. (a) Oersted's discovery of electromagnetism (https://commons.wikimedia.org/wiki/File:Oersted_discovers_electromagnetism.jpg). (b) Faraday demonstrating electromagnetic induction (https://en.wikipedia.org/wiki/Faraday%27s_law_of_induction).*

From Table 1.1, it is seen that the electrical system started with DC and later moved to AC. It is now a combination of both AC and DC in various areas of power generation, transmission and utilization. A brief discussion is made in the following sections on DC and AC systems for a better understanding of beginners to power systems.

(a) (b)

Figure 1.2 Hydro-Electric power plant at Niagara Falls. (a) Edward D. Adams Station Power Plant (https://en.wikipedia.org/wiki/Niagara_Falls_Hydraulic_Power_and_Manufacturing_Company). (b) Robert Moses Hydro-Electric Power Plant (https://en.wikipedia.org/wiki/Robert_Moses_Niagara_Power_Plant).

1.3 Development of the power system

As discussed in the previous section, power generation started with direct current (DC) and then shifted towards a polyphase AC system, finally settling with a single-phase and three-phase system with a frequency of 50 or 60 Hz. During the eighteenth century, many discoveries sought ways of using Faraday's induction principle to generate electricity. Two types of rotating electrical machines emerged. The first type was the DC machine, and the second type was the AC machine. During that time, even transmission and utilization were in DC form and later moved towards AC, considering various technical and economic merits of the latter.

1.3.1 DC generation, transmission and utilization

It is seen from Table 1.1 that the first electrical power-generating station at Pearl Street, New York City, was commissioned in 1882. It generated and transmitted power at 110 V DC for lighting purposes. Thus, initially, the power generation, transmission and utilization were at low DC voltage.

Figure 1.3 shows the waveform of 110 V DC voltage with respect to time, and the voltage (V) equation can be written as:

$$v(t) = V \tag{1.1}$$

The main problem with low-voltage DC transmission was the limitation on the distance of transmission. It is known that as the distance of the transmission line increases, the voltage drops as well and the transmission line losses increase for the same amount of power or current (I) to be transmitted through the line. This is because the resistance of the transmission line (R) depends on the length of the transmission line, which affects voltage drop (IR drop) and power loss (I^2R loss) in the line. A similar effect will result if more power is to be transmitted for the same

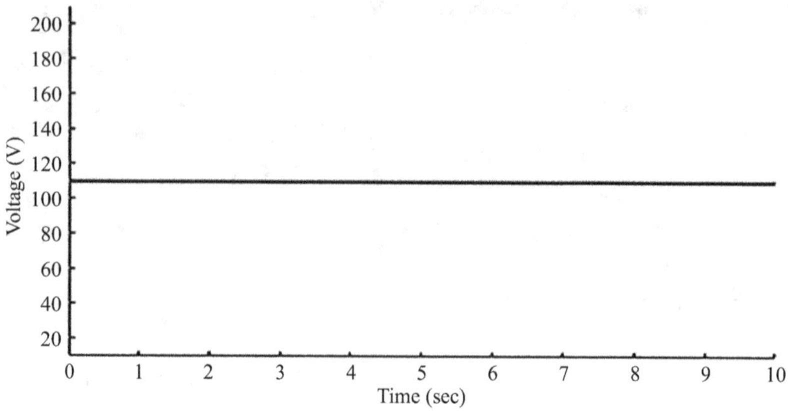

Figure 1.3 Waveform of a 110 V DC voltage

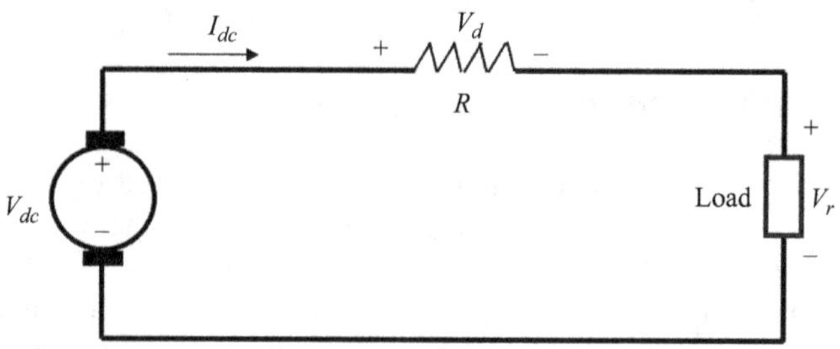

Figure 1.4 Equivalent circuit of a DC system

distance. This, in turn, affects the voltage at the receiving end of the line and the transmission power capability of the system. Thus, the distance of DC power transmission was limited at that time. This is explained through the equivalent circuit of such a DC system, as shown in Figure 1.4.

Transmission line resistance depends on the length (l), area of cross-section of the conductor (A) and type of conductor used for power transmission and is given by:

$$R = \rho \times \frac{l}{A} \tag{1.2}$$

where, ρ is the specific resistivity of the conductor. The transmitted power, P_{dc}, can be written in terms of the source voltage, V_{dc}, and the current transmitted through the transmission line, I_{dc} as

$$P_{dc} = V_{dc} \times I_{dc} \tag{1.3}$$

The main problem with the first commercial power station was the limitation on the distance of transmission due to the voltage drop across the line, which results in a reduction in the receiving end (V_r) compared to the sending-end voltage (V_{dc}) and the associated transmission line I^2R loss (T_L).

Mathematically,

$$Vr = V_S - I_{dc} \times R \tag{1.4}$$

$$T_L = {I_{dc}}^2 \times R \tag{1.5}$$

From the above equations, the voltage at the receiving end reduces, and the transmission line loss increases with the length of the line. Thus, DC power transmission was limited at that time.

One of the solutions to reduce this effect is to transmit the DC power at a higher voltage level. Under such conditions, to transmit the same power, less current will be flowing through the line, which reduces transmission line voltage drop and losses. However, generation and transmission of power at high DC voltage was not technically viable, and there was no means of stepping up the DC voltage for HVDC transmission. Hence, in the beginning, generation, transmission and distribution of power were at the same voltage level, and hence there was a limitation for distance as well as the amount of power to be transmitted.

1.3.2 AC generation, transmission and utilization

Development of a prototype transformer with a core made of laminations of soft magnetic material and coils by William Stanley (1858–1916) in 1885 paved the way for the development of transmission at high voltage and distribution at an intermediary level or utilization at a safer low voltage level [5]. Thus, with the introduction of transformers, high-voltage AC transmission is made possible so that voltage drop, transmission loss, etc., could be reduced compared to low-voltage DC transmission, and hence, more power could be transmitted from one place to another. Thus, AC power generation, transmission and utilization came into existence. Today, AC power is widely utilized in many applications, viz., household appliances, commercial, industrial and transportation. It can also easily be converted to DC using a rectifier. In AC, the instantaneous values of voltage change in magnitude and direction at every instant. Also, the phase shift between its voltage and current depends on the type of load connected to the system. Instantaneous value of AC voltage in terms of time, $v(t)$, can be written as:

$$v(t) = V_m \sin \omega t = \sqrt{2} \times V_{rms} \sin \omega t \tag{1.6}$$

where,

t is the time,

V_m is the maximum voltage, V_{rms} is the RMS value of the voltage,

ω is the angular frequency that is given by $\omega = 2 \times \pi \times f$

and f is the frequency in cycles/sec.

Waveform of a 50 Hz, 230 V single-phase supply is shown in Figure 1.5. The RMS value of the waveform is $V_{rms} = 230$ V and its peak value, $V_m = \sqrt{2} \times 230 = 325.27$ V.

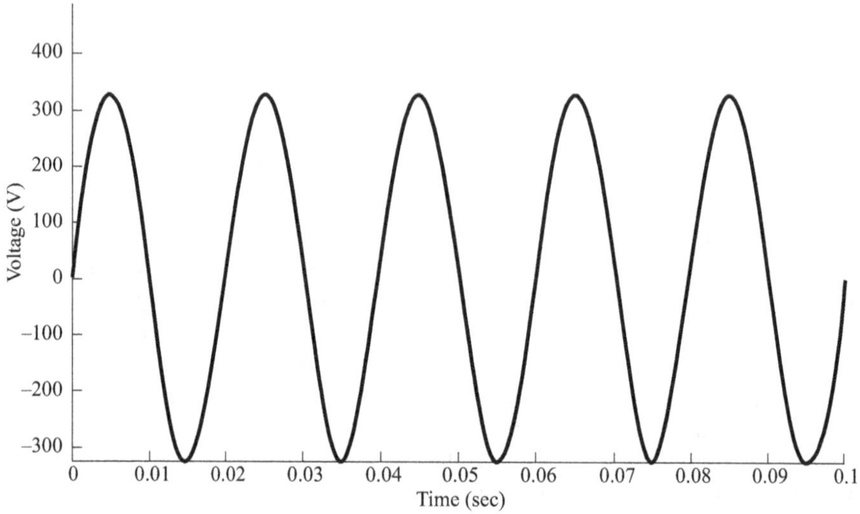

Figure 1.5 Waveform of 230 V, 50 Hz AC voltage

1.3.2.1 Single-phase and 3-phase systems

An AC power supply can be a single-phase or a polyphase system. While single-phase is generally limited to low power requirements, mainly for lighting and powering residential appliances, the three-phase system is common for commercial and industrial use. In three-phase rotating machines, there are three coils (R-Y-B) placed in space where their axes are mutually 120° apart in the stator. In the case of synchronous machines, the rotor houses a field or excitation winding that is excited by DC. This produces flux. When the rotor is driven by an external drive, the flux linkages with the stator windings become a sinusoidal function of time, and sinusoidal voltage is induced in it. The induced voltages in the three-phase windings differ in phase by 120°, as shown in Figure 1.6.

Phase voltage is written as:

$$v_R = \sqrt{2} \times V_R \sin \omega t \tag{1.7}$$

$$v_Y = \sqrt{2} \times V_Y \sin (\omega t - 120^\circ) \tag{1.8}$$

$$v_B = \sqrt{2} \times V_B \sin (\omega t - 240^\circ) \tag{1.9}$$

The magnitude of the line voltage is $\sqrt{3}$ times its phase voltage, and it leads its phase voltage by 30°. Thus, the instantaneous value of three-phase line voltage is given by:

$$v_{RY} = \sqrt{3} \times \sqrt{2} \times V_R \sin (\omega t + 30^\circ) \tag{1.10}$$

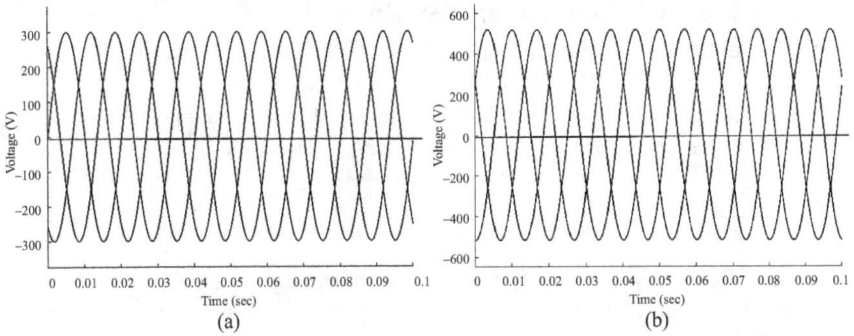

Figure 1.6 Typical 3-ϕ 50 Hz waveforms. (a) Phase – Neutral voltage (b) Phase – Phase voltage.

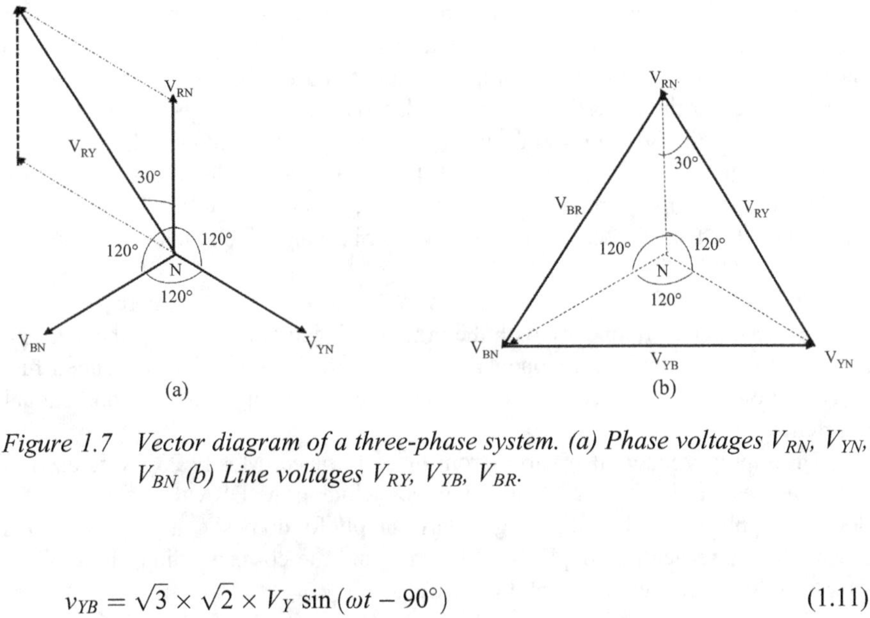

Figure 1.7 Vector diagram of a three-phase system. (a) Phase voltages V_{RN}, V_{YN}, V_{BN} (b) Line voltages V_{RY}, V_{YB}, V_{BR}.

$$v_{YB} = \sqrt{3} \times \sqrt{2} \times V_Y \sin(\omega t - 90^\circ) \tag{1.11}$$

$$v_{BR} = \sqrt{3} \times \sqrt{2} \times V_B \sin(\omega t - 210^\circ) \tag{1.12}$$

The vector diagram of the phase voltages V_{RN}, V_{YN}, and V_{BN} with respect to neutral N is shown in Figure 1.7(a). In the case of single-phase systems, loads are connected between one phase and neutral, whereas three-phase loads are connected between R-Y-B-N or R-Y-B in star or delta, respectively, depending on the requirement. In a delta connection, the line voltage is V_{RY}, V_{YB} and V_{BR}, with their magnitudes $\sqrt{3}$ times the phase voltage and leading the respective phase voltage by 30°, as shown in the vector diagram Figure 1.7(a). The vector diagram for the delta connection is shown in Figure 1.7(b).

1.3.2.2 Choice of power supply frequency

Alternating current changes its magnitude and direction periodically, as shown in Figure 1.5. For 50 Hz or 60 Hz, the direction and magnitude change 100 or 120 times per second. Frequency (f) of the generated voltage depends on the number of poles (P) and the speed of the generator (N) and is given by:

$$f = \frac{PN}{120} \tag{1.13}$$

As the frequency increases, its rate of change of flux linkages and hence induced voltages on coupled winding increases, resulting in the requirement of a reduced number of turns and hence length of winding conductor and core material for electrical machines, viz., rotating machines and transformers. This increases the inductances of the electrical equipment and transmission lines and reduces the capacitances, thereby reducing the transmission efficiency. If the frequency is too low, for the same emf, a greater number of turns or magnetic flux is required. As the maximum flux density is limited by the type of core material used, to get higher flux, a larger cross-section of core material is required. This, in turn, may result in an increase in the weight of the equipment and its cost. Also, it affects the persistence of vision or the flickering of light produced by electric lamps. Thus, there is a trade-off between high and low frequency. Globally, 50 Hz and 60 Hz power frequencies are used. Some countries use 50 Hz power grids, while other countries use 60 Hz. In Japan, supply frequency differs from region to region [6]. Eastern Japan, which includes Tokyo, uses 50 Hz, while western Japan, which includes Osaka and Kyoto, uses 60 Hz.

The choice of frequency also applies to the aviation sector, where 400 Hz is used as the standard frequency with the primary objective of reducing the size and weight of power supply components in the aircraft. This facilitates reduced fuel consumption or increased cargo handling capacity and hence operational efficiency.

The supply voltage also varies from place to place. Mostly 230 V is used to power homes, but in about 20% of countries, including the USA, 120 V is used for domestic applications. In India, single-phase supply for domestic use is 230 V. This means the phase-neutral supply is 230 V rms and the corresponding phase-phase supply is 400 V rms under normal conditions.

The rating assigned to any machine is at a specified voltage and frequency condition. Variation from rated voltage or frequency, or a combination of both, adversely affects the operating performance of all equipment in the system, including generators.

Operation of motor on different frequencies

As we have seen, globally, power supply frequencies of 50 Hz or 60 Hz are used, and electrical equipment is sourced for installation from anywhere in the world. As the majority of electrical power utilized in industry is to power an induction motor, a brief discussion on how the supply frequency affects its performance is detailed. A 50 Hertz (Hz) power supply means the voltage and corresponding current magnitude change from positive to negative and from negative to positive 50 times per second, or the changes are happening 100 times/second. For example, if the power supply frequency is 50 Hz, for a 2-pole motor, the speed is calculated as:

$$N = \frac{50 \times 120}{2} = 3{,}000 \text{ rpm} \tag{1.14}$$

Similarly, for a 4-pole motor, the speed will be 1,500 rpm. At 60 Hz, the corresponding speed for a 2-pole motor is 3,600 rpm, and for a 4-pole motor, the speed is 1,800 rpm. This indicates that operating a motor at 20% higher frequency results in 20% higher speed, and hence rotor retaining rings must bear higher centrifugal forces, and bearings must be checked for vibration. At higher frequencies, iron and eddy current losses in the stator magnetic material will also be higher.

Mechanical power, developed in a motor, is proportional to its speed N and torque T. Accordingly, for a constant torque, power output at 20% higher speed is 20% higher. The motor would be able to produce rated torque at both frequencies only if the flux, Φ, is constant, meaning that at 50 Hz the supply voltage would need to be 380 V and at 60 Hz the supply voltage would need to be 460 V. In both cases, the v/f ratio is 7.6 V/Hz. In the case of a variable torque load, such as a pump or fan, the required torque is proportional to the square of the speed and the power to the cube of the speed. Hence, a 50 Hz motor at 60 Hz will attempt to rotate at 20% higher speed, 44% higher torque and 72.8% higher power than at its design frequency. Reducing the fan or pump speed from 100% to 80% would reduce the required torque to 64% and power to 51.2%.

1.3.2.3 Development of electrical power equipment

As we have seen, electrical current is an instantaneous flow of electrical charges, driven by electromotive force. This energy is neither created nor destroyed, but it is transformed from one form to another form of energy. It is converted from primary sources such as coal, oil, natural gas, nuclear, hydro, solar, wind, etc., to electricity and hence forms a secondary form of energy. As we have seen from Table 1.1, the power system that we see today took about two centuries to develop. During this period, many devices and systems were developed to make human life easier. We shall discuss a few among them, such as dielectric material, generators, transformers, transmission and distribution systems, utilization, etc.

Dielectric material in equipment

The design of an electrical power grid is a process of optimization. It is, as best as practical, to minimize the overall size of the equipment and cost and maximize the efficiency of the system, matching the performance parameters, service life and capabilities, while maintaining mechanical, thermal, magnetic parameters, etc., within limits. Another important parameter is the rated voltage. It depends on the dielectric strength, that is, the maximum electric stress the insulation material can withstand. It is again dependent on a variety of factors – insulation medium, its imperfections, environmental parameters viz. humidity, temperature, pressure, etc.

Insulation mediums can be either solid, liquid or gaseous dielectrics that prevent the flow of current between points of different potential. There are several materials that are developed and are used either separately or as a combination to form the required temperature classification, viz., Class Y (90 °C), Class A (105 °C), Class E (120 °C), Class B (130 °C), Class F (155 °C), and Class H (180 °C) [7].

Air is the natural choice considering cost and its self-healing breaking property. At room temperature and one atmosphere pressure (1 kg/cm^2), its dielectric strength is about 3 kV/mm. Overhead transmission lines make use of this choice. At 3 kg/cm^2 pressure, its strength is about 9 kV/mm. The breakdown strength of air and hydrogen is about the same. However, hydrogen has less "windage" loss; being the lightest gas, it is often used in higher-capacity rotating machines. Another encouraging gaseous medium is SF6. It is about five times heavier than air. It is an odourless, non-toxic gas with high dielectric strength and good arc-quenching properties. SF6 is widely used as insulation for high-voltage gas-insulated substation, where all substation equipment (instrument transformers, disconnectors, circuit breakers, busbars, etc.) is housed in an enclosure, making it ideal for installation in polluted environments and costly places. Ideally, a vacuum is the best insulator and has a field strength up to about 10^7 V/cm. Air, vacuum and SF6 are generally used in circuit breakers in the increasing order of the voltage rating.

In the case of conventional substations, all conductors and equipment are placed in atmospheric air, while the electrical items are placed in SF6 gas in the case of GIS. This results in reduced electrical clearance and hence a compact layout of the substation. The space requirement of GIS is about 10% of an equivalent conventional outdoor switchyard. GIS is more popular in cities due to the scarcity of land or because the environment is polluted due to industries or coastal sites where the atmosphere is saline [8]. Figure 1.8 shows a layout view of conventional and gas-insulated substations.

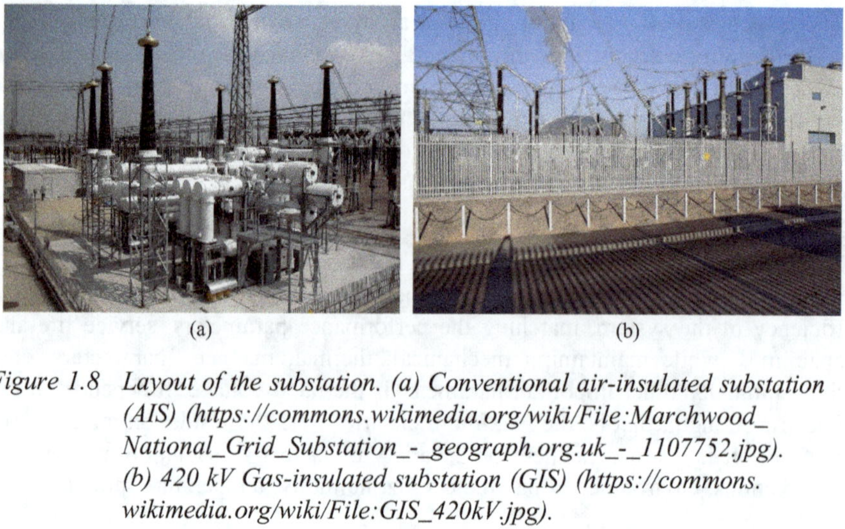

(a) (b)

Figure 1.8 Layout of the substation. (a) Conventional air-insulated substation (AIS) (https://commons.wikimedia.org/wiki/File:Marchwood_National_Grid_Substation_-_geograph.org.uk_-_1107752.jpg). (b) 420 kV Gas-insulated substation (GIS) (https://commons.wikimedia.org/wiki/File:GIS_420kV.jpg).

Table 1.2 Preferred ratings of rotating machines

Rated voltage (Un) in kV	Minimum rated output kW (or kVA)
1.0 < Un < 3.0	100
3.0 < Un < 6.0	150
6.0 < Un < 11.0	800
11.0 < Un < 15.0	2500

Rotating machines

Generators and motors are classified as rotating electrical machines, while the former convert mechanical energy to electrical energy and the latter convert electrical energy to mechanical energy. In conventional power plants, the turbine rotates the field winding of its synchronous generator to induce alternating electromotive force (emf) in its stator windings, while in the case of motors, the magnetic field and the electric current in the winding help the rotor to produce mechanical power.

As it is not practical to design and manufacture machines for every possible rating, taking into account design and manufacturing considerations IS/IEC 60034-1 [9] specifies preferred ratings as given in (Table 1.2). The standard also specifies other operating conditions like temperature, voltage and frequency variation, harmonics and waveform symmetry.

Motors and generators rated less than 100 MVA are normally cooled by air flowing over the rotor and stator. For large-capacity thermal or nuclear power plants, each unit of the plant may have a combination of steam turbines, viz., high pressure, low pressure and intermediate pressure, with a single generator rotating on a common turbine-generator shaft. Thus, the capacity of the generator becomes the sum of the capacities of all the turbines together. In recent times, the generators rated up to 1200 MW or 1400 MW are in use. Such a large turbine-generator with its overall length of more than 10 m offers a high moment of inertia. To reduce the windage losses, hydrogen gas at about 4–6 kg/cm^2 is generally used for cooling the stator core and rotor. For cooling the stator windings, pure water is passed through either its hollow stator conductor strands or other dedicated tubes placed adjacent to the stator conductors. For conductor insulation, solid insulation matching with service temperature class and environmental conditions is used. A section view of a three-phase asynchronous motor is shown in Figure 1.9.

Transformers

A transformer is another prominent piece of capital equipment in the power system. It transfers electrical power from one circuit to another by electromagnetic induction without any change in frequency. The transformer is instrumental in enabling the universal use of the alternating current system for the transmission and distribution of electrical energy. Power generated at a large capacity generating station (usually at a voltage in the range of 11–25 kV) is stepped up by a generator transformer to a higher voltage level of the order of 220 kV and above for

Figure 1.9 A section of a three-phase asynchronous motor (https://commons.wikimedia.org/wiki/File:Tesla%27s_three-phase_asynchronous_motor.jpg)

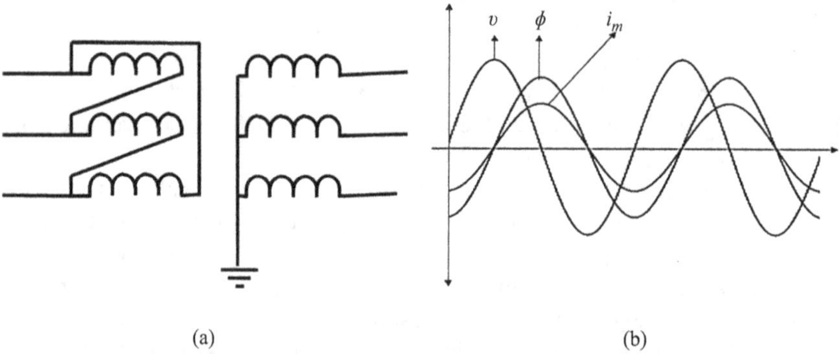

Figure 1.10 Transformer winding and waveforms. (a) Transformer windings diagram for delta-star connection (b) Waveform of voltage, fundamental component of magnetizing current and magnetic flux.

transmission. This is achieved by adjusting the turns ratio of the transformer using the relation,

$$\frac{V_2}{V_1} = \frac{N_2}{N_1} = \frac{I_1}{I_2} \tag{1.15}$$

where V_2, V_1 and N_2, N_1 are the voltage and number of turns in the secondary and primary windings, respectively. I_2 and I_1 are the secondary and primary currents flowing through the transformer.

A connection diagram of a delta-star transformer generally used in the distribution of electric energy, along with its associated waveforms for supply voltage (v), flux in the core (Φ) and magnetizing current (I_m) is shown in Figure 1.10.

Table 1.3 Preferred rating of transformers

Rated voltage (kV rms)	Capacity (MVA)	Transformer type
$(765/\sqrt{3})/(400/\sqrt{3})/33$ kV	500	1-ϕ Auto Transformer
400/220/33 kV	500/315	3-ϕ Auto Transformer
$(400/\sqrt{3})/(220/\sqrt{3})/33$ kV	167/105	1-ϕ Auto Transformer
400/132/33 kV	315/200	3-ϕ Auto Transformer
220/132 kV	200/160	3-ϕ Auto Transformer
220/66 kV	160	3-ϕ Power Transformer
220/33 kV	100	3-ϕ Power Transformer
132/33 kV	80/50/31.5	3-ϕ Power Transformer
66/11 kV	31.5/20/12.5	3-ϕ Power Transformer

Insulating materials used in transformer winding are enamelled conductors, kraft paper, mica, thermoplastic insulating tape, glass fabric, resins, etc. In the case of oil-filled transformers, mineral oil or synthetic fluid that has high dielectric strength is commonly used as an insulating medium and as a coolant along with oil-impregnated paper.

Keeping the importance of the equipment in mind, India has standardized the technical parameters of power transformers, generator transformers and shunt reactors above the 66 kV class [10]. The standardized rating of high-power transformers is tabulated in Table 1.3.

1.3.2.4 Transmission and distribution

The generator is connected to the consumer load that may often be at a far distant place from the generating station. The flow of current is governed by voltage at both ends, in addition to transmission and distribution line conductor resistance, inductance, capacitance, etc. Also, its uncontrolled flow (faults) is prevented using insulating materials and protective and controlling elements such as fuses, relays, circuit breakers, etc. The development of transmission and distribution networks is commensurate with the generation and growth of consumer demand. Thus, the network becomes complex, and power quality issues are enhanced with an increase in the network.

AC power transmission started in the late nineteenth century, from low voltage levels transmitting power over short distances, and reached a stage where huge amounts of power needed to be transmitted over larger distances. The growing power demand with localized power generation motivates engineers to find solutions by transmitting power at higher voltage levels.

The first three-phase AC transmission at 15 kV was established by connecting the hydropower generating station at Lauffen and Frankfurt in Germany with an approximately 175 km-long line (1891) [11]. Transmission voltage level increased thereafter. The first 765 kV extra-high-voltage (EHV) line put into operation was in the United States in the year 1969. The Indian power sector is also growing at an accelerated pace in generation and transmission to meet the increasing demand. In 2016, India operationalized a 1200 kV ultra-high-voltage (UHV) national test

Figure 1.11 500 kV transmission lines in Australia (https://commons.wikimedia.org/wiki/File:500kv-power-lines-australia.jpg)

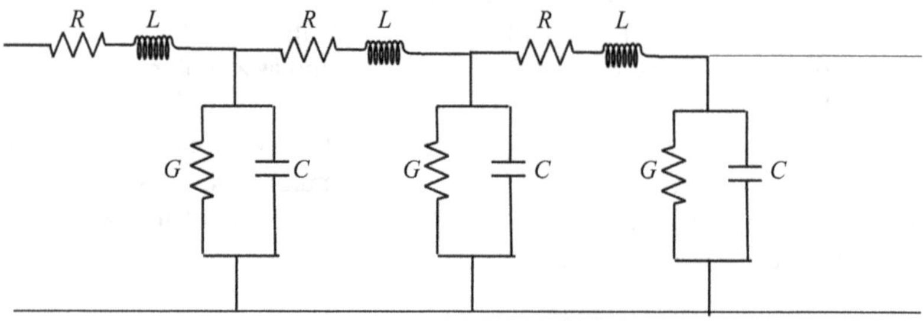

Figure 1.12 Equivalent circuit of a long transmission line

station for the development and qualification of 1200 kV equipment and transmission lines at Bina in the Indian state of Madhya Pradesh. A 550 kV transmission line in Australia is shown in Figure 1.11.

Now, let us briefly analyse the transmission line.

Power handling capacity of transmission line

One of the important characteristics of the transmission line is its power-carrying or line-loading capability. It depends on the line parameters. An AC transmission line is characterized by its distributed parameters of series resistance (R), series inductance (L), shunt capacitance (C) and shunt conductance (G), as shown in Figure 1.12.

AC resistance of the line is calculated from its DC resistance considering skin effect, temperature, spiralling of stranded conductors, bundled conductor arrangement, etc. It causes additional transmission loss (T_L) and voltage drop (V_d) across the line and is given by:

$$T_L = I_{ac}^2 \times R \tag{1.16}$$

$$V_d = I_{ac} \times R \tag{1.17}$$

where, I_{ac} is the rms value of AC current flowing through the transmission line.

G, contributes to the "dielectric loss", loss due to the leakage current through the surface of insulators due to the atmospheric contaminants such as industrial pollution, dirt, salt, moisture, etc. Typically, this leakage current is very small and hence generally neglected along with the series resistance. L, the series inductance of the line due to the alternating flux produced around the line conductor. It depends on the conductor spacing and its size. Hence, as transmission voltage increases, inductance also increases. As the induced emf is at 90° leading with respect to the current, the average active power consumed by the inductance of the line will be zero; however, it consumes reactive power. The reactive power consumed by the transmission line (Q_L) and voltage drop across the line due to the inductance of the line are given by:

$$Q_L = I_{ac}^2 \times X_L \tag{1.18}$$

$$V_d = I_{ac} \times X_L \tag{1.19}$$

where, X_L is the inductive reactance of the line.

Thus, the transmission line consumes reactive power due to its inductance. This also results in a voltage drop across the line, which in turn affects the voltage at the receiving end of the line. As this voltage drop leads the line current, the receiving-end voltage reduces from its sending-end voltage. This reduces further as the current load increases. When alternating voltage is applied between conductors separated by a dielectric medium, air in this case, charges of equal magnitude but opposite polarity will accumulate on the other side. This results in a capacitive current between the conductors and the conductor to earth and is represented as a shunt capacitance C in the equivalent circuit.

Thus, leakage current (I_C) will flow through the line and is given by:

$$I_C = \frac{V}{X_C} \tag{1.20}$$

where X_C is the capacitive reactance of the line, and V is the voltage across the capacitor, which is assumed to be the same as the system voltage.

Due to the capacitive nature, the leakage current leads the line voltage by 90°, thus generating the reactive power (Q_C) in the system.

$$Q_C = \frac{V^2}{X_C} \tag{1.21}$$

The current flowing through the series resistance and inductance of the transmission line will be the sum of I_{ac} and I_C.

$$Q_{L-TOTAL} = (I_{ac} + I_C)^2 \times X_L \tag{1.22}$$

Reactive power thus generated or absorbed by a transmission line results in deviation of the voltage profile from its ideal flat voltage profile. Under light load conditions, the reactive voltampere generated by the shunt capacitance will be more than the reactive voltampere absorbed by the series inductance. Consequently, the voltage at the receiving end is more than the sending-end voltage. This effect is known as the Ferranti effect. However, under heavily loaded conditions, reactive voltampere absorbed by the series inductance will be more than that generated by the shunt capacitor, and hence, the voltage at the receiving end will be less than the voltage at the sending end. When the reactive voltampere generated by the shunt capacitor is equal to the reactive voltampere absorbed by the inductor, the voltage at various parts of the transmission line will be the same as the sending-end voltage. Under such conditions, the transmission line is said to be surge impedance loading or in a naturally loaded condition. The power transmitted under natural, or surge impedance, conditions (P_0) is given by:

$$P_0 = \frac{V^2}{Z_0} \tag{1.23}$$

Z_0 is the surge impedance of the line, which is independent of the length of the transmission line, and is given by:

$$Z_0 = \sqrt{L/C} \tag{1.24}$$

The surge impedance loading (SIL) gives a general idea of the loading capability of the line. Because of the stability limitations, the usual practice is to load the short lines above SIL and the long lines lower than SIL. Table 1.4 shows SIL for commonly used transmission line configurations.

In fact, permissible line loading depends on many factors, such as current carrying capacity, voltage regulation, stability, geographical location of installation, etc. Again, current carrying capacity is based on thermal loading limits that depend on environmental factors such as ambient temperature, wind velocity, etc [12]. Depending on maximum permissible conductor temperature, losses, and optimization of right of way (ROW), various types of overhead conductors are developed. The ordinary hard-drawn aluminium starts annealing and losing its

Table 1.4 Surge Impedance Loading at different voltage levels[1]

Voltage (kV)	Number and size of conductor (mm^2)	Surge impedance loading (SIL), *Po* (MW)
765	4 × 686	2250
765 operated at 400	4 × 686	614
400	2 × 520	515
400	4 × 420	614
400	3 × 420	560
400 operated at 220	2 × 520	155
220	1 × 420	132
132	1 × 200	50

[1]Manual on Transmission planning criteria 2013 Annexure-II, Central Electricity Authority, India, https://cea.nic.in/wp-content/uploads/

strength at about 93 °C and is not suitable for long-term use above this temperature. A few of the commonly used overhead conductors and their maximum permitted temperatures are: Aluminium Conductor Steel Reinforced (ACSR) (85 °C), All Aluminium Alloy Conductor (AAAC) (95 °C), and High Temperature Low Sag (HTLS) conductors of different material types (150–250 °C) [13]. In addition to the above bare conductors, insulated aerial bunched cables (ABC) in congested urban areas and underground cables are also used.

Stability in the power system

At any point in time, the operation of the power system should be stable and secure. However, the system is subjected to various types of disturbance, such as small to large variations in power demand or generation and minor to major faults involving the loss of a generating unit, transmission line or large consumer demand. Thus, the system needs to remain in synchronous or equilibrium conditions under normal circumstances as well as after a disturbance and maintain voltage within acceptable limits. This is checked during the planning and design stage of the system by stability analysis. Typically, there are three types of stability analysis in power systems.

(i) Steady-state stability
(ii) Transient stability
(iii) Dynamic stability

Steady-state stability deals with the maximum power that can be transmitted through the line without losing synchronism. It indicates the capability of a power system to come back to its original state after a small disturbance. Whereas transient stability deals with the ability of a power system to come back to its original position after a large disturbance in the system. Dynamic stability deals with small disturbances, which cause oscillations in the power system. A modern generator excitation system helps to dampen this oscillation. One of the main parameters of a transmission line that affects the stability of the system is the reactance of the transmission line. The reactance of the line depends on the frequency of current flowing through the line and its inductance. The effect of transmission line reactance on power transfer is briefly explained in the next section.

Power system stability analysis. Figure 1.13 shows a typical transmission system representing generating station G and substations at the generator end with voltage E and at the load end with V and an angle δ between them. Both substations

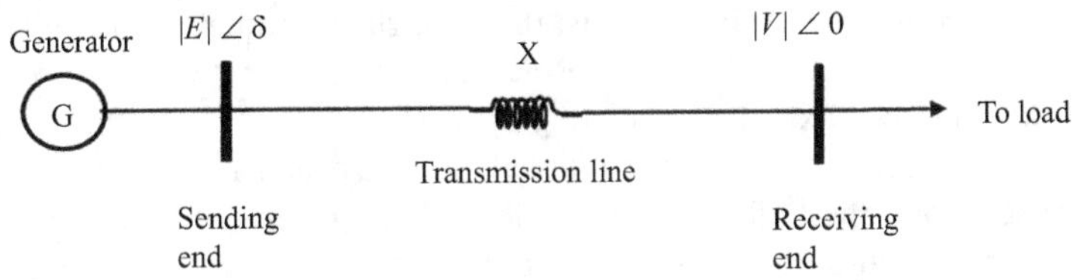

Figure 1.13 One machine model of a power system

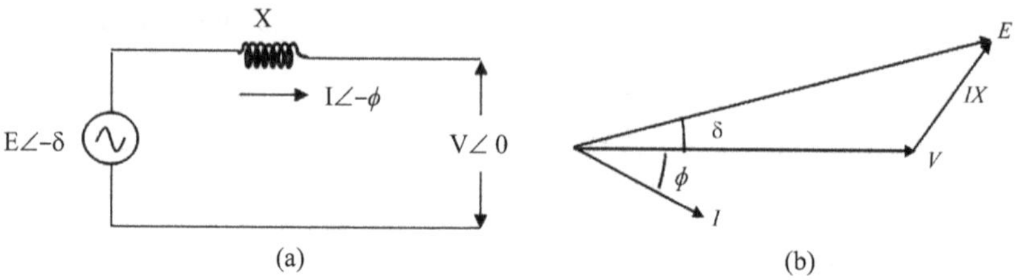

Figure 1.14 Transmission line equivalent circuit model. (a) Equivalent circuit. (b) Vector diagram.

may have many loads or interconnections, or lines. Let the line's inductive impedance of the transmission line, which connects buses, be X. Line resistance and capacitance are ignored for simplicity of analysis. The equivalent circuit of the model and its vector diagram are shown in Figure 1.14.

The apparent power transferred in the above circuit is given by:

$$S = EI^* \tag{1.25}$$

where, $I = \dfrac{E\angle\delta - V\angle 0}{jX}$ and $I^* = \dfrac{E\angle - \delta - V\angle 0}{-jX}$

Substituting I^* in S and separating imaginary and real parts, reactive power transfer from sending-end bus to the receiving-end bus is given by:

$$Q = \frac{E^2}{X} - \frac{EV}{X}\cos\delta \tag{1.26}$$

And real power transfer from the sending-end bus to the receiving-end bus is given by:

$$P = \frac{EV}{X}\sin\delta \tag{1.27}$$

And when $\delta = 90$, maximum real power given by $P_{\max} = \dfrac{EV}{X}$ is transferred.

The turbines coupled with the generator provide mechanical power to the generator rotor, and the generator converts the mechanical energy into electrical energy. This electromechanical conversion can be represented as:

$$J\frac{d^2\partial_m}{dt^2} = T_m - T_e \tag{1.28}$$

where J is the moment of inertia of the rotating machines, T_m is the mechanical input torque from the turbines, T_e is the electrical torque acting against the mechanical input torque, and ∂_m is the mechanical angle between the rotor field axis and the reference axis that is rotating synchronously.

For any sudden generation-load imbalance, say generation has a shortfall (load > generation), the higher amount of load (load-generation) will be delivered from the inertia of all the synchronously connected rotating machines in the power system, thereby reducing their speed and consequently system frequency. Hence, for large inertia systems, a change in frequency may not be appreciable for a short period of time. However, for small systems, the stability of the system may be affected.

With the increase in renewable energy generation driven by the displacement of large-capacity conventional thermal, hydel and nuclear power generation, there is a growing concern over the reduction of system inertia and hence increasing concern for frequency stability or reduction in its ability to damp out the oscillations in the system. The plot of active power P and reactive power Q transfer is given in Figure 1.15 [14].

Figure 1.15 shows that the power transmitted through the line reaches its maximum value when the load angle δ is 90°. When there is a disturbance in the system, viz., load-generation mismatches, outage of a large generation unit or transmission line, or faults in the system, viz., short circuits between two phases or phase to ground, the generator will accelerate or decelerate depending on its operating point. During the transient time, mechanical power from the turbine will remain the same as before; however, electrical output varies. As the system is continuously subjected to such transients, to ensure stability, the angle across the transmission system is usually kept below 44°. In modern power systems, the problem is minimized by the fast-acting fault-clearing protection system, generator excitation systems, etc. Phaser measurement units (PMUs) [15] are also installed at different places [16] for early warning of grid instability and blackout.

Voltage stability. Voltage instability indicates that the receiving-end voltage reduces far below its normal value and does not come back to its normal value even after a disturbance in the power system. Voltage collapse is the process by which the voltage falls to a low, unacceptable value. Voltage variation is mainly due to reactive power flowing through the line.

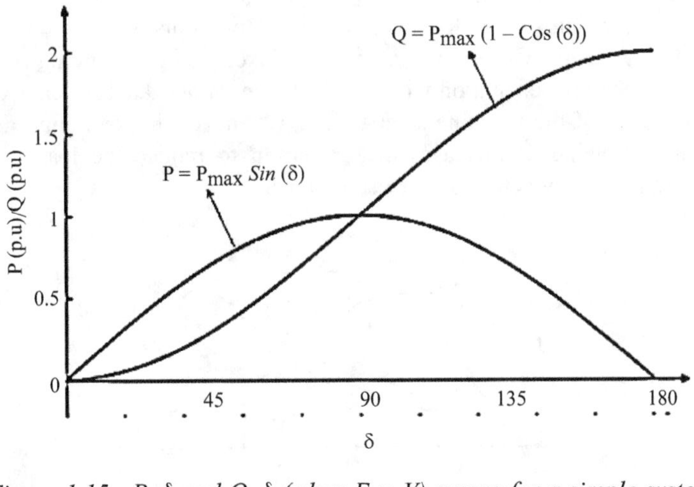

Figure 1.15 P–δ and Q–δ (when E = V) curves for a simple system

For small load angle δ, reactive power transfer can be approximated to,

$$Q = \frac{E^2}{X} - \frac{EV}{X} \tag{1.29}$$

Thus, the reactive power flow depends mainly on voltage magnitudes, and it flows from the highest voltage to the lowest voltage. This is possible with synchronous machines that are capable of injecting large amounts of reactive power during a short circuit in the system. However, with the displacement of large conventional generators by the renewable energy sources, there is a decrease in the reactive power in the system, and hence, the system damping behaviour of the system is affected.

Excess reactive power, mainly during light load on long transmission lines, gives rise to system voltage and adversely affects every piece of equipment in the power system. At the same time, inadequate reactive power from generators, transmission lines and distribution networks leads to voltage instability phenomena, in which the receiving-end voltage decreases well below its normal value and is not restored even after various fast restoring mechanisms in generator excitation systems, transformer tap settings, etc., and voltage continues to oscillate for lack of damping against the disturbances or collapse of the power grid. It is known that under surge impedance load conditions, the power system will be operating at unity power factor (UPF); under heavy-loaded conditions, it will operate at lagging power factor; and under light load conditions, it will operate at leading power factor. The margin of active power that can be transmitted through the line without losing voltage stability depends on the power factor of the system. So, for a constant power factor, a photo-voltaic (PV) diagram (nose curve) can be plotted for a typical power system as shown in Figure 1.16.

In the generator excitation system of a large generator, the automatic voltage regulator regulates the terminal voltage and the flow of reactive power within the capability of the synchronous generator by direct control of the field current using Thyristor Converters. It reduces the excitation whenever the v/f ratio is exceeded to prevent the saturation in the Generator or Generator Transformer. It also reduces the excitation when the Generator Stator current exceeds in the lagging zone of the machine and increases excitation whenever the Generator Stator current exceeds in the leading zone of the machine's capability. Whenever the load angle increases beyond the set value, it increases the excitation to reduce the load angle and dampen low-frequency active power oscillations.

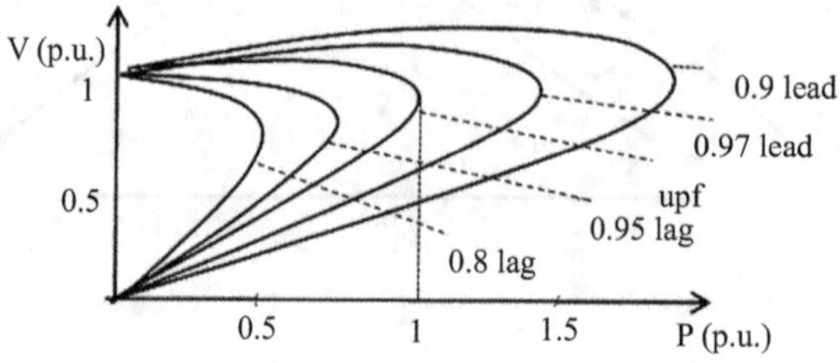

Figure 1.16 PV diagram for a radial system

From (1.26) and (1.27), it is evident that by regulating the magnitude of sending-end voltage E and/or receiving-end voltage V, its phase angle δ and inductance of transmission line X, active power and reactive power flow over the line can be controlled. Additionally, it improves the voltage profile within the limits to save the equipment from dielectric failure and maintain the stability of the system. In other words, by modifying the reactance or characteristic of the transmission line to its surge impedance loading condition, it is possible to improve the performance of the power system under any load conditions. This can be achieved using compensators in the power system. Recent advances in power electronics play an important role in providing and controlling required active and reactive power, be it lagging or leading. These are discussed in this book.

Radial power system

Figure 1.17 shows a representative simple radial system that connects the generation of electric energy to the consumers. The power-generating unit is connected to loads through a transmission line. Thus, the conventional radial system mainly consists of three parts such as generation, transmission and utilization.

One unique requirement of the electrical power supply system is to maintain power generation equal to load demand at every instant. This balance must be met instantaneously while operating the transmission system within its stability limits [17]. If power demand increases above the generation limit, the frequency of the system may go down or vice versa. This may result in tripping of the generating unit and blackout of the power system. Also, in case of failure/maintenance/unavailability of the generator or transmission line, there will be an interruption of electricity to all the consumers, and hence it is not recommended for use. Thus, the radial system shown in Figure 1.17 will not meet the requirements.

Interconnected power system

A power station may have more than one generator connected in parallel. Also, there are many power stations spread over a region. A regional power grid is created by interconnecting these stations through transmission lines. In other words, all generators, including that of the distributed generator in the grid, are in effect connected in parallel to the load through multiple transmission routes. These multiple factors contributed to a very reliable operation for the system [6].

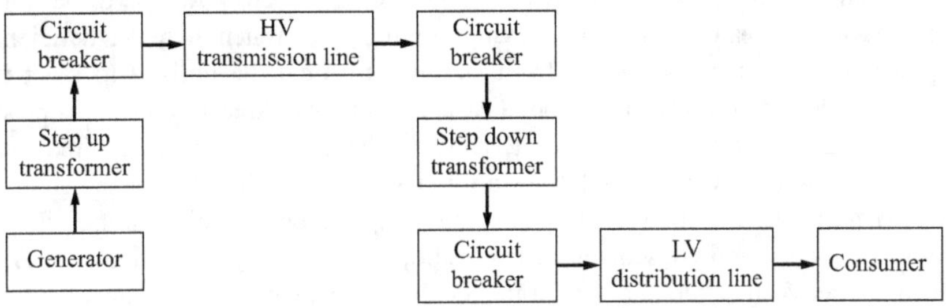

Figure 1.17 Conventional radial power system

The resulting bilateral, bi-directional movement of electricity largely goes unnoticed by the consumer. A car manufacturer in Chennai may be unaware that electricity from Rajasthan is powering its assembly line. This redundancy prevents interruptions to the customers in case of non-availability of any transmission line or generation unit due to any technical issues or periodical maintenance, or even its failure. Also, in an interconnected system with more generators, the system voltage and frequency tend to be fixed irrespective of the degree of loading present in the system. It also helps improve the affordability of supply and maximize opportunities for developing energy resources. However, it is to be controlled and monitored carefully, as a problem with one piece of equipment in one region could affect neighbouring regions and may give rise to instability or contingency conditions leading to the collapse of the system. This is avoided by enhanced system studies and efficient system monitoring and operation. Strengthening international cooperation and harmony, opening cross-border transmission projects, and pursuing joint innovation and R&D are the other benefits.

A single-line diagram of a typical interconnected system is shown in Figure 1.18. Four generating stations are connected to the transmission system through two 400 kV and 200 kV lines. If power demand increases in any one station, depending on the reactance between the interconnected regions, power will flow to the deficit area. Thus, the blackout that occurs in the radial system can be eliminated. Similarly, if maintenance is to be carried out on one piece of equipment, alternative paths can feed power to the region.

The energy transformation process from various resources for power generation to transmission is depicted in Figure 1.19.

In India, grid management on a regional basis started in the 1960s. Initially, state grids were interconnected to form regional grids, namely the Northern, Eastern, Western, Northeastern and Southern regions. These regions have now interconnected stage-wise since 1991. On 31 December 2013, with the commissioning of the 765 kV Raichur-Sholapur Transmission Line, the Southern Region was connected to the rest of the grid in synchronous mode, thereby achieving one nation-one grid-one frequency. The Indian grid is now interconnected with other neighbouring countries. Table 1.5 shows a list of interconnected power grids; this is opening a new era of international energy cooperation.

Power sharing in an interconnected system. Interconnected systems have the advantage of power sharing between different areas or regions.

Consider a two-area system (Area 1 and Area 2) where power is to be transmitted from one area to another, as shown in Figure 1.20. When there is a deficiency of power in one area, the power from surplus areas can be utilized to meet the demand in the deficit area. In the above case, assume that Area 1 has surplus power and Area 2 has a deficiency in power. Because of the deficiency in power, frequency in Area 2 may go down. The solution to this issue is load shedding or transmitting power (P_{12}) from Area 1 to Area 2. This may help the power-deficient area from a drop in frequency and consequent disconnection or blackout. Generally, there is an agreement between different areas for power sharing. However, power flow between two areas occurs when there is a change in frequency in one area due to a power imbalance. As the amount of power flow depends on frequency variation, there is no

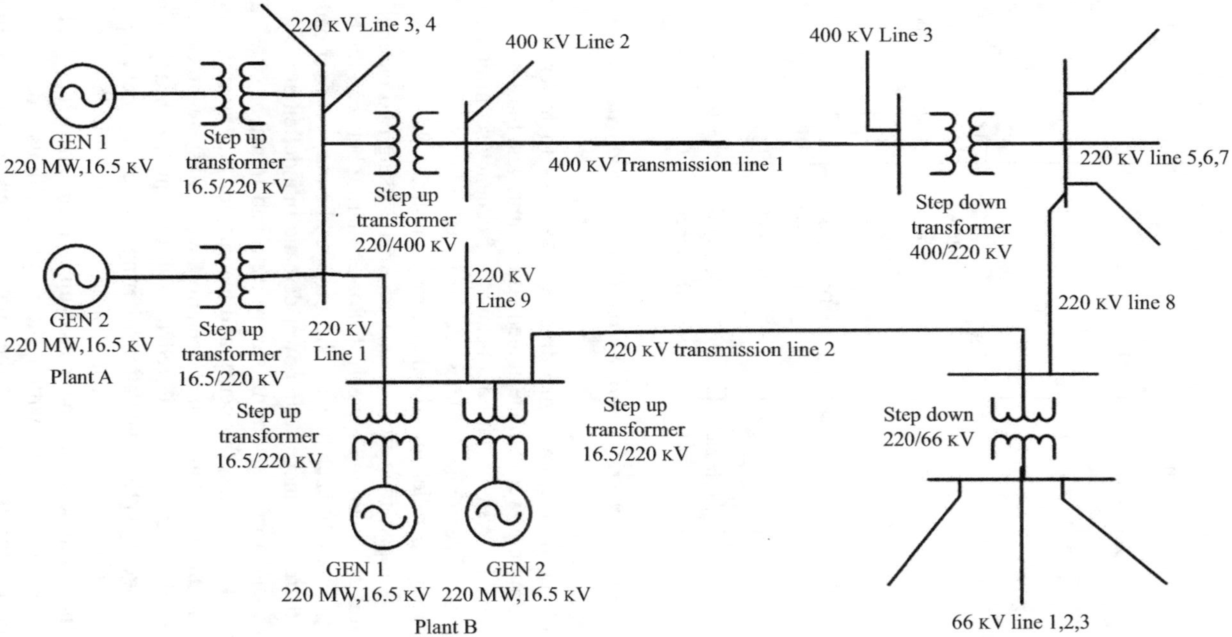

Figure 1.18 Single-line diagram of a typical interconnected system

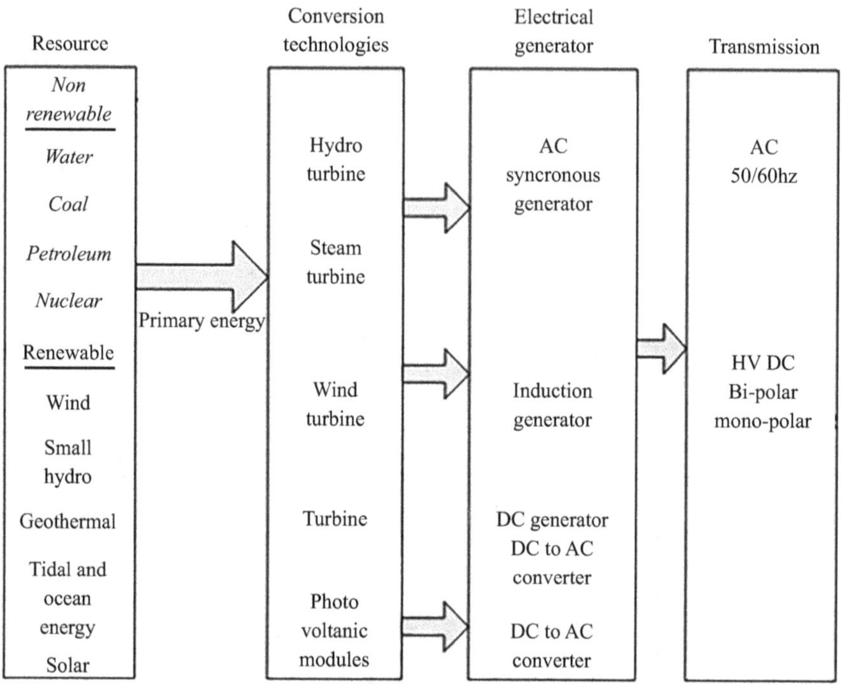

Figure 1.19 Power conversion stages from generation to transmission

control over the power flow from higher frequency regions to lower frequency regions. It may draw more power than the agreed power limit, and thus, it will affect the frequency of a healthy system. If it goes below the frequency set limit, cascaded tripping, overloading of interconnected lines, etc., may occur.

Power flow through parallel lines is also a major issue in the power system. Assume that power is transmitted from Area 1 to Area 2 using two parallel lines with different lengths as shown in Figure 1.21. Power sharing between the parallel lines depends on the length of each line, and hence, one line may get overloaded or underloaded when there is an increase in power demand in a deficit area.

In India, it is mandatory that during steady state and transient state, all the system parameters, like voltages, frequency and loadings of the transmission system, be within normal limits during normal operation, when all system elements are available and when the system is subjected to a disturbance with a loss of an element (single contingency) [13]. Even after suffering one contingency, the grid is still vulnerable to experiencing a second contingency, though less probable, wherein some of the equipment may be loaded up to their emergency limits. To bring the system parameters back within their normal limits, load shedding/ re-scheduling of generation may have to be applied either manually or through automatic system protection schemes (SPS). Such measures are generally applied within one and a half hours (1½) after the disturbance.

Table 1.5 Examples of interconnected power grids

Bay of Bengal Initiative for Multi-Sectoral Technical and Economic Cooperation (BIMSTEC) for Grid Interconnection [32] Low-voltage power supply: 230/400 V, 50 Hz			
Bangladesh	Bhutan	India	Myanmar
Nepal			
Gulf Cooperation (GCC) Power Grid Interconnection [33,34] Low-voltage power supply: 230/400 V, 50 Hz			
Bahrain	Kuwait	Saudi Arabia	Oman
Qatar	United Arab Emirates (Abu Dhabi, Dubai, Sharjah, Ajman, Umm al-Qalwain, Ra's al-Khaimah, Fujairah)		
Interconnected Network ENTSO-E [35] (European Transmission System Operators (TSOs) – 36 countries) Low-voltage power supply: 230/400 V, 50 Hz			
Albania	Estonia	Latvia	Portugal
Austria	Finland	Lithuania	Romania
Belgium	France	Luxembourg	Serbia
Bosnia and Herzegovina	Germany	Macedonia	Slovakia
Bulgaria	Greece	Montenegro	Slovenia
Croatia	Hungary	Netherlands	Spain
Cyprus	Iceland	Northern Ireland	Sweden
Czech Republic	Ireland	Norway	Switzerland
Denmark	Italy	Poland	Ukraine
North American Electric Reliability Corporation (NERC) [36] Low-voltage power supply: 120/240 V, 60 Hz			
Canada	United States of America (USA)	Northern portion of Baja California	Mexico

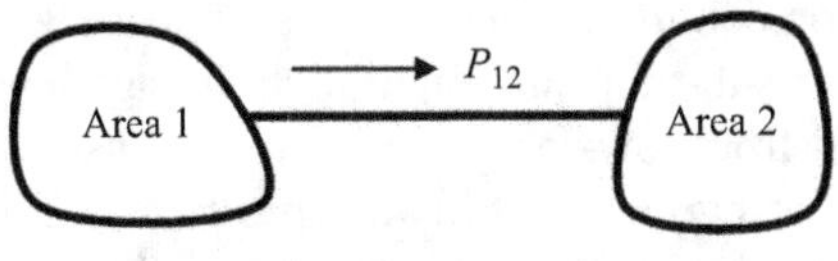

Figure 1.20 Representation of power transfer in a two-area system

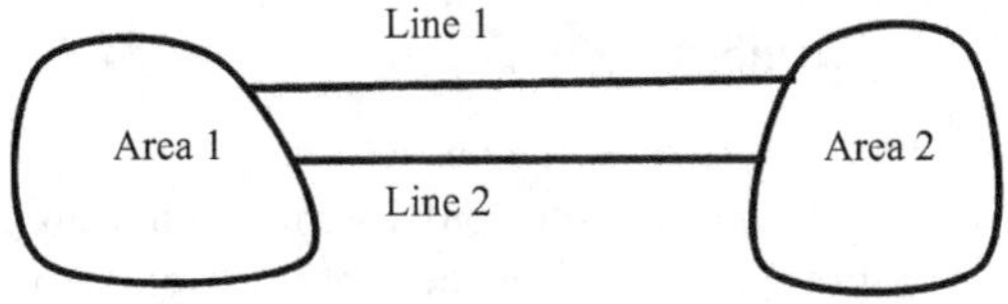

Figure 1.21 Representation of power transfer in a two-area system using two parallel lines

Frequency and voltage control in an interconnected. Frequency is determined by the rotational speed of the generator, while voltage depends on the changing magnetic flux of the generator's stator windings as the generator's rotor rotates. In a stand-alone synchronous generator, frequency is kept constant by the governor, and voltage is varied by increasing or decreasing the rotor magnetic field with an automatic voltage regulator (AVR). In the grid, frequency is stabilized by the rotational inertia of large turbogenerators. When connected to the grid, small generators rotate at the grid frequency. Thus, a grid-connected small generator makes no attempt to regulate frequency; it just injects power in step with the grid's frequency. Regulation of voltage by a grid-connected generator often depends on the preference of the grid operator. Voltage varies throughout the system and depends on generation, distribution of rating and types of loads, and power factor. In some cases, a power producer regulates its generator to keep a constant power factor. In other cases, the distributed generator may regulate voltage.

Availability-based tariff. Frequency-linked pricing of power, or tariff for the unscheduled interchange of power, popularly called Availability-Based Tariff (ABT), is a way to enforce grid discipline or control of frequency in the grid. In this system, if the grid has surplus power, i.e., the frequency is above 50.0 Hz, the rate would be lower. If the excess generation takes place at the time of generation shortage in the system (in which condition the frequency would be below 50.0 cycles), the payment for extra generation would be at a higher rate. Accordingly, the tariff structure for power generation comprises three components: (a) a capacity charge towards reimbursement of the fixed cost of the plant, linked to the plant's declared capacity to supply MWs; (b) an energy charge to reimburse the fuel cost for scheduled generation; and (c) a payment for deviations from schedule, at a rate dependent on system conditions.

1.3.2.5 Utilization of electric energy

On the consumer side, utilization equipment, depending on its constructive characteristics and for safety reasons, must operate at lower voltage levels. Thus, the power transmission line operates at a high-voltage level to reduce the transmission losses using transformers that step the voltage up from generation to transmission and step it down to the distribution or utilization voltage level.

Large industrial or commercial consumers or loads are often fed from AC transmission systems or sub-transmission at high-voltage levels. Whereas residential loads are often fed from low- or medium-voltage distribution levels. Broadly, all these loads can be categorized into two types based on how they modify AC waveforms while utilizing power, viz., linear loads and non-linear loads. Resistive, inductive and capacitive loads are linear, while most of the electronic loads exhibit non-linear characteristics. Or the other way around; linear loads obey Ohm's law, while non-linear loads do not obey Ohm's law. A brief discussion on these loads is detailed in the following sections.

Resistive loads

This type of load draws current in a sinusoidal variation and in phase with supply voltage, as shown in Figure 1.22, and hence these are of the unity power factor type. The vector diagram of such loads is given in Figure 1.22(b). These include

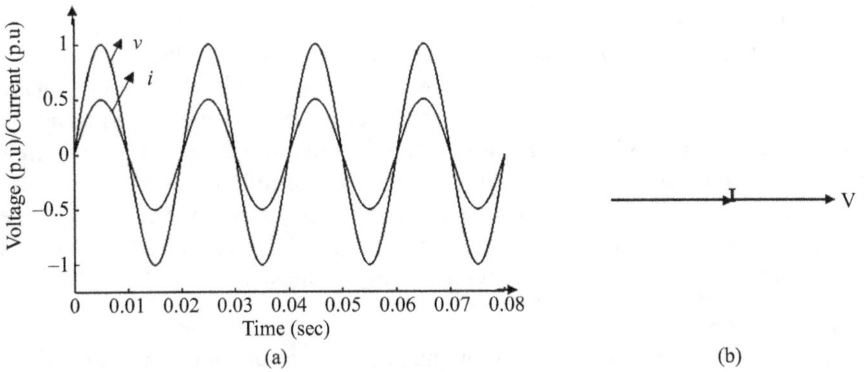

Figure 1.22 Voltage and current waveform in a resistive load. (a) Waveform of voltage and current. (b) Phasor diagram.

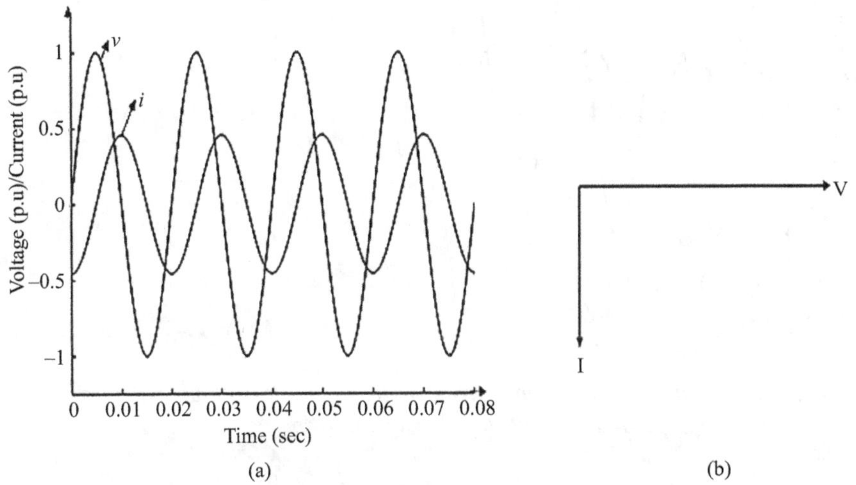

Figure 1.23 Voltage and current in an inductive load. (a) Waveform of voltage and current. (b) Phasor diagram.

incandescent lights (being phased out) and resistive heating loads such as toasters, ovens, water and space heaters, etc.

Inductive loads

In contrast to resistive loads, these loads draw current in a sinusoidal pattern that peaks after the voltage peaks, so the zero, maximum and minimum values are out of phase as shown in Figure 1.23. The majority of the loads are of this type, and hence the power factor is lagging. Single-phase and polyphase induction motors used in residential, agricultural and industrial sectors are of this type. Such loads are also found in a variety of household devices, such as fans, vacuum cleaners, dishwashers, washing machines and compressors in refrigerators and air conditioners.

Capacitive loads

In a capacitive load, current and voltage are out of phase as with an inductive load; however, the current reaches its maximum and minimum value before the voltage does, as in Figure 1.24. In this case, the current leads the voltage waveform, and the power factor is said to be leading. Capacitors and the line capacitance between a high-voltage transmission line and ground are examples of these types.

In all the above cases, as voltage increases, current also increases and vice versa. Hence, the *V–I* characteristics of a linear type of load are linear in nature, as shown in Figure 1.25. Slope varies with $\frac{V}{I}$ ratio, in ohms (Ω).

Non-linear loads

Most of the electronic loads have front-end AC–DC power converters using diode rectifiers with capacitors or controlled rectifiers with SCR or some other switching

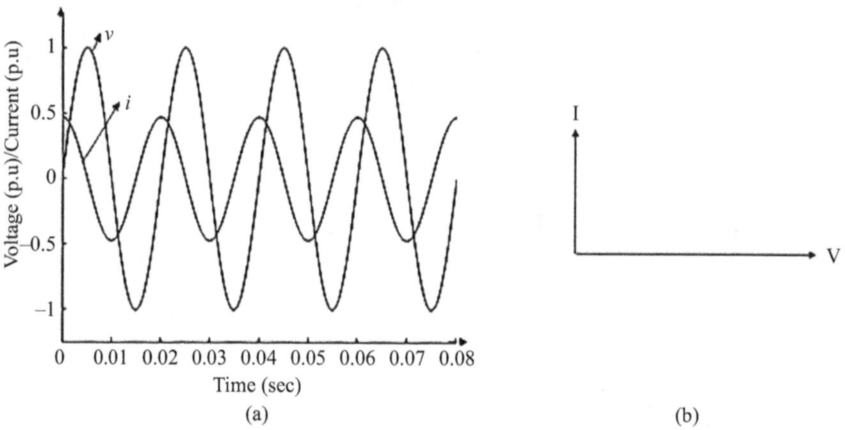

Figure 1.24 Voltage and current waveform in a capacitive load. (a) Waveform of voltage and current. (b) Phasor diagram.

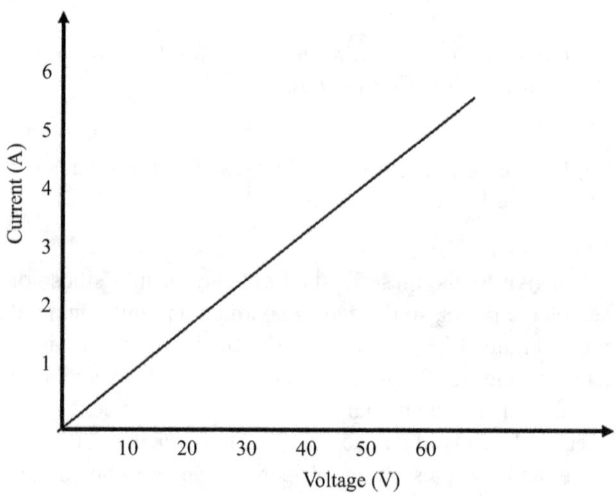

Figure 1.25 V–I characteristic of a linear load

device. These converters draw practically no current until the rectifiers begin to conduct, and thereafter as required by the load. This is repeated in every cycle of the input voltage waveform. Thus, the input current has pulses with absences and peaks in its waveform. That is, the load draw current is not in proportion to the input voltage waveform and has harmonics in its waveform. Thus, it draws a distorted current even if the input voltage is sinusoidal. Thus, non-linear loads produce harmonics, switching transients, notches, spikes, etc., in the network. In general, these power quality issues decrease the efficiency and life of the distribution system components, viz., transformers, capacitors, induction machines, etc. Chapters 2 and 3 discuss it in detail.

1.4 Modern power system

Recent developments in power systems are mainly due to the application of power electronics in power systems. Power electronics helped in various areas of power systems like generation, transmission and distribution. A brief application of power electronics in various sections of the power system is explained in this section, and detailed analysis is provided in other chapters of this book.

1.4.1 Distributed generation

Distributed Generation (DG) [18] is an electrical power source connected to the network, serving a customer on-site or providing support to the network. DG technologies may comprise small- or medium-sized modular energy conversion units which are generally placed near the consumer. The energy source may be for limited or intermittent generation (solar, wind) or dispatchable generation, such as diesel generators, combined heat and power (CHP), etc. It helps to improve the voltage profile and increase energy security and reliability of supply. It also permits islanding mechanisms that protect customers against power outages in networks. DG technology can increase the reliability of supply by providing network support, reduce congestion on the upstream transmission grid, and reduce transmission and distribution losses. It can lower fossil fuel consumption and stimulate competition in supply, allowing more players to enter the electricity market. The use of distributed energy resources is increasingly being practiced as a supplement and an alternative to large conventional central power stations.

1.4.2 Renewable energy sources

In India, the Electricity Act 2003 has charted provisions for the promotion of renewable energy. Section 3(1) of the act instructs: Government to prepare the National Electricity and tariff policy from time to time, for the expansion of the power systems based on optimal utilization of resources. India has up-scaled the target of renewable energy capacity to 175 GW by 2022 [17,19], which includes 100 GW from solar, 60 GW from wind, 10 GW from bio-resources and 5 GW from small hydropower. With this ambitious target, India will become one of the largest green energy producers in the world. This transformation will significantly alter India's CO_2 emissions and climate targets.

Table 1.6 Installed generation capacity in India

Installed generation capacity in India as of April 2024[1]		
Category	**Installed capacity (MW)**	**% share**
	Thermal	
Coal	210969	47.64
Lignite	6,620	1.5
Gas	24,818	5.6
Diesel	589	0.1
Total thermal	2,42,996	54.87
Nuclear	8,180	1.8
Renewable		
Hydro	46,928	10.60
Wind	46,161	10.42
Solar	82,634	18.66
BM power/cogen	10,355	2.34
Waste to energy	591	0.13
Small hydropower (<25 MW)	5,005	1.13
Total (renewable)	191679	43.28
Total installed capacity	442856	

[1]Central Electricity Authority, India Executive summery on power sector April 2024 https://cea.nic.in/wp-content/uploads/executive/2024/04/Executive_Summary_April_2024.pdf

Solar and wind energy generation is also not uniform throughout the country. Some states are richer in renewable energy compared to others. With the increase in renewables, the renewable-rich states will experience periods when wind and solar make up most of the generation, and it will then become imperative to monitor local system and inertia requirements [20].

Solar energy has become the other most popular renewable distributed energy source, wherein energy is extracted directly from sunlight using PV modules. The generation pattern from PV modules gradually increases from morning and decreases to zero during the night. It depends on the season, clouds or their unpredictable movement, rain, dust and other geographical conditions. Hence, other conventional generating units, such as thermal or hydro stations, must ramp up and ramp down their generation depending on the availability of solar energy to meet the daily load. This is a technically challenging task for stations and grid operators. Installed generation capacity of renewable and conventional energy sources in India is shown in before Table 1.6. As per the National Electricity Plan [21] 2018, of India, the projected daily generation pattern from solar PV and wind during December–January is shown in Figure 1.26.

1.4.2.1 Challenges in grid-connected renewable energy sources

As indicated in the previous section, grid-connected renewable energy systems pose a lot of challenges to the operation of power systems. Flexible operation of the existing conventional generation to integrate the wind and solar, ramping up its

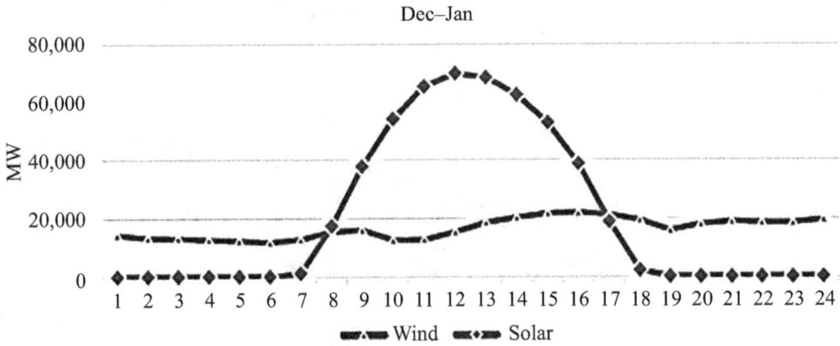

Figure 1.26 Projected power generation curve of solar and wind energy (https://cea.nic.in/wp-content/uploads/2020/04/nep_jan_2018.pdf, page 6.15)

capacity, and start/stop duration are to be met, apart from issues related to power quality [22]. In the case of linear loads, the issue is with reactive power compensation; power factor and consequently, voltage regulations are the issues.

Solar PV generates DC power and uses inverters to convert the DC to AC for synchronization with the grid [23]. Thus, traditionally, the systems run at unity power factor. Hence, with the solar PV, the grid supplies the balanced active and full reactive power requirements of the loads. Thus, the power factor at the main generating station or the system is reduced. Hence, other compensating ways or systems need to be installed.

1.4.3 Flexible alternating current transmission system

It is evident from an earlier section that to achieve stable operation of the system, the system should deliver the active power as well as the reactive power to the fluctuating loads instantaneously. Generally, passive compensators with energy storage components, inductors and/or capacitors connected in series or parallel, are used to control reactive power requirements. Though these are simple for implementation, they are specific to the load. To cope with the changing load-generation scenario, a more adaptive compensation is required. This is achieved by using power electronic devices and power system control known as Flexible AC Transmission System (FACTS) [24]. This technological revolution has opened opportunities for controlling and enhancing the power transfer capability, also on existing transmission systems.

FACTS is, in fact, a power electronic-based system that controls one or more AC system parameters to improve controllability and to change the power transfer capability of the line. FACTS technology was incorporated in the transmission system by replacing mechanical switches with thyristors, which reduces switching losses and helps in faster operation and control. However, static var compensation (SVC) with thyristors injects harmonics into the power system. Further advancement of the technology replaces thyristors with converter-based controllers, which

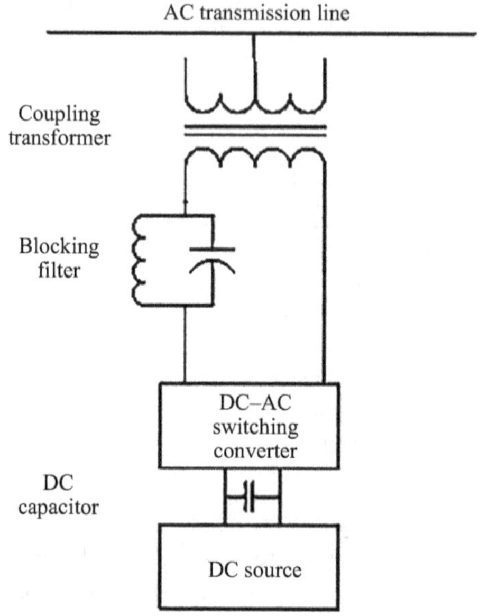

Figure 1.27 VSC-based shunt compensation

eliminate the drawback of thyristor-controlled FACTS systems. One such system is a Static Synchronous Compensator (STATCOM). It is a voltage source converter (VSC)-based system, with the voltage source behind a reactor as shown in Figure 1.27. A DC capacitor acts as the source of voltage, and therefore, a STATCOM has very little active power capability. The reactive power supplied by the STATCOM depends on the amplitude of the voltage at its terminal. For example, if the terminal voltage of the VSC is higher than the grid voltage at the point of connection, the STATCOM generates reactive current; on the other hand, when the amplitude of the terminal voltage is lower than the grid voltage, it absorbs reactive power.

For dynamic compensation, STATCOM is preferred over SVCs due to its faster response, while the mechanically switched reactors or capacitors would be preferred for reactive compensation under steady state. With increased penetration of renewable energy sources, the use of STATCOMs is expected to provide dynamic reactive power support and reliable power evacuation. Table 1.7 shows a few of the key STATCOM installations in India.

More on thyristor-based SVC systems and converter-based VAR compensation techniques are detailed in Chapters 6 and 7, respectively.

1.4.4 High-voltage direct current (HVDC) transmission

The first power station at Pearl Street in Lower Manhattan, for providing light and power to customers, built by Thomas Alva Edison in the late nineteenth century, was a DC system. However, because DC could not be readily transformed to higher

Table 1.7 Key STATCOM installations in India

Location	Dynamic compensation (STATCOM) MVAR	Mechanically switched compensation (MVAR)	
		Reactor	**Capacitor**
Hyderabad	±200	2 × 125	1 × 125
Udumalpet	±200	2 × 125	1 × 125
Trichy	±200	2 × 125	1 × 125
Rourkela (Siemens Press release 2018)	600	2 × 125	
Fatehgarh-& Bhadla-II Rajasthan [Solar energy zone]	±600	2 × 125	4 × 125
NP Kunta, Andhra Pradesh	±100	–	–

voltages for long-distance transmission, AC systems became the standard because it was possible to use transformers with AC systems. Later, the HVDC transmission system was developed to transmit huge power over long distances through overhead lines and cables, either submarine or underground, and for asynchronous interconnection between grids. In HVDC, at the sending end, power is converted from AC to direct current (DC) before being transmitted and is then converted back to AC at the receiving system. This conversion involves rapid switching, which generates harmonics that affect power quality and also interfere with telecommunications facilities. AC harmonic filters are needed, especially to eliminate such harmonics.

This system is used for transmitting electricity between two grids that are running at different frequencies, making the overall system more efficient and resilient to failures. DC transfers active power. Hence, losses are lower than with AC, which transfers reactive power also. Unlike long-distance AC transmission, the HVDC system has no stability constraint. Overhead HVDC transmission lines can usually be easily identified because they have two conductors per circuit, rather than three as in AC. Further, it requires less space (right of way) due to the reduction in the number of conductors and hence the cost of construction. HVDC technology is used to integrate energy generated by long-distance offshore wind turbines, connecting small electrical networks to the main grid.

The reactive power consumed by the transmission line, which is the root cause of various problems with high-voltage AC transmission, is eliminated by HVDC systems. This is because the reactive power consumed by the DC line is zero, as there is no inductive and capacitive reactance present in the DC line under steady-state conditions. Also, control of power transmission in HVDC systems can be done easily by controlling the firing angle of the thyristors used in power converters.

Figure 1.28 shows an HVDC transmission line connected between a 220 kV AC grid in an interconnected system with four generating stations connected to the grid. The HVDC system mainly consists of AC–DC and DC–AC converters at the sending end and receiving end, respectively, and a smoothing reactor and filter at the DC transmission line.

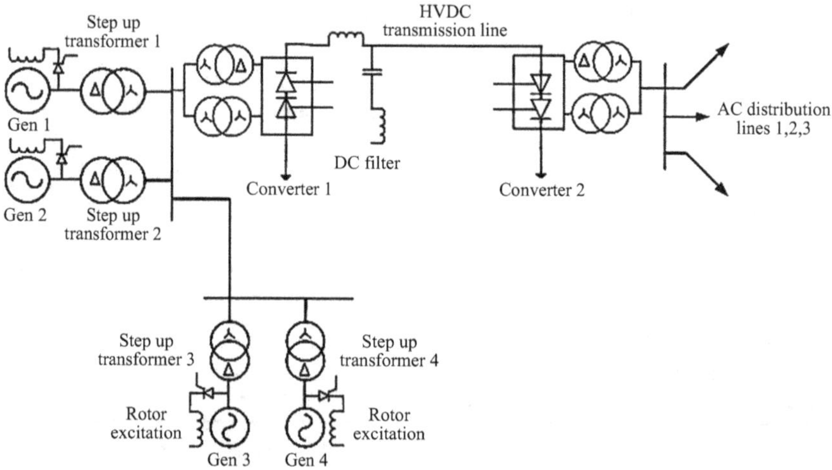

Figure 1.28 A typical HVDC transmission system

HVDC systems started with natural or line commutation for switching between converter devices using the zero crossing point of the AC waveform. The system started in 1945 with mercury arc rectifiers that used liquid mercury in a glass tube. With the invention of thyristors, HVDC systems moved to thyristor-based systems during the 1970s. A thyristor, being a gate-controlled switching device, needs an additional gate control circuit for generating a triggering current (electrically triggered thyristor – ETT) pulse on every half cycle of the AC waveform. With the light-triggered thyristor (LTT) introduced in 2000, the number of components in the system and hence losses reduced significantly. Thus, it improves system performance.

Converters using thyristors for HVDC transmission give proper control; however, they inject harmonics into the system. This issue can be mitigated using the latest technology, using VSC-based HVDC technology, where thyristors are replaced by semiconductors with turn-off-capable IGBTs. By controlling the switching operation of the device, active and reactive power transfer, as well as distortion power taken by the converters, can be controlled. Figure 1.29 shows waveform of gate pulse generation technique using PWM control.

The VSC performs conversion at very high frequencies, using pulse-width modulation. This gives a very high degree of control to maintain high power quality.

To provide an uninterruptible power supply, India has installed an extensive network of transmission systems and has commissioned a ±320 kV VSC-based HVDC system between Pugalur in Tamil Nadu and Thrissur in Kerala as a part of the 800 kV HVDC link between Western India and Southern India in June 2021. Transmission lines above 220 kV, along with their lengths in circuit km of line in India, are given in Table 1.8 [25].

Recently, residential and industrial consumers added more power electronic equipment to make their systems more efficient. However, it may result in certain

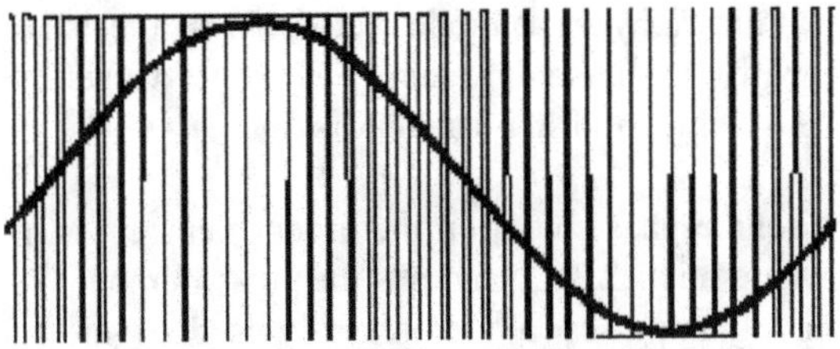

Figure 1.29 Pulse-width modulation (PWM) control used with VSC

Table 1.8 Transmission lines above 220 kV in India

Transmission line	**±800 kV DC**	**±500 kV DC**	**±320 kV DC**	**765 kV AC**	**400 kV AC**	**220 kV AC**
Length in km	9655	9432	288	54,797	203,890	207,661

power quality issues. A few examples of such loads are explained in the next section.

1.4.5 Development in utilization sector

In recent times, we find increased use of electronic instruments in every area of life. In domestic applications, microwave ovens, washing machines, dishwashers, air fryers, light-emitting diode (LED) lighting, smart TVs, DVD players, personal computers, rooftop solar PV generation units, etc., are increasingly used more than ever before. In the industrial sector, we use rectifiers, switched-mode power supply (SMPS), uninterruptible power supplies (UPS), variable frequency drives (VFD), video display terminals (VDT), and electric vehicle battery charging. In the commercial sector, we find wide application of personal computers (PCs), fax machines, copiers and various electronic communications.

It is estimated that the majority of electrical energy is processed by electronic circuits and power converters. This percentage is growing at a faster rate than before. As discussed earlier, unlike linear loads, the above non-linear loads draw a current from the system that is not proportional to the voltage applied to it. In other words, it draws a non-sinusoidal or distorted current from the system. The basic Ohm's law cannot be applied to find out the current in such cases. The impedance of the non-linear load is not constant with respect to supply, as in the case of a linear load. Its *V–I* characteristic will be non-linear. A few such loads are explained in the next section.

1.4.5.1 AC–DC power converters

In general, all electronic devices or equipment, be smartphones, computers, televisions, refrigerators, air conditioners, battery chargers, emergency power supplies, etc., require a DC power supply. Rectifiers are used to convert AC to DC. A typical waveform of current drawn by a 1-ϕ diode rectifier with a resistive load and a filter capacitor, along with a sinusoidal supply voltage waveform, is shown in Figure 1.30. As the value of filter capacitors increases, the current waveform becomes peakier.

In the case of a three-phase rectifier, the order of current harmonics injected is given by $np \pm 1$, where n is 1,2,3 ... and p is the pulse number. To limit the current harmonic injection, higher pulses (n) are generally used. Figure 1.31 shows a block diagram of a 3-phase, 12-pulse rectifier where the order of the harmonic is $h = 12n \pm 1$, and the magnitude of each order of harmonic component of current (I_h) is $I_h = \frac{I_1}{h}$, where I_1 is the fundamental component of current.

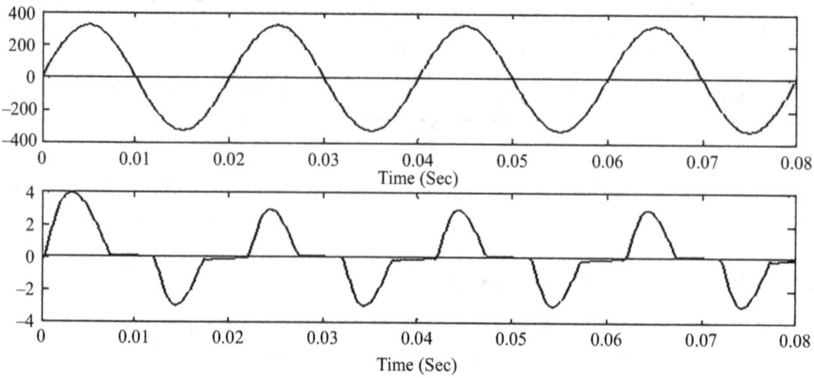

Figure 1.30 Current drawn by a 1-ϕ diode rectifier with a resistive load and filter capacitor along with a sinusoidal supply

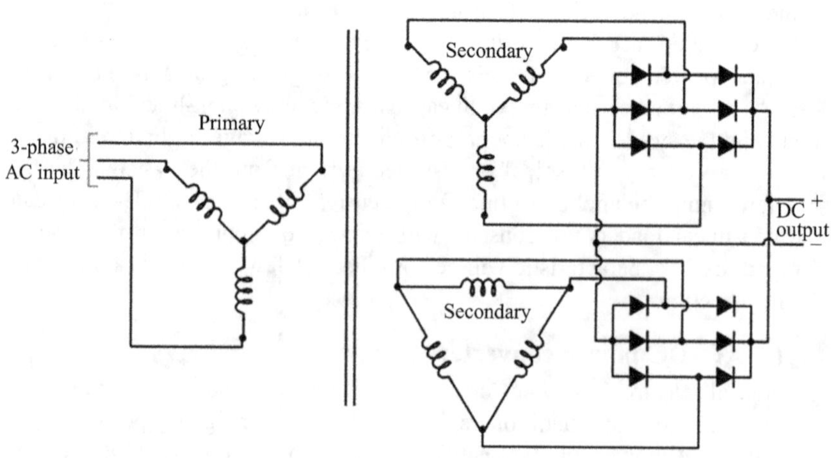

Figure 1.31 Block diagram of 3-phase, 12-pulse rectifier

1.4.5.2 Uninterruptible power supply

It is common practice to supply critical loads such as banking, transportation and information technology computers; SCADA and telecommunication systems; medical equipment, etc., with uninterruptible power supplies. Figure 1.32 shows a block diagram of a double conversion UPS. The 12-pulse rectifier circuit at the input converts AC to DC and supplies DC power to a battery bank to keep it charged and then to an inverter. The inverter converts DC back to AC. Under normal conditions, the rectifier feeds the inverter directly from the AC supply. In case of loss of input AC supply, the batteries feed the inverter. In this way, UPS isolates utility power quality issues, and there is no interruption of the power flow to the load when a power outage takes place. The waveform of current drawn by the UPS at the input sinusoidal supply is like that of AC–DC power converters, as the first input device of the UPS is a rectifier.

1.4.5.3 Soft starters

Soft starters reduce motor voltages during start-up and increase them to match the start-up requirement of the motor load. Limiting the voltage supplied to the motor during start-up limits the starting or inrush currents and torque. Thus, it provides a gentle acceleration up to full speed. The reduced inrush currents improve the system voltage profile and avoid other problems associated with the starting/running of other critical motors on relatively small sources, especially on UPS supply. A single-line diagram of the power circuit of the soft starter is shown in Figure 1.33. As can be seen from the figure, the power circuit has thyristors, and the motor voltage is controlled by the gate control circuit.

1.4.5.4 Lighting loads

An LED emits visible light when current passes through it. It is very compact and offers high luminous efficacy accompanied by energy efficiency. In recent times, it has been widely accepted as a replacement for incandescent and fluorescent lighting fixtures. It is also widely used in traffic lights and decorative lighting. As indicated, an LED requires an appropriate DC current. Hence, to operate the LED luminaire from a low-voltage AC supply, a driver circuit is required to convert the low-voltage AC into DC , matching with the luminaire. The driver circuit in its simplest form is a single-phase AC–DC converter with a capacitor on the DC side.

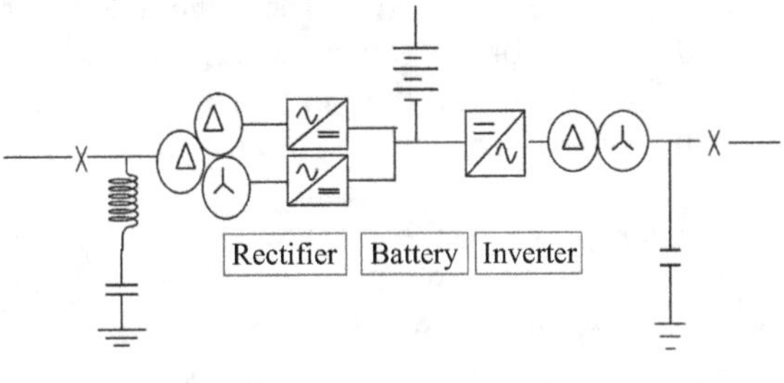

Figure 1.32 Block diagram of a double conversion UPS

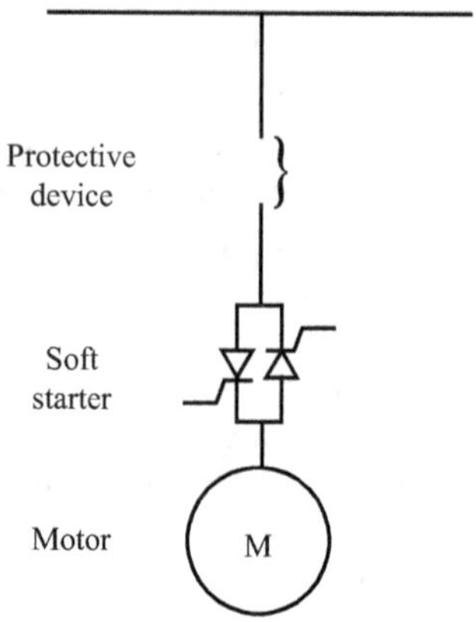

Figure 1.33 Single-line diagram of a soft starter

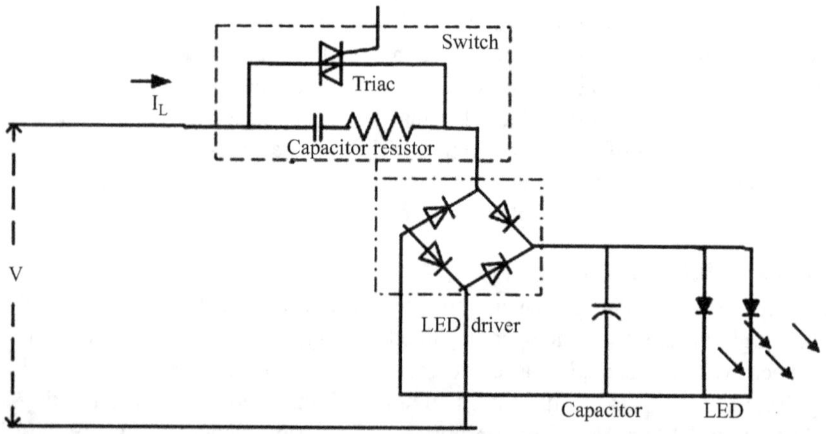

Figure 1.34 Lighting LED circuit diagram

Hence, it draws high inrush current due to the driver capacitor during starting, which later stabilizes to the LED load current. The magnitude of inrush current also varies depending on the instant of the voltage waveform. This non-linear single-phase load primarily injects significant third-order harmonics that are added together in the neutral of the system. Though the current drawn by an individual LED fixture is less, lighting contributes a significant portion of the distribution system load. This creates a major power quality issue apart from the reduction in power factor. Thus, it may overload the neutral conductor and lighting or distribution transformers in the connected circuit. A detailed analysis is done in Chapter 3 on harmonics. Loads of Figure 1.34 indicate their components, and Figure 1.35 shows the resulting current waveform.

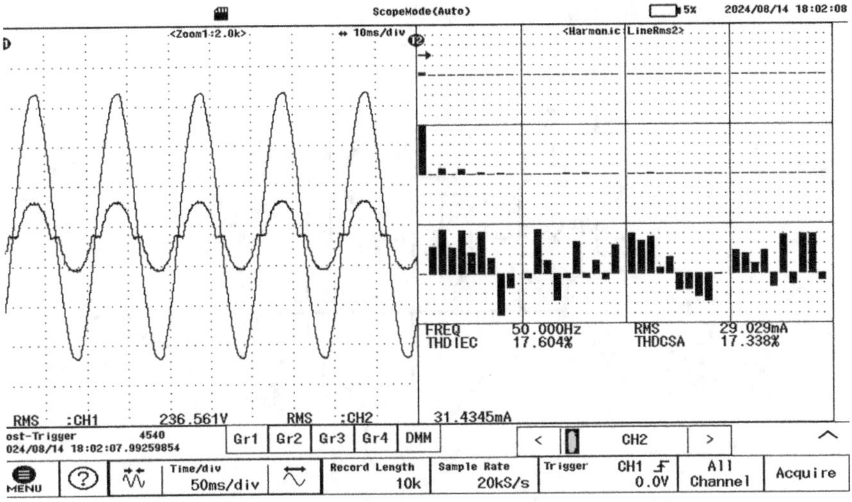

Figure 1.35 LED current waveform

1.4.5.5 Electric motor drives

Electric motors form the major industrial load. With the advancement of power electronics, motor drives are increasingly used in industry due to their higher efficiency, better performance and higher productivity. In DC motor applications, these drives are used for starting, stopping, speed control and reversing the rotation of DC motors. In the case of common AC applications, such as fans or pumps, motor shaft power (p) is proportional to the cube of its speed (N). Since the power of a motor is proportional to its torque (T) and speed (N), torque will be proportional to the square of speed. So, reducing the speed from 100% to 80% would reduce the required torque to 64% and power to nearly 50%. Nowadays, VFDs are increasingly used for adjusting the V/Hz pattern of the motor according to its load requirement. Figure 1.36 shows a schematic diagram of the inverter side of the drive. If the V/Hz ratio is too low, the motor torque may not be sufficient, and the motor may experience a pull-out condition. If the ratio is too high, although the achievable torque of the motor is increased, it results in saturation of the stator core and hence higher magnetizing current and damage to the motor. To avoid motor pull-out, the V/Hz ratio must be high enough to provide enough torque for the application. Depending on the type of drive, microprocessor-based control adjusts the firing angle of the switch for the desired output voltage form. These power electronics drives distort the sinusoidal nature of the power supply waveform and may create fast rise time surges per second. When the rise time is very fast, the motor insulation becomes overstressed. Excessively high dv/dt can cause premature breakdown of standard motor insulation.

However, during a power failure, VFDs disconnect the motors from the power supply and thus prevent access to the inertia stored in the motors that otherwise will help in slowing down falling grid frequency.

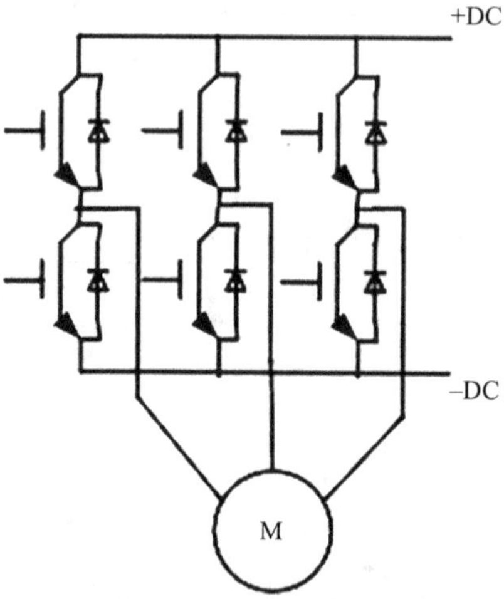

Figure 1.36 Schematic diagram of the output side of a VFD

1.4.5.6 Electric vehicle charging station

Increasing demands for energy and the finite nature of fossil fuels, accompanied by global warming and climate change, force countries to replace conventional vehicles with electric vehicles (EV). This necessitates the development and deployment of the battery charging station for transportation. These high charging loads of fast EV charging stations adversely affect the operating parameters of the distribution network. The electrical charging station is a non-linear load to the grid and hence affects the voltage profile, increases the peak load, and can cause harmonic distortion, DC offset, phase imbalance and voltage deviations in the distribution network.

1.5 Recent developments in the power system

Typical electricity supply chain showing generation, transmission and utilization of electricity is shown in Table 1.9. Over the past decade, electronic systems have been integrated into all areas of power systems, starting with the generation, transmission, distribution and utilization of electrical energy. Thus, conventional systems are increasingly replaced by electronic switching devices with more information technology (IT) and automation systems [26]. Though the switching devices help in increasing the production process, they cause current interruption due to their on-off operation under steady-state conditions. Thus, the shape of the current drawn by modern power electronic equipment deviates from its traditional sinusoidal current. If we analyse the current drawn by such equipment, it is seen that the current consists of several harmonic components that are multiples of the fundamental frequency of the system. The amplitude and phase angle of a harmonic component are dependent on

Table 1.9 Typical electricity supply chain

Generation (source of energy)		Transmission	Utilization
Conventional sources/ Non-Renewable (generally, up to 21 kV)	Non-conventional energy/ renewables	AC transmission (69–765 kV)	Residential 120/240 V (60 Hz) 230/400 V (50 Hz)
Coal	Wind energy	HVDC	Agricultural
Natural gas,	Solar thermal		Hospitals EV Charging Commercial
Petroleum liquids	Solar photo-voltaic (SPV)		Industrial (>230/400 V)
	Wave and tidal energy Salinity gradient energy Ocean thermal energy conversion		Traction systems DC: 750/1500/ 3000 V AC: 2.5 kV
Nuclear	Bioenergy biomass such as wood, animal waste and traditional charcoal Biofuels from bagasse and other plants		
Hydro-electric	Micro hydel ≤ 0.1 MW Mini hydel >0.10 MW to ≤ 2.00 MW Small hydel >2.00 MW to ≤ 25.00 MW Geothermal energy		

the circuit, switching frequency and the load it drives. For example, a six-pulse converter using thyristors will inject 5,7,9,11 ... order of harmonics into the system, while the order of harmonics present in a 12-pulse converter is 11, 13, 15, etc., as discussed earlier.

A typical modern power system showing distributed generation, HVDC and HVAC transmission, FACTS devices, EV charging stations, etc., is shown in Figure 1.37. Microgrids, smart grids, numerical relays for protection, PMU [27], real-time assessment of transmission line capability, data communication, etc., are a few of the emerging fields in modern power systems, which are explained in the next section.

1.5.1 Numerical relays in protection

Protection systems are an extremely important part of any power system. Their primary function is to detect inadvertent electrical connections – that is, short circuits between different voltages in the system. When faults occur, very large currents, typically 10–20 times normal load currents. Since power is proportional to the square

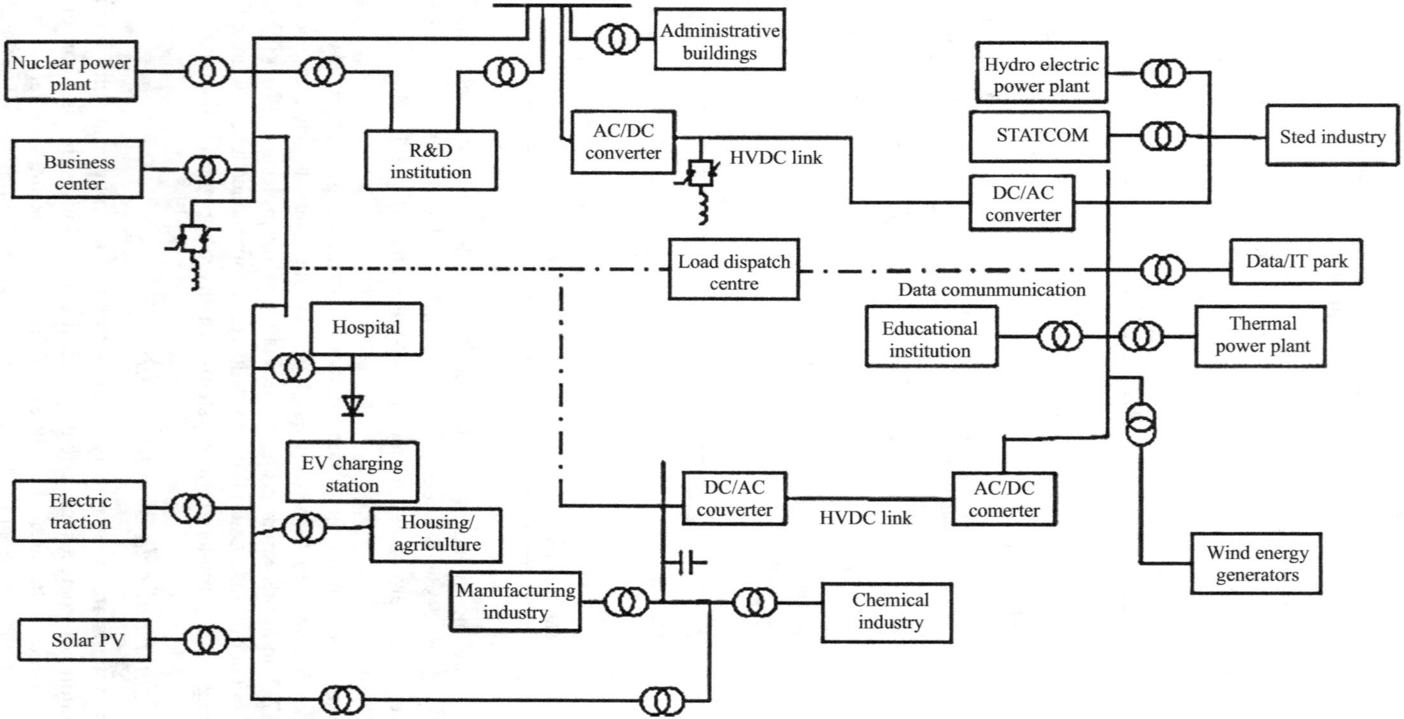

Figure 1.37 A typical modern power system

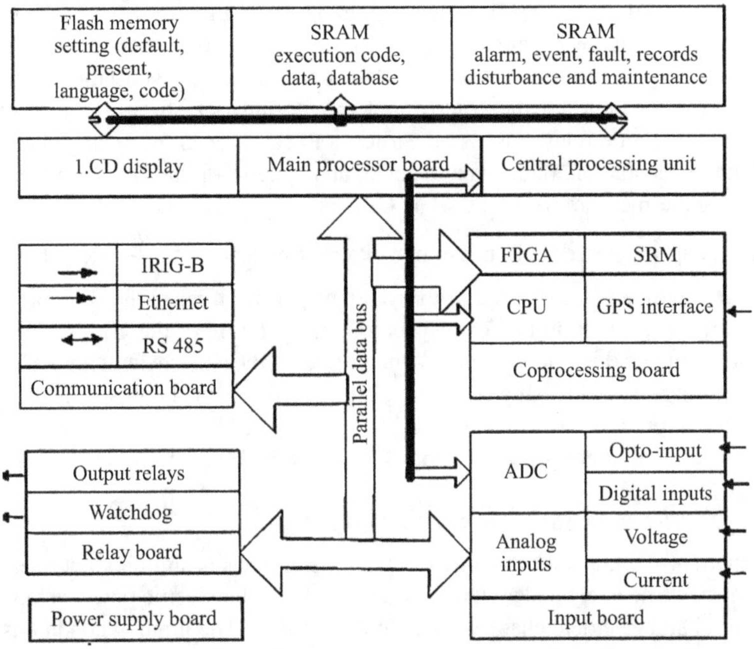

Figure 1.38 Modern numerical relays (intelligent electronic devices)

of the fault current, a large quantity of energy can be delivered to an unintended path in a very short time, which can damage the equipment or system. The goal of protection systems is to isolate and de-energize faults before they can harm or cause serious damage. Protective relays are used to detect these abnormal conditions in a very short time, typically in milliseconds, and operate circuit breakers to isolate the malfunctioning system components, preventing damage to the system components. Originally, protective relays were electromechanical; each one performed a specific task, and thus installation required several different types of relays. Many relays nowadays use solid-state electronic components controlled by microprocessors. A major advantage of the microprocessor-based design is that the functions of many discrete relays can be incorporated in a single relay. These relays, often called intelligent electronic devices (IED), are communicating with other relays at faraway places in real time. Figure 1.38 shows the modern numerical relays (DSP-based).

These relays are affected by power quality issues and could trip the feeder or lines, causing loss of supply to many consumers. Hence, it needs to be qualified for the PQ environment.

1.5.2 Phasor measurement unit

The transition from an isolated grid to a national or international grid was aided by the development of protections, control and monitoring devices and systems. The invention of microprocessor-based relays, or IED, as an alternative to electromechanical relays; open-source technologies as an alternative to proprietary software; the adaptation of interoperability as per IEC 61850; and optical fibre ground wire (OPGW) for communication and data transfer among substations and their control centres simplified the control and monitoring operation of the grid. Like numerical protection relays discussed earlier, a PMU collects the analogue electrical signal at the installation, digitizes it using an analogue-to-digital (A/D) converter and passes it to the measurement CPU. The CPU processes this digitized signal into the corresponding magnitude and phase of voltage, frequency (f), $\frac{df}{dt}$, etc. These parameters are stamped in real time with a GPS signal to a central synchrophasor for taking corrective action to prevent asynchronous operation of the grid.

The Indian grid is implementing synchrophasor measurement using PMUs [28] for monitoring the electrical grid on a real-time basis at substations of 220 kV with generating stations, substations of 400 kV and above, HVDC stations and pooling points of renewable energy generating stations of 50 MW and more [29].

1.5.3 Real-time assessment of line capacity

Assessment of the transfer capability of transmission lines is an important operational requirement of the integrated system. The ability to transfer electric power is limited by the physical and electrical characteristics of the systems. The limiting conditions, such as the temperature limitation of line conductors, voltage limits and stability at some portions of the transmission network, can shift as the network operating conditions change over time. Accordingly, transfer capability is to be worked out in real time so that neither the security of the grid nor the economy of interchange is compromised. Calculations must also consider the network topology, planning criteria, operating philosophy, operating standards, generation (including that from renewable sources), distribution of load, behaviour of the utilities, meteorological information, etc.

1.5.4 Data communication

Communication is the process of conveying information from a transmitter to a receiver through a channel or a link connected between them. Historically, dedicated pilot wires have been used in power systems, followed by power line carrier communications (PLCC). In recent times, fibre-optic systems have been the choice due to their availability and immunity from electromagnetic interference. Also, in the area of instruments for sensing, measuring, computation, etc., similar developments have taken place. In protection, the traditional electromechanical relays are increasingly replaced by numerical relays employing digital signal processors (DSPs). Thus, with the advancement of communication and computing power, operation and control of the power system today is automated by a set of interconnected IED, PMUs for advanced inter-regional instability monitoring [27], under-voltage/frequency load shedding (UVLS/ULS) and islanding schemes, weather monitoring schemes, automatic generation control (AGC), etc. In other words, the power grid is evolving into a big Internet of Things (IoT) in a Wide Area Management System (WAMS), as shown in Figure 1.39.

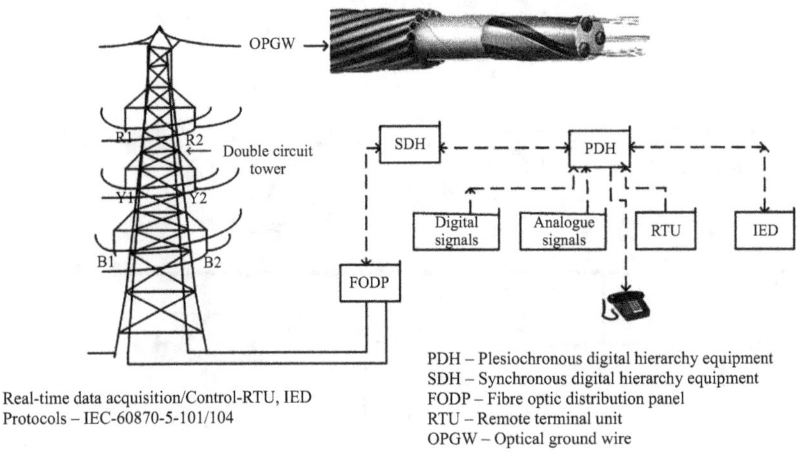

Figure 1.39 Typical data communication in power systems

1.5.5 Microgrid

A microgrid [30] is a localized group of interconnected loads and distributed energy resources of renewable and non-renewable types within a defined electrical boundary, like cities, communities, etc., that act as a single controllable unit with respect to the main electric grid. It can be operated in grid-connected synchronous mode or isolated from the main grid (islanded) mode as per the requirement. It is established primarily for local access to electricity at remote places (social needs), operational economics and reliability and better control. A forecast issued by Navigant Research identifies "several new trends emerging in the global microgrid market as it moves toward becoming a $30.9 billion industry by 2027".[4]

A block diagram of a typical microgrid with AC/DC converters, communication modules and links, and a grid management system is shown in Figure 1.40.

1.5.6 Smart grid

A smart grid, in simple terms, is an electricity distribution network that combines a bi-directional power flow network integrated with a bi-directional information flow (communication) in such a manner as to facilitate various optimization and control features. The vision of India on smart grids is to "Transform the Indian power sector into a secure, adaptive, sustainable and digitally enabled ecosystem that provides reliable and quality energy for all with active participation of stakeholders". India has a few pilot smart grid projects at various locations such as: Panipat and Manesar in Haryana, Kala Amb in Himachal Pradesh, Siliguri in West Bengal, Agartala in Tripura, Guwahati in Assam, Naroda in Gujarat, Mysore in Karnataka, IIT Kanpur, etc. Also, projects at Chandīgarh and Rajasthan are under

[4]https://www.microgridknowledge.com/editors-choice/article/11430269

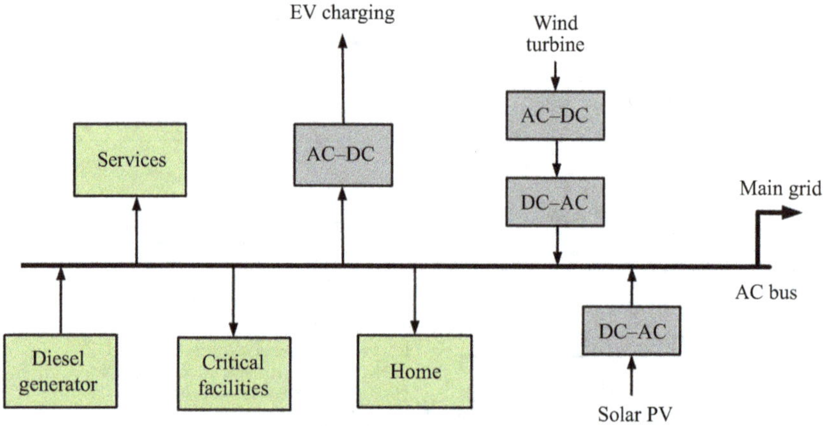

Figure 1.40 Power electronics interfaces in a microgrid

implementation.[5] Smart grid technology involves the implementation of smart meters that enable peak load management, development of DG, smart grid management of power quality, etc. [31].

1.6 Internet of Things

Given the extensive communication and controls, more and more devices are being upgraded to be able to communicate intelligently over the Internet and thereby form a so-called "Internet-of-Things" (IoT), as shown in Figure 1.41. Thus, there is vulnerability to cybersecurity threats wherein a virus or any intrusion with malicious intent or compromise onto the network may render the grid operation insecure, thereby affecting the entire system or even causing a cascade tripping of the units or a complete blackout.

1.7 Challenges in power system

Recent development in power systems, which involves more power electronic devices in generation, transmission and utilization, coupled with microprocessor-based devices for protection and control. This results in a few challenges in the modern power system. Equipment containing former devices in general causes power supply disturbances that affect the latter.

1.7.1 Power quality

The term "power quality" describes the preservation of distribution bus voltage and current waveforms that are almost sinusoidal at their rated magnitude and frequency.

[5]https://www.nsgm.gov.in/en/sg-projects

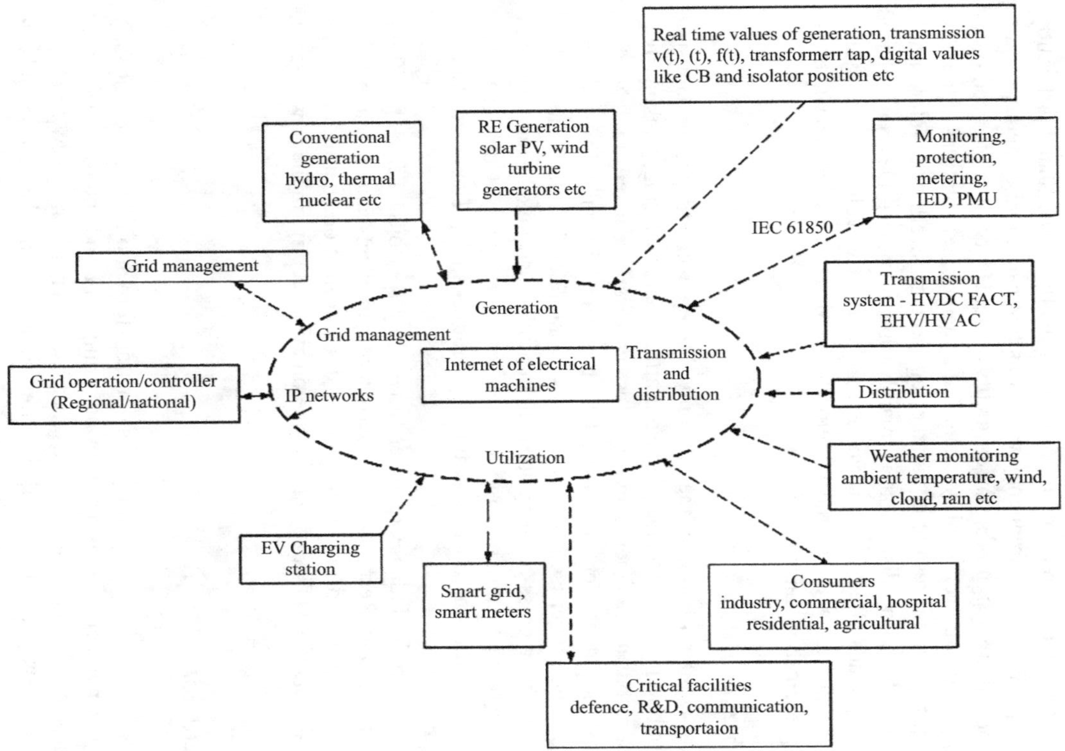

Figure 1.41 Internet of Things (IoT) in power system

But, as we have seen from the preceding sections, there is a widespread application of power electronics in power systems such as AC–DC converters, which are front-end equipment in almost all the electronic equipment that draws non-sinusoidal current from the system. Such non-linear loads are increasingly added from generation to utilization in every area of energy transformation. These non-linear devices affect the power quality and performance of various equipment in the system. This issue is referred to as a power quality issue. Stated differently, abnormal voltage, current, or frequency variations that cause end-use equipment to malfunction or fail are a sign of a power quality issue. Hence, it is the need of the hour to analyse the power quality issues and determine solutions for them. Disturbance in the system results in voltage sag, voltage swell, brief interruption, unbalance, etc. However, distortion is mainly due to power electronics-based equipment. Such loads affect both the voltage and current waveform of the system and, hence, the power factor of the system. A detailed discussion on waveform distortion is presented in Chapter 2.

1.7.2 Cybersecurity

Electric power is the backbone of the economy. It empowers the people. Housing, agriculture, public health, education, transportation, manufacturing, banking, telecommunication, business, emergency services, etc., represent only a few of its critical downstream dependent activities. For its enhanced availability and economy, the grid is no longer isolated; rather, it is synchronously interconnected and even connected with a cross-border grid, with multiple, distributed energy sources, including fluctuating generation from wind and solar PV sources. Consequently, it is vulnerable to security threats or so-called cyberattacks [16]. Hence, the need to provide adequate security.

India has set up the Indian Computer Emergency Response Team (CERT-In) for Early Warning and Response to cybersecurity incidents and to have collaboration at the national and international level for information sharing on mitigation of cyber threats [25,31]. Also, a guideline on cybersecurity in the power sector is in place.

1.8 Conclusion

The electric power grid is undergoing a major transformation. Over the last forty years or so, the number of power electronic equipment and microprocessors or DSP-based protection-control systems has increased dramatically. New and smart technologies are being added to the generation, transmission, distribution and utilization components of electricity. Though these intelligent automation technologies give many benefits to power systems, they can also do harmful activities, such as the isolation of a circuit, leading to load shedding or even an entire blackout.

The technological changes that are taking place from traditional rotating machines to power electronic-based renewable energy systems, particularly solar photo-voltaic generation, transmission and utilization, from pure AC systems to hybrid AC/DC systems, are bringing out major challenges, notably on voltage and frequency stability, decreased system strength and increased distortion of the very fundamental nature of its sinusoidal waveform of the alternating current. Although the complexity is increasing, it is bringing new possibilities and avenues. Chapter 2 presents power quality in power supply, and Chapter 3 presents harmonics in power supply systems, their sources, effects and regulation. Chapter 4 details harmonics and reliability indices, and Chapter 5 is on power factor and THD under sinusoidal and non-sinusoidal conditions. Mitigating technologies are presented in Chapter 6 using passive components and Chapter 7 on emerging converter-based technology and active compensation.

1.9 Problems and solutions

Problem 1.1 Consider a generator rating of 12 kW; 120 V DC is feeding 30 numbers of 40 W, 120 V lamps, which are placed 10 km away from the generating station. Resistance of the transmission line is assumed to be 0.1 Ω per km. A single-line diagram of the system is given in Figure 1.42. Voltage variation allowed is 5% of the nominal voltage, and the limit for transmission loss is 10%.

Calculate:

1. Total power demand
2. Current through the transmission line
3. Voltage drops across the line
4. Voltage at the load centre or receiving end of the transmission line
5. Transmission line loss
6. Power delivered by the source
7. Check whether voltage variation and transmission loss are within the limit
8. How many lamps can be connected to the system for its safe operation?
9. What is the safe distance that the power can be transmitted to meet the same power demand?
10. If the limit is exceeded, suggest solutions

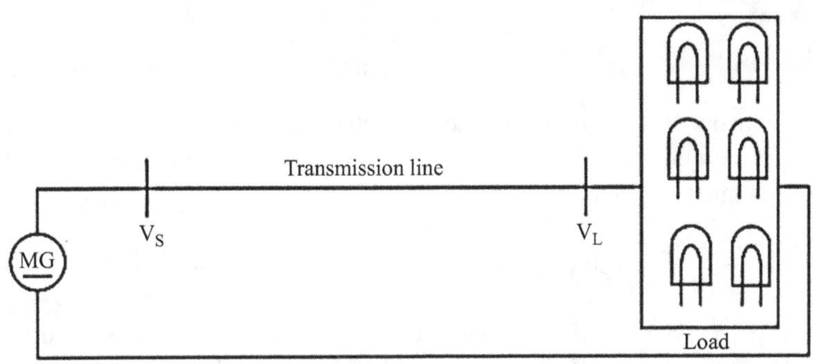

Figure 1.42 Single-line diagram of the system of problem 1.1

Solution:

1. Considering rated condition, power demand (P_D)

$$P_D = 40 \times 30 = 1200 \text{ W}$$

2. Current through the transmission line (assuming ideal situation $V_s = V_r$)

$$I = \frac{P}{V} = \frac{1200}{120} = 10 \text{ A}$$

3. Voltage drop across the line (V_d)

Total resistance of transmission line $(R) = 0.1 \times 10 = 1\ \Omega$

$$V_d = I \times R = 10 \times 1 = 10 \text{ V}$$

4. Voltage at the load centre or receiving end of the transmission line (V_r)

$$V_r = V_s - V_d = 120 - 10 = 110 \text{ V}$$

5. Transmission line loss (T_L)

$$T_L = I^2R = 10^2 \times 1 = 100 \text{ W}$$

6. Power delivered by the source (P_s)

$$P_s = P_D + P_L = 1300 \text{ W}$$

7. Check whether voltage variation and transmission loss are within the limit. Ideally, $V_r = V_s$

$$\text{Percentage variation of receiving end voltage} = \frac{V_s - V_r}{V_s} \times 100$$

$$= \frac{120 - 110}{120} \times 100 = 8.3\%$$

Hence, in the given system, the receiving-end voltage exceeds its limit of 5%.

$$\text{Percentage transmission loss} = \frac{T_L}{P_s} \times 100 = \frac{100}{1300} \times 100$$

$$= 7.692\%, < 10\%, \text{ hence with in limit}$$

8. How many lamps can be connected to the system for its safe operation?
   To maintain a voltage variation of 5%, V_r should be limited to satisfy the following condition.

$$\frac{V_s - V_r}{V_s} \times 100 = 5$$

So, V_r should be 114 V, and the current through the line should be 6 A. As current/lamp = 0.33 A, the number of lamps for meeting 5% voltage variation is 18 nos.

9. What is the safe distance that the power can be transmitted to meet the same power demand?

   As $V_r = 114$ V, $V_d = 6$ V, R should be 0.6 Ω, and the length of the line for safe operation of the system to meet the same power demand is 0.16 km.

10. If the limit is exceeded, suggest solutions

If the limit is exceeded, the solution may be (i) load shedding, (ii) reducing the distance of transmission, (iii) generating power at a higher voltage level, (iv) transmitting the voltage at a higher voltage level.

Problem 1.2 Consider the same load as in problem 1.1, and assume that a single-phase AC generator of 12 kW, 120 V is used to feed the load. A step-up transformer of 120/440 V is connected at the sending end, and a step-down transformer of 440/120 V is connected at the receiving end of the line. The resistance of the transmission line is the same as in problem 1.1, and the inductive reactance of the line is 0.02 Ω/km. A single-line diagram is shown in Figure 1.43.

Calculate:

1. Current through the transmission line
2. Voltage drops across the line
3. Voltage at the load centre or receiving end of the transmission line
4. Transmission loss
5. Actual power delivered by the source
6. Check whether voltage variation and transmission loss are within the limit

Solution:

1. Current through the transmission line

$$P_D = 1200 \text{ W}$$

$$\text{Load current,} \quad I_L = \frac{P}{V} = \frac{1200}{120} = 10 \text{ A}$$

Current through the transmission line is equal to the primary current flowing through the load-end transformer.

$$\text{Current through } 440/120 \text{ V transformer secondary, } I_s = I_L = 10 \text{ A}$$

$$\text{Current through } 440/120 \text{ V transformer primary, } I_P = \frac{V_2}{V_1} \times I_s = 2.72 \text{ A}$$

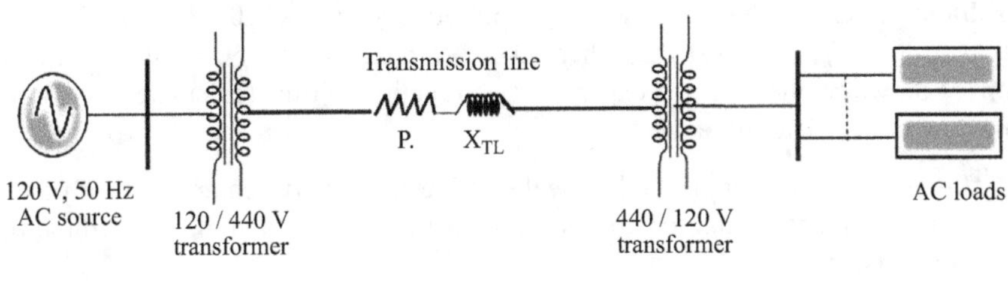

Figure 1.43 Single-line diagram of the system of problem 1.2

So, instead of 10 A in the case of the DC system, the transmission line current is 2.72 A because of the higher transmission voltage compared to the voltage in problem 1.1.

2. Voltage drops across the line

$$\text{Total impedance of transmission line } (Z) = (0.1 + J0.02) \times 10$$

$$= (1 + j0.2)\ \Omega$$

$$V_d = I_P \times Z = 2.72 \times 1.02 = 2.77\ \text{V}$$

Note that V_d is reduced to 2.77 from 10 V compared to problem 1.1

3. Voltage at the load or receiving end of the transmission line

Receiving end voltage at primary side of 440/120 V transformer, V_{rp}

$$= 440 - 2.77 = 437.23V$$

Receiving end voltage at secondary side of 440/120 V transformer, V_{rs}

$$= \frac{120}{440} \times 437.23 = 119.24\ \text{V}$$

4. Transmission line loss (T_L)

$$T_L = I^2R = 2.72^2 \times 1 = 7.4\ \text{W}$$

5. Power delivered by the source (P_s)

$$P_s = P_D + P_L = 1207.4\ \text{W}$$

6. Check whether voltage variation and transmission loss are within the limit
   Ideally, $V_r = V_s$

$$\text{Percentage variation of receiving end voltage} = \frac{V_s - V_r}{V_s} \times 100$$

$$= \frac{120 - 119.2}{120} \times 100 = 0.66\%$$

Transmission line loss = TL/Ps = 7.4/1207.4 = 0.613%
Hence within limit.

Problem 1.3 Given that a sinusoidal supply voltage $v(t) = 230\sqrt{2}\ \sin\omega t$ V is supplying a load $Z_L = 10 + j15\ \Omega$, where $\omega = 2\pi f$ rad/s and $f = 50$ Hz. Calculate the current drawn by the load and express it as a function of time (calculate $i(t)$). Based on $v(t)$ and $i(t)$, determine the following:

1. Instantaneous power $p(t)$ and show the active and reactive components of power.
2. The real, reactive and apparent power for this loading and also determine the power factor.
3. Repeat parts 1 and 2 for loads (a) $Z_L = 10 - j15\ \Omega$, (b) $Z_L = 10\ \Omega$, and (c) $Z_L = j15\ \Omega$ and comment on the nature of the current drawn by the load and power.

Solution:

Given data: $v(t) = 230\sqrt{2}\ \sin\omega t$ V, $\omega = 2\pi f$ rad/s and $f = 50$ Hz, $Z_L = 10 + j15\ \Omega$ (RL load)

1. Instantaneous power calculation:
   Phasor form of voltage:

   $$V = 230\angle 0^\circ\ \text{V}$$

   Load impedance:

   $$Z_L = 10 + j15\ \Omega = \sqrt{10^2 + 15^2}\angle\tan^{-1}\left(\frac{15}{10}\right) = 18.03\angle 56.31^\circ\ \Omega$$

   Phasor current:

   $$I = \frac{V}{Z_L} = \frac{230\angle 0^\circ}{18.03\angle 56.31^\circ} = 12.706\angle - 56.31^\circ\ \text{A}$$

   Thus, the time-domain current is:

   $$i(t) = 12.76\sqrt{2}\sin(\omega t - 56.31^\circ) = 18.04\sin(100\pi t - 56.31^\circ)\ \text{A}$$

   Instantaneous power $p(t)$:

   $$p(t) = v(t)\cdot i(t) = V_m \sin(\omega t)\cdot I_m \sin(\omega t - \phi)$$

   where:

   $$V_m = 230\sqrt{2}$$

   $$I_m = 12.76\sqrt{2} = 18.03$$

   $$\phi = 56.31^\circ$$

   Using the identity:

   $$\sin A \sin B = \frac{1}{2}[\cos(A - B) - \cos(A + B)]$$

   $$p(t) = V_m I_m \cdot \frac{1}{2}[\cos(\phi) - \cos(2\omega t - \phi)]$$

   $$p(t) = VI\cdot\cos(\phi)\cdot 1 - \cos(2\omega t) - VI\cdot\sin(\phi)\cdot\sin(2\omega t)$$

   $$p(t) = 230 \times 12.76 \times \cos(56.31^\circ)\cdot\{1 - \cos(2\omega t)\} - 230 \times 12.76 \times \sin(56.31^\circ)\cdot\sin(2\omega t)$$

   $$p(t) = \underbrace{1627.31\{1 - \cos(2\omega t)\}}_{\text{Active component}} - \underbrace{2441.90\sin(2\omega t)}_{\text{Reactive component}}\ \text{W}$$

2. Power calculations:
   Real power (P):

   $$P = V_{rms} I_{rms}\cdot\cos(\phi) = 230 \times 12.76\cdot\cos(56.31^\circ) = 1627.93\ \text{W}$$

Reactive power (Q):

$$Q = V_{rms}I_{rms} \cdot \sin(\phi) = 230 \times 12.76 \cdot \sin(56.31°) = 2441.90 \text{ VAR}$$

Apparent power (S):

$$S = P = V_{rms}I_{rms} = 230 \times 12.76 = 2934.8 \text{ VA}$$

Power factor (PF):

$$PF = \cos(\phi) = \cos(56.31°) = 0.555 \text{ (lagging)}$$

3. For different loads:

(a) $Z_L = 10 - j15\ \Omega$

The load given in this case will have the same magnitude, but the phase angle will now be $-56.31°$, meaning the current is now leading.

$$i(t) = 12.76\sqrt{2}\sin(\omega t + 56.31°) = 18.03 \sin(100\pi t + 56.31°) \text{ A}$$

P, Q, S and PF are same as the above, however PF will be leading.

(b) This load is purely resistive, which means that the reactive component of power will be zero, and voltage and current waveforms will be in phase.

$$I = \frac{V}{Z_L} = \frac{230\angle 0°}{10} = 23\angle 0° \text{ A}$$

$$i(t) = 32.53 \sin(100\pi t) \text{ A}$$

$$p(t) = 5290\{1 - \cos(2\omega t)\} \text{ W}$$

$$P = 5290 \text{ W}$$

$$Q = 0 \text{ VAR}$$

$$S = 5290 \text{ VA}$$

$$PF = 1(\text{in} - \text{phase/UPF})$$

(c) $Z_L = j15\ \Omega$

This load is purely inductive, which means that the active component of power will be zero, and the current will lag the voltage waveform by 90°.

$$I = \frac{V}{Z_L} = \frac{230\angle 0°}{15\angle 90°} = 15.33\angle -90° \text{ A}$$

$$i(t) = 21.68 \sin(100\pi t - 90°) \text{ A}$$

$$p(t) = -3525.9\{\sin(2\omega t)\} \text{ W}$$

$$P = 0 \text{ W}$$

$$Q = 3525.9 \text{ VAR}$$

$$S = 3525.9 \text{ VA}$$

$$PF = 0 \text{ (Purely reactive)}$$

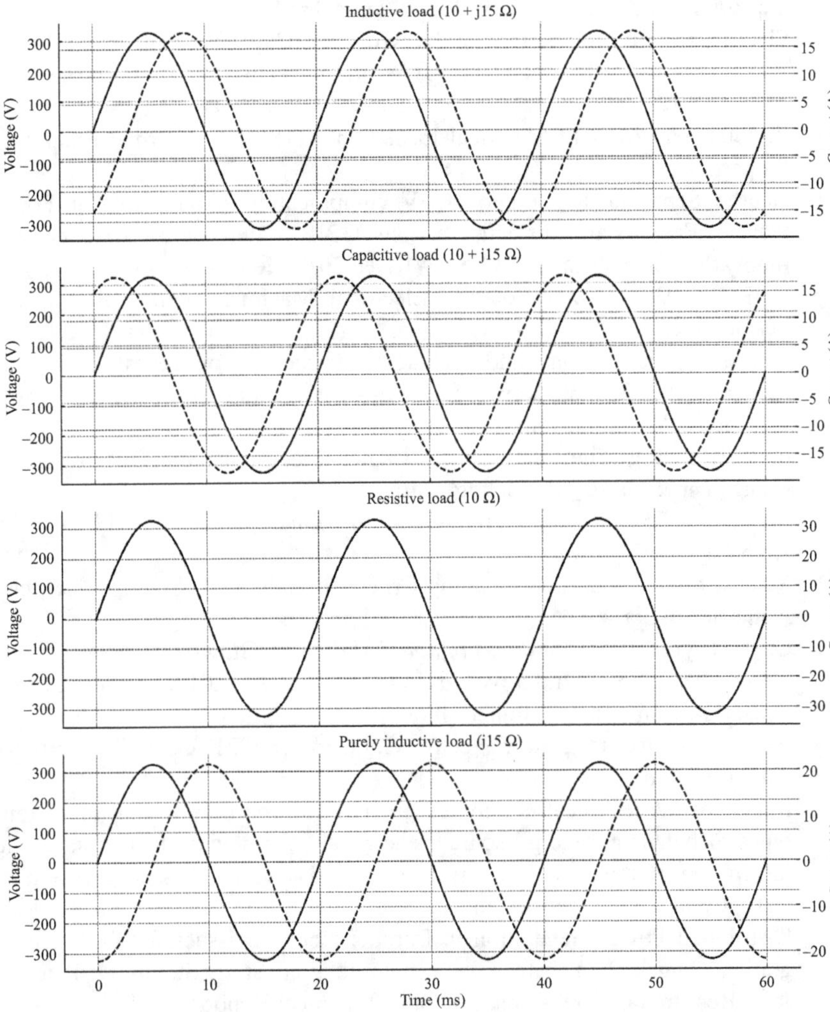

Figure 1.44 Voltage and current waveform for problem 1.3

The relationship between the voltage and current waveforms for different loading conditions is shown in Figure 1.44.

References

[1] https://www.britannica.com/biography/Alessandro-Volta
[2] https://en.wikipedia.org/wiki/Faraday%27s_law_of_induction
[3] https://ethw.org/Milestones:Pearl_Street_Station,_1882
[4] https://ethw.org/Milestones:Adams_Hydroelectric_Generating_Plant,_1895

[5] https://edisontechcenter.org/WilliamStanley.html

[6] Electric Power distribution in the World: Today and Tomorrow: EPRG Working Paper 1826, Cambridge Working Paper in Economics 1846, August 2018.

[7] IS-8504 (part 6) 2012 Electrical Insulation Materials – Thermal Endurance Properties.

[8] George Sebastian and others, EHV equipment in Indian Nuclear Power Plants – Failures and corrective actions, CBIP conference on Advances and Innovations in Substation, 30 Nov–1 Dec 2017 New Delhi, India.

[9] IS/IEC 60034-1, 2004 Rotating Electrical Machines, Part 1 Rating and Performance.

[10] Standard specifications and technical parameters for Transformers and Reactors (66 kV & above Voltage class), India April 2021 Central Electricity Authority, India.

[11] https://ethw.org/Milestones:Long_Distance_Electric_Power_Transmission_Using_Three-Phase_Alternating_Current,_1891

[12] Manual on Transmission Planning Criteria January 2013, Central Electricity Authority, India.

[13] Manual on Transmission Planning Criteria March 2023, Central Electricity Authority, India.

[14] C. W. Taylor, *Power System Voltage Stability*, McGraw-Hill, 1994.

[15] Synchro phasors - Initiative in India, December 2013: Power System Operation Corporation Limited, India.

[16] Cyber Security in Power Sector Guidelines, 2021 Central Electricity Authority, India.

[17] Report of the technical committee on study of optimal location of various types of balancing energy sources/energy storage devices to facilitate grid integration of renewable energy sources and associated issues, December 2017, Central Electricity Authority, India.

[18] Distributed Power Generation in Europe: technical issues for further integration, Angelo L'Abbate, Gianluca Fulli, Fred Starr, Stathis D. Peteves. Joint Research Centre – Scientific and Technical Reports.

[19] Renewable Energy Certificate Mechanism – Key learnings, Data analysis and way forward Report dated 31-7-2018 – Power System Operation Corporation Limited, India.

[20] Report on Assessment of Inertia in Indian Power System JANUARY 2022 – Power System Operation Corporation Limited in collaboration with Indian Institute of Technology Bombay.

[21] National Electricity Plan, Volume I, Generation, January 2018, Central Electricity Authority, India.

[22] Concept paper on the assessment of solar PV (SPV) system for power quality and safety issues 2018, Central Electricity Authority, India.

[23] Technical Standards for Connectivity to the Grid Regulations, 2019 Central Electricity Authority, India.

[24] Narain G. Hingorani, and Laszlo Gyugyi, Understanding FACTS Concepts and Technology of Flexible AC Transmission Systems, IEEE Press Standard Publishers Distributors; first edition, 1999.
[25] Central Electricity Authority, India, Executive summery on power sector April 2024 (https://cea.nic.in/wpcontent/uploads/executive/2024/04/Executive_Summary_April_2024.pdf)
[26] Guidelines on introduction of automation in distribution sector, 2018, Central Electricity Authority India.
[27] Synchro phasor Applications for Wide Area Monitoring and Control, ISGAN Discussion Paper, Philip Overholt, U.S. Department of Energy, Kjetil Uhlen, Norwegian University of Science and Technology, Brian Marchionini and Olivia Valentine, Energetics Incorporated.
[28] Unified Real Time Dynamic State Measurement (URTDSM) – A Report 2012, Power Grid Corporation of India Ltd.
[29] Central Electricity Authority – (Technical Standards for Construction of Electrical Plants and Electric Lines) Regulations 2019.
[30] https://microgridknowledge.com/
[31] India Country Report: Research Development Demonstration and Deployment of Smart Grid in India, Government of India, Ministry of science and Technology New Delhi, June 2017 https://dst.gov.in/sites/default/files/India%20Country%20Report%20on%20Smart%20Grids.pdf
[32] https://powermin.gov.in/en/content/international-cooperation-electricity
[33] Hassan K. Al-Asaad and Ahmed A. Ebrahim, The GCC Power Grid: Benefits & Beyond.
[34] https://www.gccia.com.sa
[35] https://www.entsoe.eu/about/inside-entsoe/members/
[36] https://www.nerc.com

Chapter 2
Power quality issues in power supply systems

Power quality: "The characteristics of voltage or current wave on an electrical power supply, assessed with respect to a set of reference values."

Goals of the chapter

After going through the lesson, you shall get a broad idea of the power quality issues in modern power supply systems, various terms pertaining to power quality, causes and effects of each issue, etc. Recommendations contained in various international standards, such as those developed by the Institute of Electrical and Electronics Engineers (IEEE) and International Electrotechnical Commission (IEC) on power quality, are also briefly covered to make it more application-oriented.

2.1 Introduction

As discussed in Chapter 1, the need to provide continuous electrical power with steady voltage and frequency has been recognized since the inception of the power system. In the early stages, power stations were of hydel, thermal or nuclear, wherein generators were of synchronous machines. These generators are further interconnected with loads that were mostly of the linear type, thus forming a grid. Thus, by virtue of linear generation-load characteristics, waveforms at various parts of the system were pure sine waves. Some of the early concerns were variation in the luminescence of incandescent bulbs and changes in the rotational speed of motors. This was mainly due to load variations resulting in changes in the supply voltage and frequency. These changes can be broadly categorized based on time as short-duration issues, such as voltage sag, voltage swell, brief interruption, etc., and long-duration issues such as undervoltage, overvoltage, sustained interruption, etc. Additionally, the system has been subjected to transients from external sources or internal switching.

The widespread use of non-linear items, viz., power electronic-based equipment in the transmission, distribution, and utilization sectors, has resulted in distortions – such as DC offset, harmonics, inter-harmonics, notches and noise – in the supply waveform. More recently, renewable energy in the generation sector, particularly wind turbines and solar PV, has been added to the system. Integration of these variable and intermittent (diurnal, seasonal and weather-dependent) generations, which are also based on power electronics, together with the existing system, has compounded the challenges to the rated sinusoidal waveform. These deviations from rated sinusoidal conditions are termed under power quality. These variations adversely affect the performance of the electrical equipment, components and the power system. Hence, it is very important to analyse these power quality issues [1–3] in detail and find solutions to them.

2.2 Propagation of waveform distortion in power system

Consider a simple system with a conventional load connected to a synchronous generator through a transmission line having equivalent series impedance (Z_L) that is composed of resistance (R_L), inductance (X_L) and capacitance (X_C) as shown in Figure 2.1. When current 'i_s' flows through the line, the receiving-end bus voltage v_r differs in magnitude from the sending-end bus voltage v_s by the drop across the transmission line impedance (i_sZ_L).

Let us consider the load as a large agricultural or industrial pump driven by an induction motor with a direct online (DOL) or star-delta starter. With the supply voltage as in Figure 2.2(a), the motor draws a lagging current as shown in Figure 2.2(b). With a large motor, due to the high starting current of about six times the full load current (i_s), the voltage at the motor terminal drops by about six times i_sZ_L and improves thereafter to its running condition, with a drop of i_sZ_L. Thus, the receiving-end voltage sag or undervoltage occurs during starting. This may affect the performance of the motor itself and other loads connected to the system. Now, let us consider the motor with a modern drive composed of an AC–DC and DC–AC converter to limit the current. Converters draw non-sinusoidal current as shown in Figure 2.2(c). We may note that the active component of current in Figure 2.2(b) contributes to the power drawn by the motor and is equal in both cases. However, the current drawn in the second case is distorted from the ideal sinusoidal waveform.

The system may also be subjected to another form of transient disturbance, such as lightning, switching or different types of faults. Current resulting from lightning causes an impulse voltage, which results in flashover through transmission line

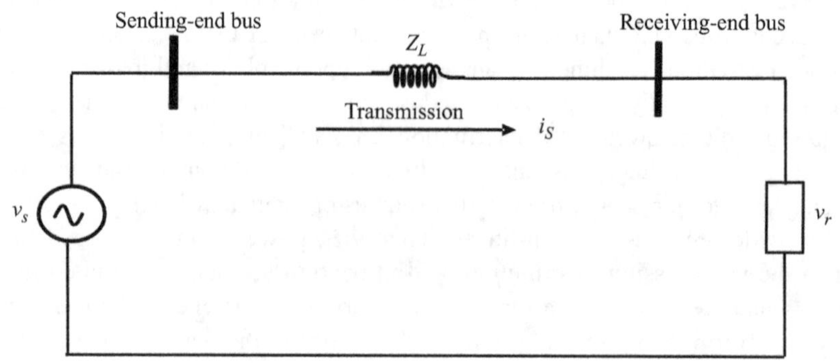

Figure 2.1 Simple power system with linear load

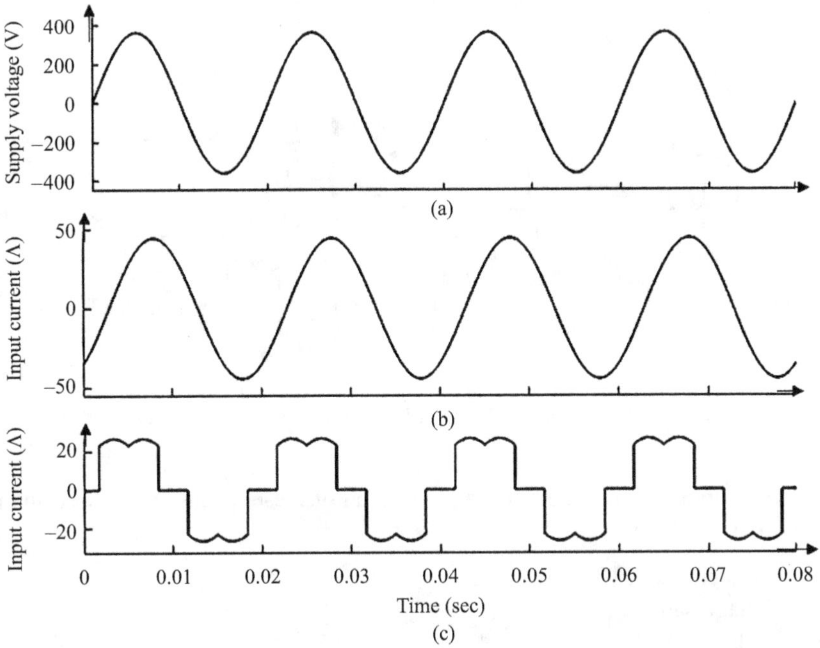

Figure 2.2 Typical waveform of supply and line current drawn by a 3-ϕ induction motor. (a) Supply voltage, (b) current with DOL starter and (c) current with motor drive.

insulators. This, in turn, may result in a short circuit and failure of high voltage (HV) equipment like transformers. The current resulting from a short circuit causes the voltage v_r to sag or disappear completely. Similarly, the voltage rises above the normal operating level due to the disconnection of the load or under no-load conditions. All these disturbances cause the voltage to change its magnitude as well as its shape.

The issue of supply voltage and current deviation is increasing in recent times with the wide application of non-linear equipment and the switching of higher-capacity loads in power generation, transmission, distribution and utilization. As discussed, with non-linear equipment, the current i_s drawn by it, causing non-sinusoidal voltage across the transmission line and transformer impedances (v_L). Thus, the voltage (v_{PCC}) at the point of common coupling (PCC) deviates from its rated sinusoidal form. This deviated power supply affects other linear or non-linear loads as well. A block diagram of the system with loads and interrupting devices, such as circuit breakers (CB), is shown in Figure 2.3.

2.3 Power quality terms, causes and effects

As discussed in the previous section, the power system is subjected to various types of disturbances that affect its waveform or power quality at the utilization point.

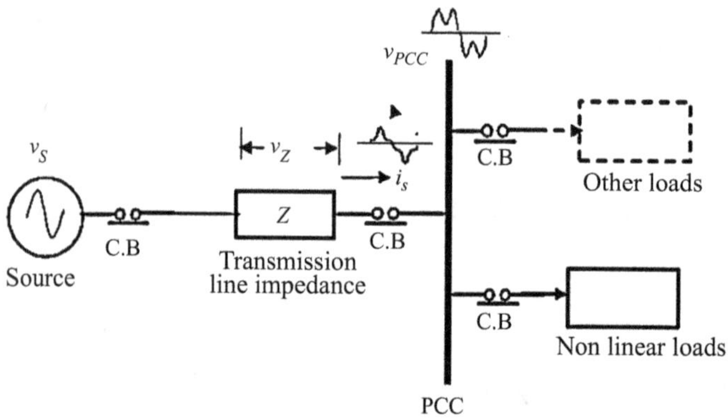

Figure 2.3 Propagation of waveform distortion in a power system

Depending upon its characteristics and duration of these deviations, power quality is classified [4–7] under

(i) Transient
(ii) Voltage sag
(iii) Undervoltage
(iv) Voltage swell
(v) Overvoltage
(vi) Voltage interruption
(vii) Voltage fluctuation
(viii) Frequency variation
(ix) Unbalance
(x) Harmonics
(xi) Inter-harmonics
(xii) Notches
(xiii) Noise
(xiv) DC offset

The typical waveforms of these deviations from the ideal sinusoidal wave are illustrated in Figure 2.4.

Various power quality terms listed above, along with their causes and effects, are discussed in the ongoing sections.

2.3.1 Transients

A transient is always a part of the electrical power supply system and is a discontinuity in alternating current or voltage waveforms characterized by a sharp rise and fall in the magnitude for a brief period. It may be of either polarity and hence may be additive or subtractive from the nominal waveform. Transients can be of two types: (i) impulsive transients and (ii) oscillatory transients.

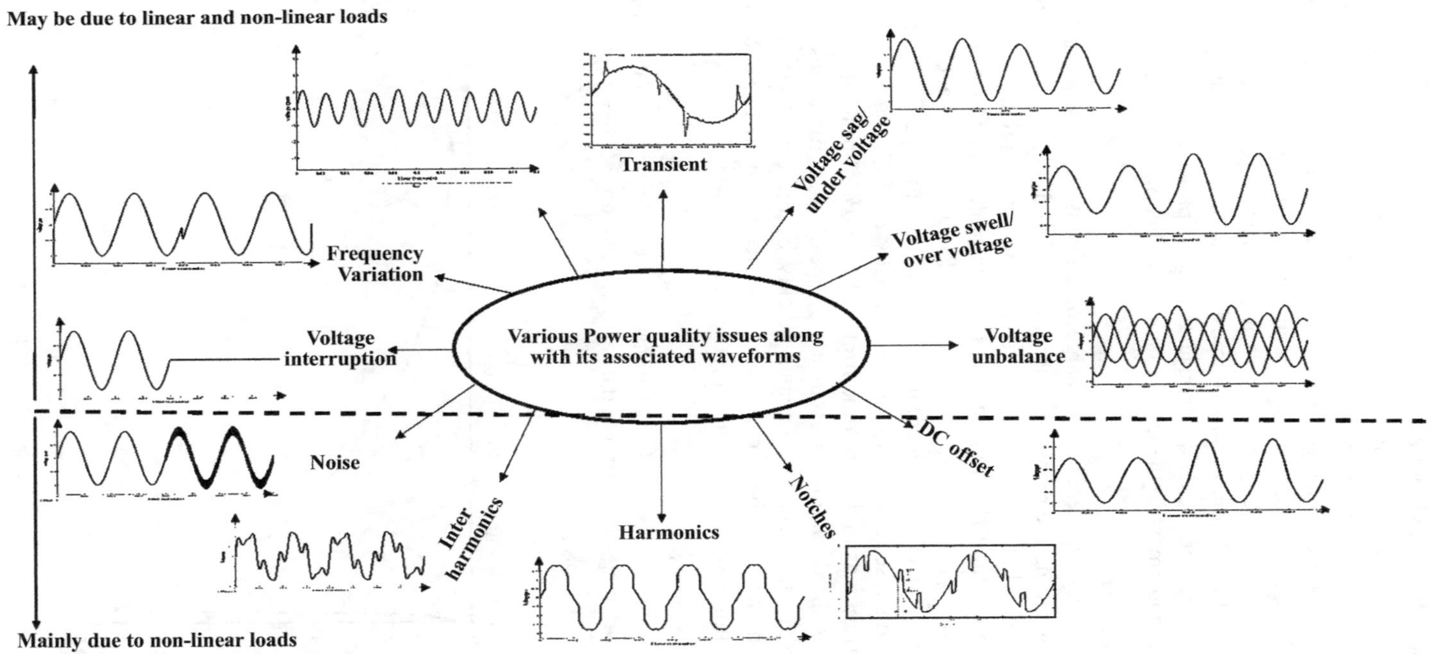

Figure 2.4 Power quality issues, along with typical waveforms

2.3.1.1 Impulsive transients

An electrical power system is often subjected to overvoltage originating from the atmosphere – lightning – and produced within the system due to switching. These types of transients are a change of state of the waveform from its steady-state condition and are associated with very high potential energy. Also, it is unidirectional with either positive or negative polarity, and hence it changes the waveshape from the nominal condition. Impulsive transients can also excite the power circuit to resonate and produce oscillatory transients. Hence, protective measures are to be considered for equipment in the transmission, distribution and utilization system.

Lightning transients

Lightning is a very high-energy discharge of millions of volts, produced by the charged clouds during the monsoon season. It is often discharged on the overhead extra high voltage (EHV) transmission system. Shielding or ground conductors are provided on top of the EHV transmission tower to protect the phase conductors of the transmission system against direct strikes. The lightning waveform is characterized by a high rise time to its peak value and decay to about 50% peak value in a very short (microseconds) duration. It is usually described as having a 1.2/50 wave. 1.2 corresponds to the rise time from 10% to 90% of peak in microseconds, and 50 denotes the duration from start to 50% peak in microseconds. Due to the high frequencies, transients are damped quickly by the circuit element's impedance, but equipment is to be protected against dielectric failure due to overvoltage. A typical lightning wave is shown in Figure 2.5.

Switching transients

Power systems are also subjected to transient overvoltages generated within the system by switching. It may be during energization of the capacitor bank,

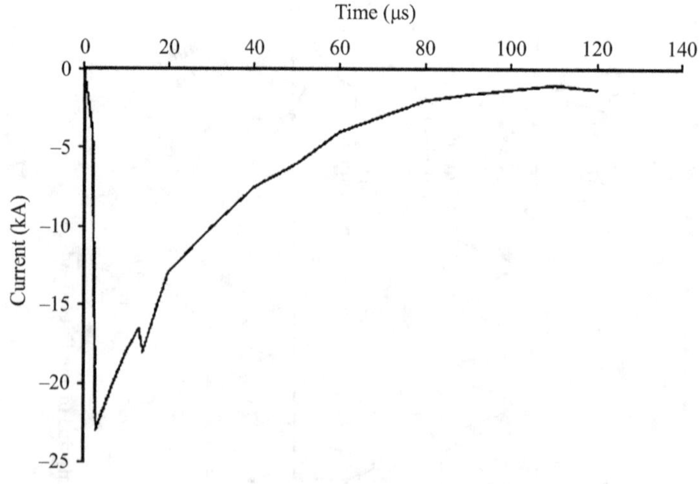

Figure 2.5 Impulsive transient due to lightning

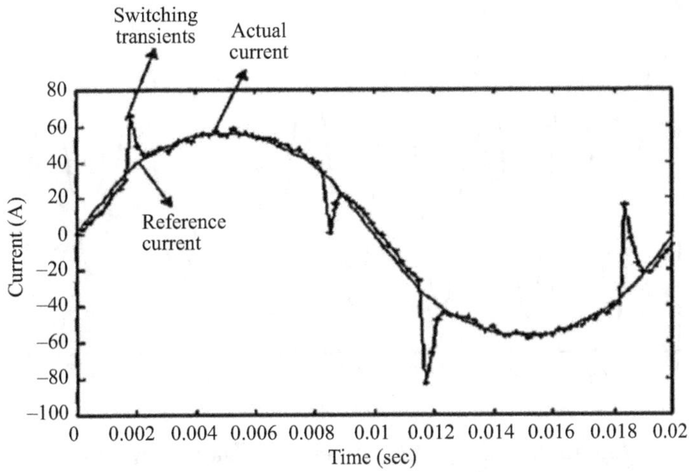

Figure 2.6 Transient due to the switching of three rectifiers

ferroresonance of the transformer, etc., due to harmonic components in the inrush current. This is due to the formation of a resonating inductive-capacitive circuit at these harmonic frequencies. It is also produced by power electronic switching devices like SCR, TRIAC, etc., which force current to zero suddenly or cause a change in the magnetic field. A typical switching transient from a simulated input current waveform of closed-loop control of a converter circuit is shown in Figure 2.6. It is seen that the actual current tracks the reference waveform. However, during the switching transition, a pulse occurs, as shown in the figure, which is known as a switching transient.

2.3.1.2 Oscillatory transients

It is produced in the transmission and distribution system during load rejection, capacitor switching, ferroresonance, fault clearing, etc. It may be of relatively long duration, but undamped or slightly damped. An oscillatory transient consists of a voltage or current whose instantaneous value changes in polarity rapidly, as shown in Figure 2.7. It is described by its spectral content (predominant frequency), duration, and magnitude. Transients of this nature may result in excitation of power system resonance circuits. It can also cause overvoltage stress on the insulation material and cause its failure.

A summary of both types of transient issues discussed is presented in Table 2.1.

2.3.1.3 Protection against transient overvoltage

It is observed that on EHV systems above 220 kV, switching overvoltage has a higher amplitude than that of lightning. However, at a lower level, switching overvoltage has a lesser amplitude than lightning. In both cases, switching overvoltage can be of longer duration and have more energy.

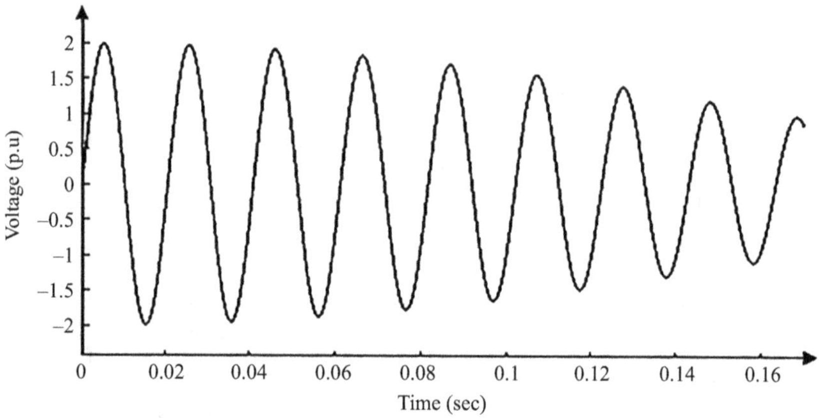

Figure 2.7 A typical transient oscillatory waveform

Table 2.1 Overview of transient disturbances

Disturbance type	**Description**	**Possible causes**	**Impact**
Impulsive	Characterized by peak value, rise time, span	Lightning, excitation of transformers, switching of capacitor banks	High voltage stress may lead to dielectric breakdown
Oscillatory	Characterized by the peak value, frequency, span	Switching operations involving capacitor banks or abrupt load changes	Undesired resonance phenomenon in the power network resulting in insulation deterioration

Overvoltage of any type can overstress insulation and possibly lead to equipment failure. Hence, equipment needs to be protected. Surge suppression/diverter devices are commonly used to protect the system. It presents variable impedance according to the transient voltage applied. At rated voltage, it offers very high impedance, whereas during an overvoltage, it has reduced impedance for diverting the energy associated with the overvoltage and keeping the system's operation within the safe range [8].

Metal oxides of zinc (ZnO) exhibit very non-linear behaviour ($I = kV^{\alpha}$, where α varies from 10 to 50, depending on the formation of the disc). Hence, it is commonly used for limiting surge voltages on equipment by discharging or bypassing surge current through it. It also prevents the continued flow of power frequency current to ground once the surge energy gets discharged. It can also repeat these functions multiple times without degradation. Figure 2.8(a) shows its connection diagram in the system, and Figure 2.8(b) shows a few ZnO discs of different current ratings [9]. In low-voltage systems, protection against transient

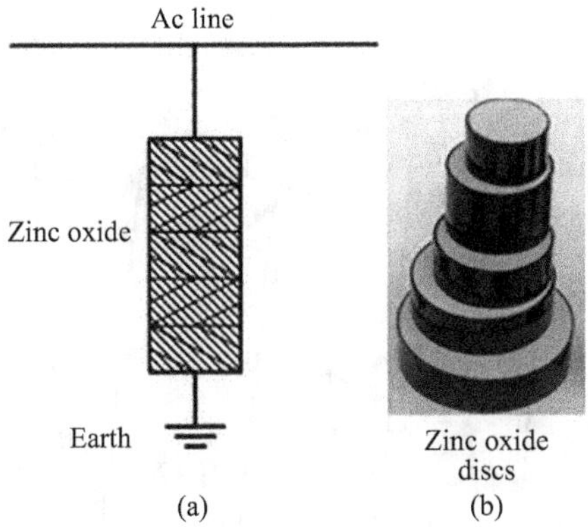

Figure 2.8 Surge diverter. (a) ZnO discs connected to a system. (b) ZnO discs of different current ratings.

over voltages is provided by the installation of surge protection devices (SPD) that contain a non-linear component to limit or divert surge currents to ground, or to protect sensitive utilization equipment in LV system against transferred surge voltages [8,10].

2.3.2 Voltage variation (short and long duration)

Voltage variation is classified as short and long duration, which depends on the duration of variation of voltage. There are the following terms under this category: (i) voltage sag, (ii) under voltage, (iii) voltage swell, (iv) overvoltage, (v) voltage interruptions and (vi) voltage fluctuation and flicker [11], which are described below.

2.3.2.1 Voltage sag or dip

Voltage sag is a condition in which there is a reduction in the root mean square (rms) value of voltage to 10%–90% of the nominal voltage (0.1–0.9 pu) for a short duration of about half a cycle to several seconds. Accordingly, it is further classified as:

(i) 0.5 cycles to 30 cycles, it is considered an instantaneous voltage sag
(ii) 30 cycles to 3 s, it is considered a momentary voltage sag
(iii) 3 seconds to 1 min, it is considered a temporary voltage sag
(iv) More than 1 min is termed as sustained voltage sag or dip

Typical voltage waveform during a sag is shown in Figure 2.9.

In power systems, voltage sag is usually caused by the starting of large induction motors, switching of other large inductive loads, disconnection of capacitor banks or system faults, etc.

Voltage sag affects many components of the system, depending upon its magnitude and duration. It causes malfunctioning of sensitive components and control devices such as programmable logic controllers (PLC), information technology equipment (ITE), Internet of Things (IoT), computers and communication systems.

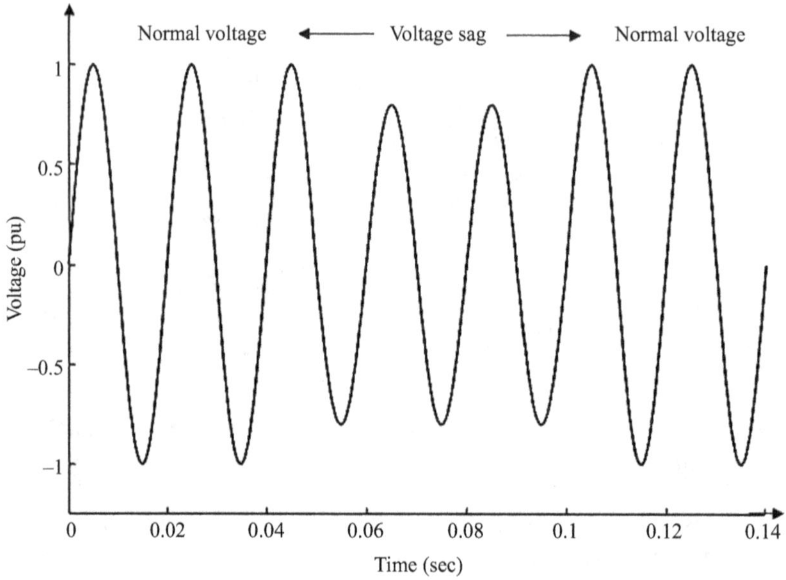

Figure 2.9 Supply voltage waveform with voltage sag

2.3.2.2 Undervoltage

Undervoltage is a long-duration dip in rms voltage to about 80%–90% (0.8–0.9 pu) of the rated value for a period of greater than one minute. It may be due to running large inductive loads, such as induction motors on a weak system or switching off large capacitor banks, resulting in inadequate capacitive compensation. Undervoltage also occurs during overloading of transmission or distribution lines/ feeder circuits. Integration of renewable energy generation like solar PV (SPV) with the electrical grid also contributes to voltage variation near the SPV plant occasionally [12]. This is detailed below.

Consider a system with SPV with a conventional DC–AC converter connected to the grid and generating active power of P_{DG} as shown in Figure 2.10. With SPV generation, the active power demand of the load P_D from the generator G is reduced by P_{DG}. Hence, the total load on generator G is corrected to $P_D - P_{DG} + jQ_D$. Since the reactive power demand on generator G has remained Q_D, inductive current continues to flow from generator G to the load, and hence its power factor reduces. This may cause a reduction in bus voltage at PCC.

Performance of most of the equipment connected to the power system, viz., incandescent lamps, heating elements, power supplies, capacitors, electromagnetic contactors, data processing equipment, line commutated converters, electromagnetic contactors, etc., is affected by undervoltage. In induction motors, it results in a reduction of output torque and deteriorates their starting/running performance. Figure 2.11 indicates the variation of torque slip characteristics of a typical induction motor for 110%–80% variation of supply voltage. It may be seen

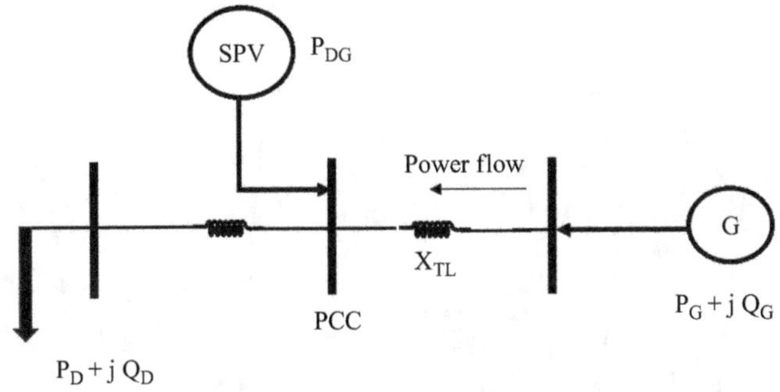

Figure 2.10 An SPV-connected grid

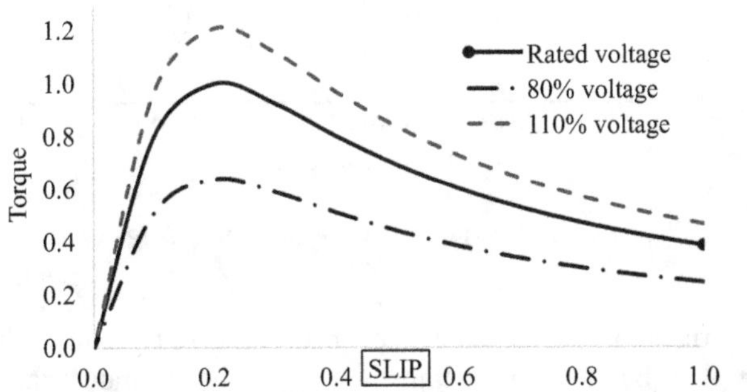

Figure 2.11 Typical variation of induction motor torque speed with supply voltage

that the developed torque varies as the square of the voltage, and the motor may pull out of step if the developed torque is less than the torques required by the pump/fan.

Equipment that cannot tolerate prolonged undervoltage is generally isolated from the system using undervoltage tripping of relays or provided with an uninterruptible power supply (UPS). At the transmission end, transformers are generally provided with either on-load tap changers (OLTC) or off-load tap changers to correct undervoltage in the system.

2.3.2.3 Voltage swell

A momentary increase in the power frequency voltage, outside the normal tolerance level of 110%–180% (1.1–1.8 pu), for a duration of about half a cycle to several seconds, is called a voltage swell. It is also classified similarly to voltage sag as instantaneous, momentary, temporary and sustained voltage sag. A sample waveform with voltage swell is shown in Figure 2.12.

Long high-voltage transmission lines experience voltage swell during lightly loaded conditions. This effect, called the Ferranti effect, arises due to line

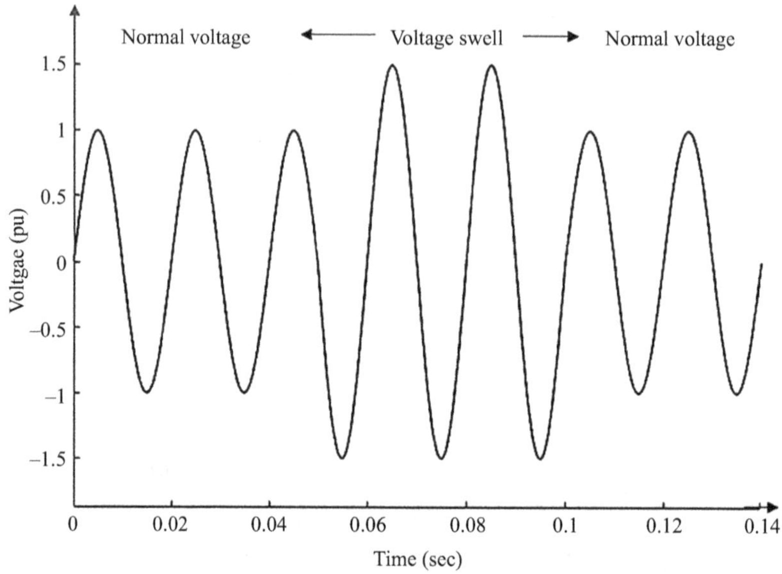

Figure 2.12 A typical waveform of voltage swell

capacitance. A rise in voltage can also occur during load rejection. This can happen at generating plant bus terminals, transmission, or heavy industrial units during a load shedding or disconnection of major loads on the system. Voltage can also rise during the switching on of a large capacitor bank. Voltage swells also occur during ground fault conditions. A line-to-ground fault can cause a voltage to rise on the healthy phases and fall on the faulty phase. This is because during a line-to-ground fault on an ungrounded system or isolated system, voltages on the healthy phases may be $\sqrt{3}$ times the line-to-ground pre-fault voltage. For a grounded system, the rise will be much less than $\sqrt{3}$ times and will depend on the system sequence impedances.

Voltage swells may adversely affect the insulation system of equipment, which ultimately causes its premature failure. When voltage swells beyond the set points, voltage-operated relays and contactors may cause their operation, resulting in system isolation.

2.3.2.4 Overvoltage

Overvoltage is a long-duration rise in rms voltage between 110% and 120% (1.1–1.2 pu) for a period of greater than 1 minute. The cause of overvoltage is like that of voltage swell. It can adversely affect equipment performance due to its dielectric failure. In transformers, overvoltage causes magnetic flux (Φ) to increase its magnitude ($\Phi \propto \frac{v}{f}$) and causes the core to saturate, resulting in a higher temperature rise and failure. To prevent this condition, overvoltage protection and overflux protection are installed in the electrical protection system.

2.3.2.5 Voltage interruptions

A condition in which the voltage falls lower than 10% (0.1 pu) of the nominal voltage is called voltage interruption. It is a complete loss of voltage in one or more phase conductors, and based on its duration, voltage interruption [4] is classified as:

(i) Momentary: if voltage interruption is for a period of 0.5 cycles to 3 seconds.
(ii) Temporary: if voltage interruption is for a period of 3 seconds to 1 minute.
(iii) Sustained: if voltage interruption is for a period of more than 1 minute.

A sample voltage interruption waveform is given in Figure 2.13. Reasons behind interruptions may be power system faults, equipment failures, load shedding, etc.

The majority of the faults on overhead transmission lines are transient in nature, usually caused by lightning or temporary contact with trees, birds, monkeys, etc. (overhead lines in forest areas are prone to this type of fault). Immediate tripping of CB may clear the fault. Subsequent re-energization of the affected line is usually successful. Auto-reclosing (detailed in Chapter 4) allows CB to be reclosed within a few seconds of a fault. On the occurrence of such a fault, single-phase auto-reclose schemes, if implemented on line, enable a three-phase circuit breaker to trip and reclose only the faulted phase of the line. The overall effect of a voltage interruption would be loss of supply for a very short time to allow for fault-arc de-ionization, but it would affect a larger number of consumers. The remaining faults are of either the semi-permanent or permanent type. With a fast-acting protection system, the circuit breaker disconnects the faulty part from the system, which results in the interruption of voltage that affects all the users connected to the system. A detailed summary of various types of short- and long-duration voltage variations discussed above has been provided in Table 2.2.

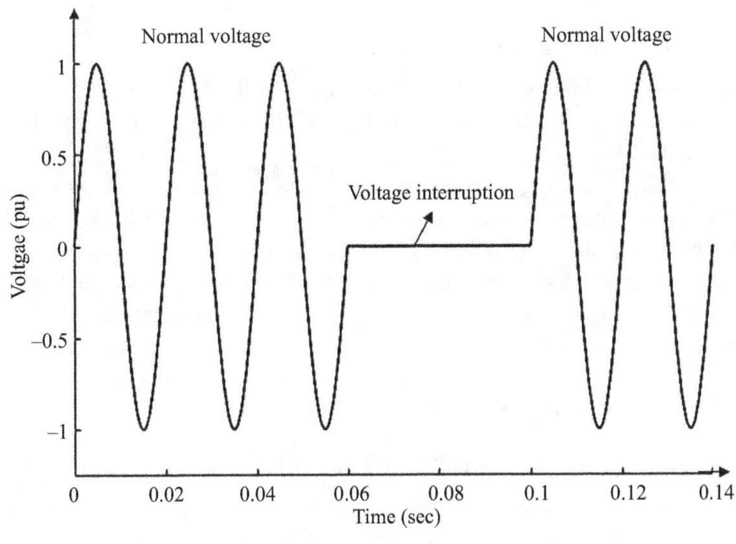

Figure 2.13 Typical waveform of voltage interruption

Table 2.2 Disturbances on account of voltage variation

Disturbance type	Description	Possible causes	Impact
Voltage sag	Short-term reduction in RMS voltage level characterized by its magnitude and duration	Starting of large motors, single line-to-ground faults in the network	Improper operation of protection systems and interruption of industrial processes and production
Voltage swell	Short-term increase in RMS voltage level characterized by its magnitude and duration	Switching of capacitor banks and/or heavy loads, system faults	Increased voltage stress on equipment, malfunction of protection systems
Supply Interruption	Complete loss of voltage supply, for a brief period of time	Short duration faults in the system	Shutdown of ongoing processes and production and triggering of false alarms
Sustained interruption	Long-duration loss of supply voltage	Permanent faults in the electrical network requiring system repair	Total power outage leading to stoppage of operations and production
Undervoltage	Prolonged reduction of voltage magnitude below the nominal level	Sudden connection of loads or disconnection of capacitor banks	Higher system losses and possible overheating of electrical equipment
Overvoltage	Extended increase in voltage magnitude above the rated value over a certain period of time	Disconnection of loads or energizing of capacitor banks	Potential damage to insulation and electrical devices

2.3.3 Voltage fluctuation

It is a series of voltage variations or sag and swell changes in a cyclic order. Figure 2.14 shows a typical cyclic variation of the voltage waveform that could result in flicker [4,11].

In case of fast switching of industrial and process load such as rolling mills, cement and petrochemical plants, rock crushers, arc furnaces, elevators, cranes, space heaters controlled by thermostats, arc welding, driers, etc., voltage at the load terminal also rises and drops or fluctuates depending on the system strength, impedance and magnitude of load current. The rhythmic or periodic change in the voltage magnitude that results in frequent changes in the brightness of the incandescent light is termed voltage flicker.

Voltage fluctuation affects the loads connected to the system, depending on their sensitivity. For example, in the case of an incandescent bulb, brightness or its luminous intensity, which is proportional to $\frac{V^2}{R}$, changes rapidly or fluctuates with the variation in supply voltage. When it is sudden and large enough (typically 7–8 dips per second) to be noticeable to the eye, the effect of light is called flicker. Thus, a varying current results in voltage fluctuation due to the series impedance of

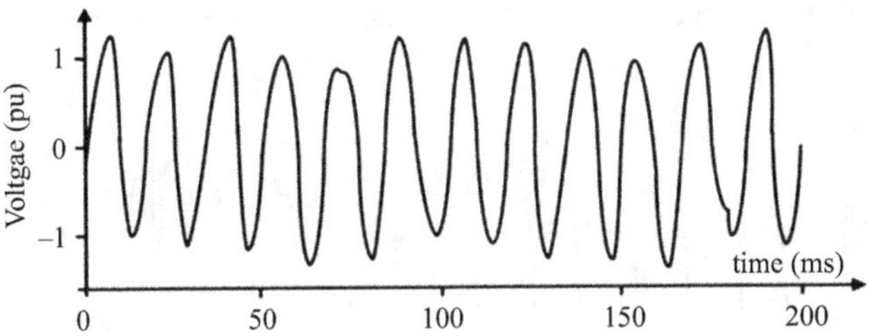

Figure 2.14 Typical fluctuating voltage waveform

Table 2.3 Disturbances on account of voltage fluctuations

Disturbance type	Description	Possible causes	Impact
Voltage fluctuations	Continuous or intermittent voltage variation/ changes	Rapid changes in heavy loads	Light intensity variation, nuisance tripping, health hazards

the system. If it is of sufficient magnitude, it can cause the human eye to perceive flicker in the light output of the lamp.

Excessive flicker can cause affected persons to experience weakness or fatigue, headaches or dizziness, lower concentration levels, etc. Lamps with ballast, such as fluorescent or sodium vapour, the ballast stabilizes the power, and therefore the light output from it is less sensitive to the supply voltage within a range. The performance of the light output from the LED is not an exception. Considering the adverse effects of LED lighting issues and to mitigate them, IEEE 1789–2015 formulated recommended practices for modulating current in high-brightness LEDs. The problems caused by voltage fluctuation are noticeable in computers, PLC, rotating machines, by their abnormal noise and fluctuating torque, etc. It results in maloperation of relays and contactors, unwanted switching of UPS units, etc. In general, all loads draw variations in load current due to voltage variations, and their performance is affected. Table 2.3 gives a summary of voltage fluctuation phenomenon discussed here.

Voltage variation in the transmission system is a rare event and is controlled at various points generally by providing automatic voltage regulators (AVR) at the generator side and OLTC at the transformers on the transmission side [13]. Thus, the rms voltage may remain within the nominal values; however, it may fluctuate within a range and hence rarely causes equipment failure or supply interruptions.

2.3.4 Power frequency variation

The deviation of the power system fundamental frequency from its nominal value is defined as a frequency deviation. A 50 Hz sinusoidal waveform with variation in frequency is shown in Figure 2.15.

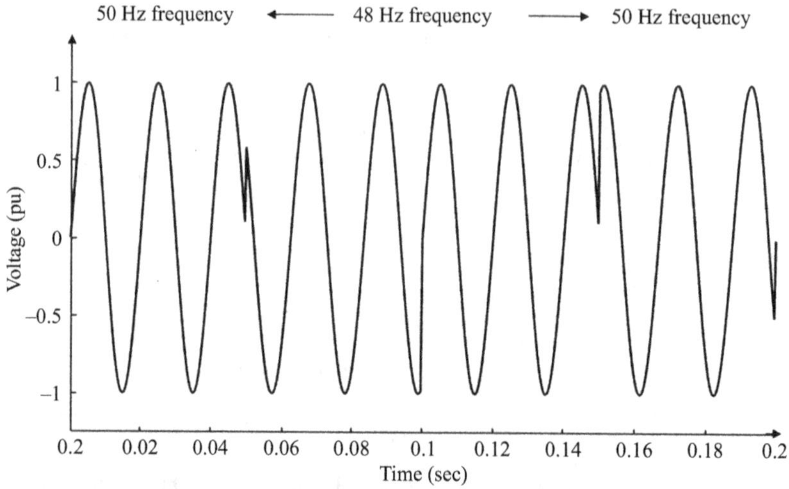

Figure 2.15 Type waveform of frequency variation

Frequency variations from the normal steady-state value can be caused by load-generation mismatch, faults in generation, transmission and distribution systems, load shedding, etc. Disturbances due to load-generation imbalance occur continuously, and the system adjusts itself accordingly without much impact on the stability because of the large inertia of the interconnected system. However, for large power imbalances, the system frequency can deviate at a much faster rate from the nominal operating frequency range. Large frequency variations result in long-term effects on all components of the power system, starting from the generator to the end utilization equipment. It may also affect system stability and lead to blackout of the grid or a large-scale power interruption.

Maintaining frequency stability is critical for the operation of an interconnected power system. The magnitude of the frequency shift and its duration depend on the load-generation balance, inertia of the combined turbine-generator system, rotating motors, and response of the generator-governor system (load frequency controller). Small frequency variation occurs almost continuously due to the continuous change of load. As the electricity generation mix today evolves with a larger contribution from renewable energy such as solar PV, the electricity system's "natural" inertia is decreasing. And hence the higher the rate of change of frequency (RoCoF) following a contingency in the system. Thus, large imbalances between generation and utilization, higher deviations in frequency. These differences can make the system unstable or, in extreme cases, cause a risk of loss of synchronism. Therefore, the system protection schemes need to cope with extreme events of RoCoF. A comprehensive summary of the power frequency variation issue has been provided in Table 2.4.

In an electrical system, a steep fall in the frequency is controlled by system inertia and further by automatic under-frequency load shedding (AUFLS). For example, with a nominal frequency of 50 Hz, AUFLS may trigger at 49.4 Hz for

Table 2.4 Disturbances on account of power frequency variations

Disturbance type	Description	Possible causes	Impact
Power Frequency Variation	Characterized by "Rate of change of system frequency (ROCOF)"	Large load changes, system faults and islanding etc.	Can cause mechanical stresses and damage to generator rotors and turbine shafts, adversely affect the performance of frequency dependent loads

load shedding to stabilize the grid frequency. If frequency is found to be still dropping, further loads will be isolated for grid stabilization.

2.3.5 Unbalance

Unbalance (*Synonymously* called imbalance [4]) is a condition in which the rms value of the 3 line-to-line voltages or currents is not equal, or the phase angles between them are not equal to 120°. Unbalance in the system may be due to improper transposition of transmission lines, mismatch in transformer turn ratios, problems with tap changers, unequal distribution of single-phase loads among the three phases, or to asymmetrical faults, or the absence of any phase voltage due to the operation of protective devices such as a fuse. Assume that the current i_a, i_b, and ic in a three-phase system are unbalanced as shown in Figure 2.16.

This unbalance is often analysed using symmetrical components comprising (a) balanced positive sequence components represented with subscript p, (b) balanced negative sequence components represented with subscript n and (c) zero sequence components represented with subscript 0. A vector diagram of unbalanced current and sequence components is shown in Figure 2.17.

Thus, rms value of current in each phase I_a, I_b, I_c can be expressed in terms of the sequence component as:

$$\begin{aligned} I_a &= I_{ap} + I_{an} + I_{a0} \\ I_b &= I_{bp} + I_{bn} + I_{b0} \\ I_c &= I_{cp} + I_{cn} + I_{c0} \end{aligned} \tag{2.1}$$

The current in each phase can therefore be written in matrix form:

$$\begin{bmatrix} I_a \\ I_b \\ I_c \end{bmatrix} = \begin{bmatrix} 1 & 1 & 1 \\ 1 & a^2 & a \\ 1 & a & a^2 \end{bmatrix} \begin{bmatrix} I_{a0} \\ I_{ap} \\ I_{an} \end{bmatrix} \tag{2.2}$$

where $a = 1 < 120°$, $a^2 = 1 < 240°$ or $1 < -120°$ and hence $1 + a + a^2 = 0$.

Similarly, symmetrical component can be derived from three-phase current as:

$$\begin{bmatrix} I_{a0} \\ I_{ap} \\ I_{an} \end{bmatrix} = \frac{1}{3} \begin{bmatrix} 1 & 1 & 1 \\ 1 & a & a^2 \\ 1 & a^2 & a \end{bmatrix} \begin{bmatrix} I_a \\ I_b \\ I_c \end{bmatrix} \tag{2.3}$$

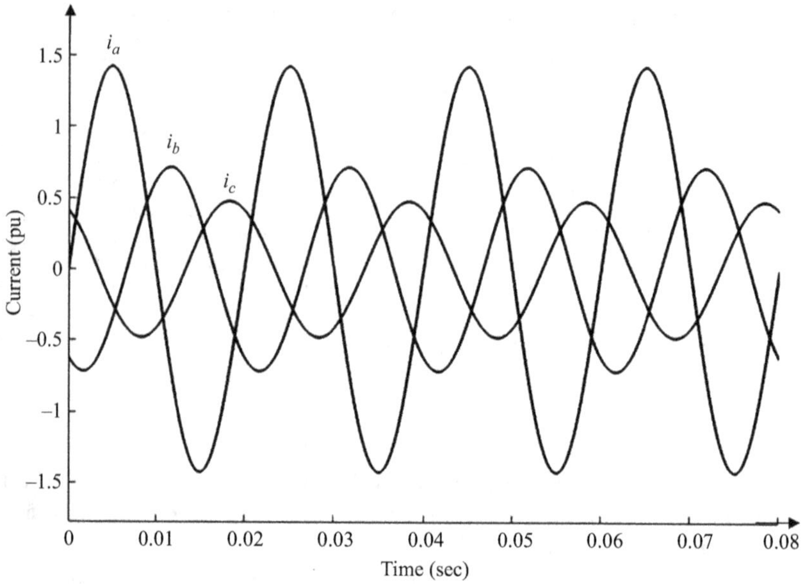

Figure 2.16 Waveform of current unbalance

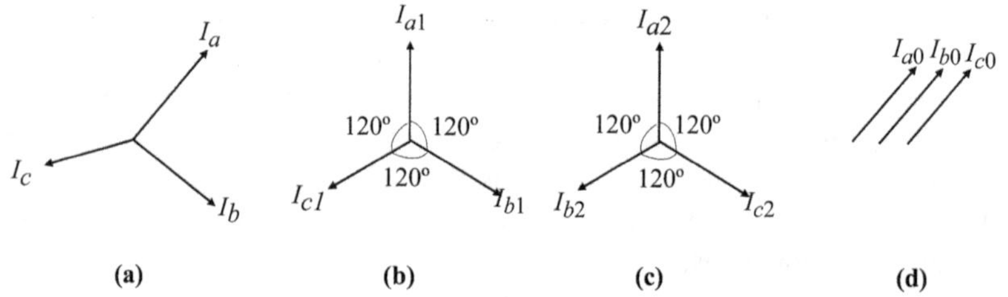

Figure 2.17 Vector diagram of unbalanced current and its sequence component. (a) Unbalanced current, (b) positive sequence component, (c) negative sequence component and (d) zero sequence component.

Neutral current (I_N) is given by:

$$I_N = I_a + I_b + I_c = 3I_{a0} = 3I_{b0} = 3I_{c0} \tag{2.4}$$

From (2.3.), if I_a, I_b, I_c balanced, the zero sequence current I_{a0}, I_{b0}, I_{c0} and negative sequence current I_{an}, I_{bn}, I_{cn} I_{an} are zero and only a positive sequence current I_{ap}, I_{bp}, I_{cp} will be present. Otherwise, there will be a negative sequence and a zero sequence current in the system with a neutral conductor carrying $3I_{a0}$. Hence, a neutral conductor needs to be rated for the increased

Table 2.5 Disturbances on account of voltage imbalance

Disturbance type	Description	Possible causes	Impact
Voltage imbalance	It is characterized by unequal voltage magnitude and/or phase angle among the three phases, analysed using symmetrical component theory	Highly uneven loading of the 3 phases of a 3-ϕ system, loss of one phase, open conductor, or unbalanced distribution network conditions	Motor overheating, increased losses, torque pulsations, and reduced lifetime of electrical equipment

current [8]. Similar equations are applicable to the voltage unbalance case too. Ratio of the negative sequence component to the positive sequence component in percentage is defined as the imbalance factor (UF) [4]. For voltage imbalance, it is expressed as the

$$\%VUF = \frac{V_n}{V_p} \times 100 \tag{2.5}$$

where V_p and V_n are the rms values of positive and negative sequence components of the voltage. A similar equation is applicable for the current unbalance too.

Unbalance in voltage and/or current is harmful to all components of the power system. When an induction motor is supplied with an unbalanced supply voltage, positive sequence flux is produced due to the positive sequence of current rotating in the same direction as the fundamental component of current, while negative sequence flux is produced from the negative sequence of current rotating in the opposite direction. This results in pulsating torque. Also, it results in a reduction in motor locked, pull-up and breakdown torque values, and the motor may no longer be adequate for its intended application. As per the IEC 60034-26 standard, voltage unbalances at the motor terminals shall not exceed 1%, and if they exceed 1%, the nameplate power rating of the motor is required to be de-rated to prevent long-term damage. The standard also mandates a derating of up to 5% for voltage imbalance [14].

From utility to users, all the single-phase loads should be distributed on the three-phase system such that they present nearly equal loads to the three phases to reduce the unbalance. Also, proper testing to identify the disturbing equipment having unbalanced reactance, and replacing it with a balanced one, is suggested to resolve the problems. A summary about voltage imbalance is tabulated in Table 2.5.

2.3.6 Waveform distortion

Waveform distortion due to non-linear loads, which cause harmonics, inter-harmonics, voltage notches, noise, DC offset, etc., is discussed in detail below.

2.3.6.1 Harmonics

Harmonics are sinusoidal quantities having frequencies that are integer multiples of the nominal frequency at which the supply system is designed to operate. Current harmonics are injected into the power system mainly due to non-linear power electronic loads, which may result in voltage harmonics too in the power system. Unlike disturbances caused during switching or faults that occur for a short period of time, distortion due to harmonics load is continuous as long as non-linear loads are connected in the system [15].

Simulated sample waveforms of a power supply consisting of fundamental, second-order and third-order harmonic components are shown in Figure 2.18(a), which result in an unsymmetrical waveform as shown.

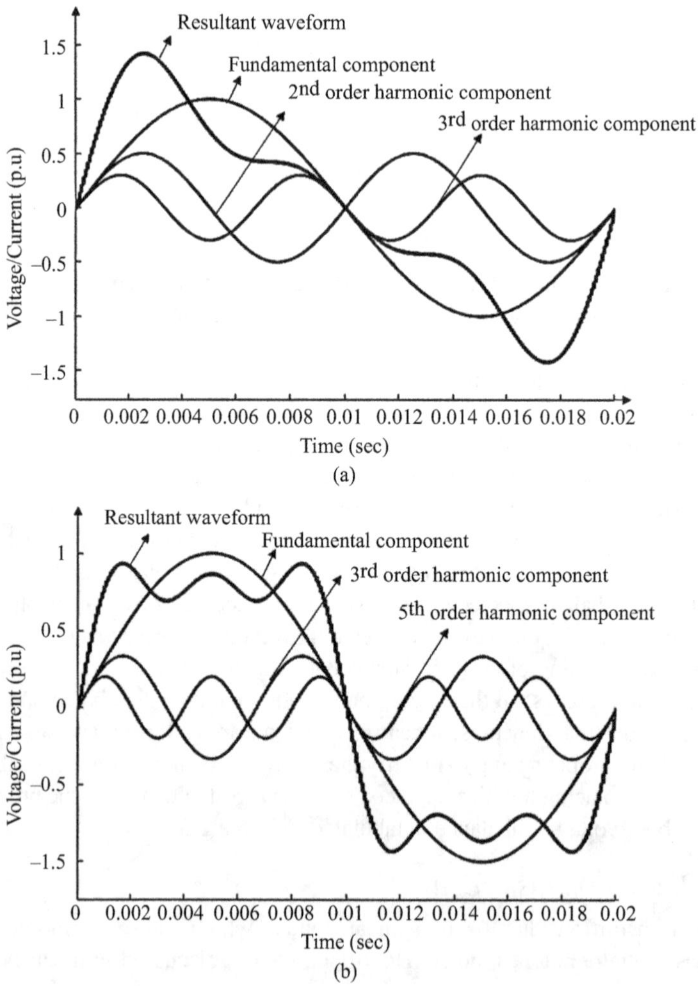

Figure 2.18 Waveform distortion due to harmonics. (a) Unsymmetrical resultant waveform and (b) symmetrical resultant waveform.

Generally, even harmonics are not present in most of the loads. However, loads such as arc welding and half-wave rectifiers produce even harmonic current, due to which the resultant current waveform becomes unsymmetrical. Most of the loads consist of odd harmonics, which result in a symmetrical waveform as shown in Figure 2.18(b). This waveform is simulated with fundamental, 3rd and 5th order harmonic waves.

The characteristics or the behaviour of each of the harmonics depend on its order of harmonics. Table 2.6 summarizes the rotational sequence of harmonics up to the seventh order. It can be seen from the table that harmonics of the orders 4, 7, 10, etc. have properties of the positive sequence symmetrical components and rotate in the same direction as that of the fundamental components. On the other hand, harmonics of the orders 2, 5, 8, etc., possess properties of negative sequence symmetrical components and rotate in the opposite direction. Harmonics of the order 3, 6, 9... (multiples of 3) are called triple harmonics and have properties of zero sequence components. Zero sequence component of current will get added up and flow through the neutral current.

Harmonics are mainly produced on the downstream of the network and are transmitted to the upstream of the network in proportion to the strength of the

Table 2.6 Rotational sequences of harmonic frequencies

Frequency	**Phases (Phase sequence a-b-c)**			**Rotational sequence**
Fundamental	*a-1* *0°*	*b-1* *120°*	*c-1* *240°*	Positive
Second harmonic	*a-2* *2 × 0°* *= 0°*	*b-2* *2 × 120° = 240°* *= −120°*	*c-2* *2 × 240° = 480°* *= −240°*	Negative
Third harmonic	*a-3* *3 × 0°* *= 0°*	*b-3* *3 × 120° = 360°* *= 0°*	*c-3* *3 × 240° = 720* *= 0°*	Nil
Fourth harmonic	*a-4* *4 × 0°* *= 0°*	*b-4* *4 × 120° = 480°* *= 120°*	*c-4* *4 × 240° = 960* *= 240°*	Positive
Fifth harmonic	*a-5* *54 × 0 °* *= 0°*	*b-5* *5 × 120° = 600°* *= −120°*	*c-5* *5 × 240° = 1200* *= −240°*	Negative
Sixth harmonic	*a-6* *6 × 0°* *= 0°*	*b-6* *6 × 120° = 720°* *= 0°*	*c-6* *6 × 240° = 1440* *= 0°*	Nil
Seventh harmonic	*a-7* *7 × 0°* *= 0°*	*b-7* *7 × 120° = 840°* *= 120°*	*c-7* *7 × 240° = 1680* *=24 0°*	Positive

network, and they cause problems for all power users and utilities. Voltage at various parts of the network depends on the flow of current and impedance of the network from the source to the specified point. Strength of the network is decided by the short-circuit impedance of the network. If it is low, the network will be strong, whereas the network will be weak if the value is high. In the upstream of the power system network, the short-circuit impedance is low. So, the effect of harmonics on voltage at the upstream will be less compared to downstream for the same magnitude of harmonic loads. Hence, the current harmonics limit at the upstream will be at a higher level compared to downstream.

The rms value of the harmonics, expressed as a percentage of the fundamental component, is a popular means of quantifying harmonics and is termed as total harmonic distortion (THD). For example, THD for current harmonics is given by:

$$THD_i = \frac{\sqrt{\sum_{h=2}^{\infty} I_h{}^2}}{I_1} \tag{2.6}$$

where I_h represents the rms value of h order harmonic component of current.

Several methods are employed while designing the system to reduce harmonics. For example, transformers for non-linear loads are selected such that their windings share a common core to nullify the effect of harmonics. Delta-connected transformer primaries are widely used for balancing triple harmonic currents generated on the secondary side by non-linear electronic load equipment. Apart from these, passive and active harmonic filters are developed for mitigating these harmonics. These are detailed in subsequent chapters.

2.3.6.2 Inter-harmonics

Inter-harmonics [4] are those frequency components of voltage or current having non-integer multiples of fundamental frequency. Non-integer harmonics with an order less than one are termed as subharmonics. For example, if a waveform with fundamental components of 50 Hz contains a frequency component of 175 Hz and 25 Hz, the order of inter-harmonics and subharmonics is 3.5 and 0.5, respectively. The main sources of inter-harmonics are static frequency converters, HVDC back-to-back systems, cycloconverters, induction furnaces, arcing devices, etc. Figure 2.19 indicates a typical distorted waveform of 3.5 and 5.5 inter-harmonics to a 50 Hz sinusoidal waveform.

Flow of Inter-harmonics adversely affects power line carrier signalling, causes telephone interference and brightness of incandescent lamps and flickers in display devices such as cathode ray tubes (CRTs). Other problems include possible resonance between the parasitic inductive and capacitive elements in the network and possible overloading of capacitors, leading to malfunctioning and premature ageing. Inter-harmonics also affect generators, motors, transformers, other equipment such as filters and process such as signalling, communication systems, etc.

Inter-harmonics can be eliminated using active or passive power filters. It can also be reduced with the help of improved equipment design.

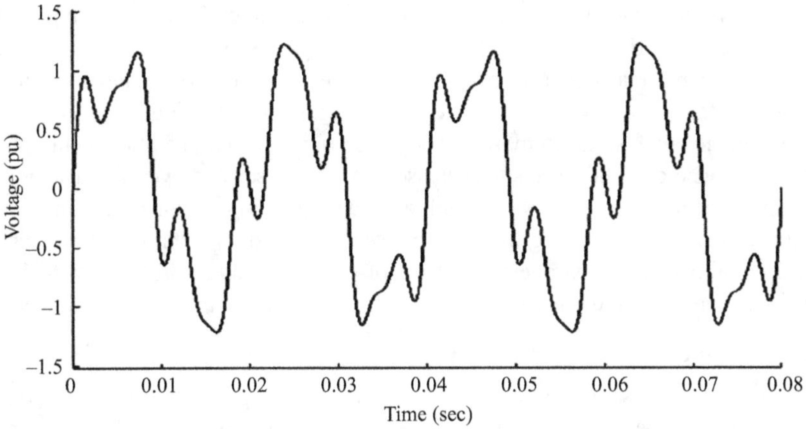

Figure 2.19 Typical waveform with inter-harmonic

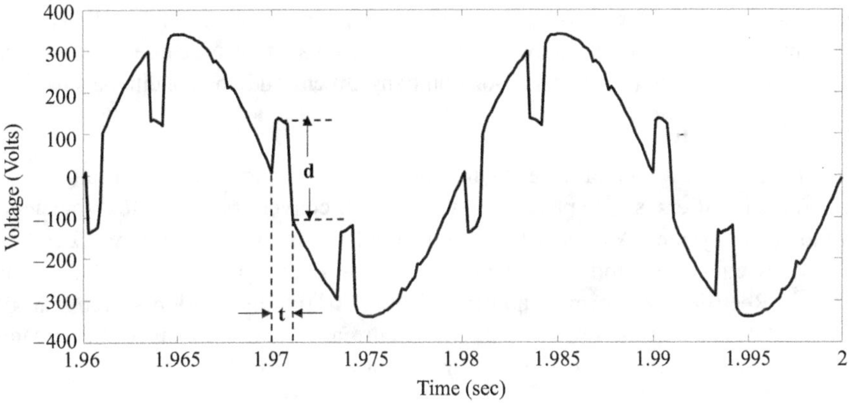

Figure 2.20 Typical waveform of input voltage notches by a six-pulse converter

2.3.6.3 Voltage notches

It is a periodic disturbance caused by the normal operation of power electronics switching devices when current is commutated from one phase to another. AC to DC converters that produce a steady DC output are the most important source of voltage notches in the system. The source inductance is the main culprit for the notches. The switching of the converter results in a short circuit across input supply terminals during the commutation period. Though such short circuits last only a few microseconds, they produce spikes or dips in the supply voltage waveform as shown in Figure 2.20, which introduces harmonics of varying frequency in the supply voltage. This is generally known as voltage notches, which are produced at the input terminal voltage of a six-pulse converter. The severity of the notch is specified in terms of width (t) in microseconds, depth (d) in volts and area (t × d) in microsecond-volts. The width and depth of the notches vary with the trigger delay angle and source inductance.

It can be seen from the voltage waveform that it can cross the x-axis multiple times in a cycle due to notches. Hence, it may cause frequency or timing errors on power electronic circuits that count zero crossings for their control operation. The power system can form a resonating circuit, causing ringing with the high frequency associated with the notch and system inductance or capacitance. This additional current due to the resonance condition leads to increased heating losses and even faster ageing of the capacitors used in the power factor compensation. This resonating condition is generally avoided by changing the system resonance conditions through harmonic filters. By minimizing the sharpness and depth of voltage notches using soft switching or better snubber circuits, the harmonics present in the voltage notch can be reduced.

2.3.6.4 Noise

Noise in a power system is unwanted electrical signals typically lower than 200 kHz superimposed on the power system phase or neutral system conductors. Noise in power systems can be caused by power electronic devices, control circuits and arcing equipment. It can be enhanced by improper grounding. A typical noise signal in power voltage is shown in Figure 2.21. Noise disturbs electronic devices such as microcomputers and PLC, which are used for the control of electrical equipment and thus the malfunction of the equipment. It also results in overloading of capacitors. Problems caused by noise can often be mitigated by using filters, isolation transformers, and line conditioners.

2.3.6.5 DC offset

DC offset in the AC waveform is mainly due to power electronic converters such as half-wave rectifiers, single phasing, asymmetry of converters and system transients in the power system. A typical DC offset in voltage is shown in Figure 2.22.

Many equipment and systems are susceptible to DC present in the AC supply voltage. Rectifier transformers and transformers in the distribution system can be damaged because of the DC current. Direct current flow in the transformer creates magnetic core saturation that results in increased losses, thus overheating of the

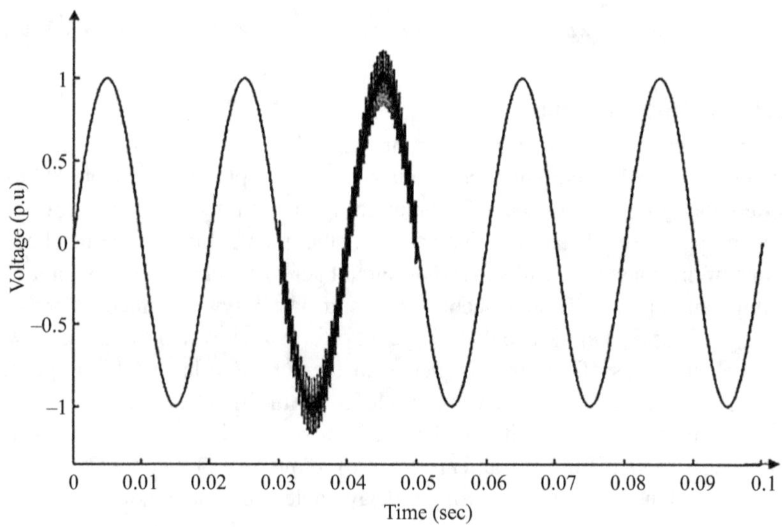

Figure 2.21 A typical voltage waveform with a noise signal

transformer beyond its design capacity. Consequently, DC offset may lead to overheating of the dielectric system, whether it is of solid (dry type or cast resin type) or oil-filled transformers, causing accelerated ageing of the insulation system, leading to its failure. By using proper mitigation techniques such as DC blocking capacitors, transformer and converter design modification or by using filters, etc., DC offset can be minimized. Table 2.7 gives a detailed summary of various power quality issues due to waveform distortion in power systems.

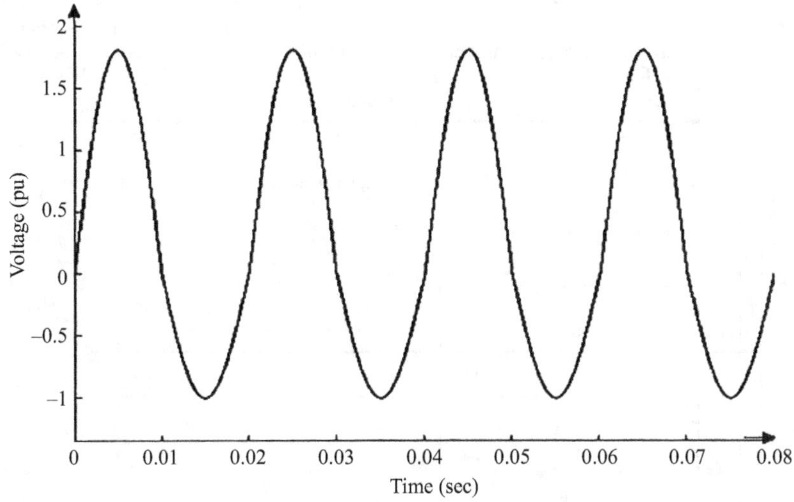

Figure 2.22 A typical voltage waveform with DC offset due to a half-wave rectifier

Table 2.7 Disturbances on account of waveform distortion

Disturbance type	**Description**	**Possible causes**	**Impact**
Harmonics	Distortion in voltage or current waveforms characterized by Total Harmonic Distortion (THD) and the presence of frequency components that are integer multiples of the fundamental frequency	Nonlinear electrical loads such as power electronic converters, rectifiers, and variable speed drives	Additional power losses, overheating of equipment, and degradation of the system power factor
Interharmonics	Frequency components that occur between the integer multiples of the fundamental frequency in the spectral content of the waveform	High-frequency switching converters, HVDC transmission systems, and industrial heating equipment such as arc furnaces	Generation of audible noise, mechanical vibrations, and possible interference with nearby electrical devices
Notches	Small periodic disturbances in the voltage waveform, typically caused during the switching or commutation process of power electronic devices	Switching actions in power electronic converters and controlled rectifier circuits	Accelerated degradation of capacitor banks and malfunction of sensitive electronic equipment

(Continues)

Table 2.7 (Continued)

Disturbance type	Description	Possible causes	Impact
Noise	Unwanted high-frequency signals superimposed on the normal power waveform	Industrial equipment such as arc furnaces, arc lamps, and various power electronic converters, SMPS etc.	Disturbances in electronic systems and interference with control or communication signals
DC offset	Presence of a direct current component in an otherwise alternating voltage or current waveform	Geomagnetic disturbances, asymmetrical system faults, or rectification effects in converters	Magnetic core saturation in transformers, leading to increased losses and overheating

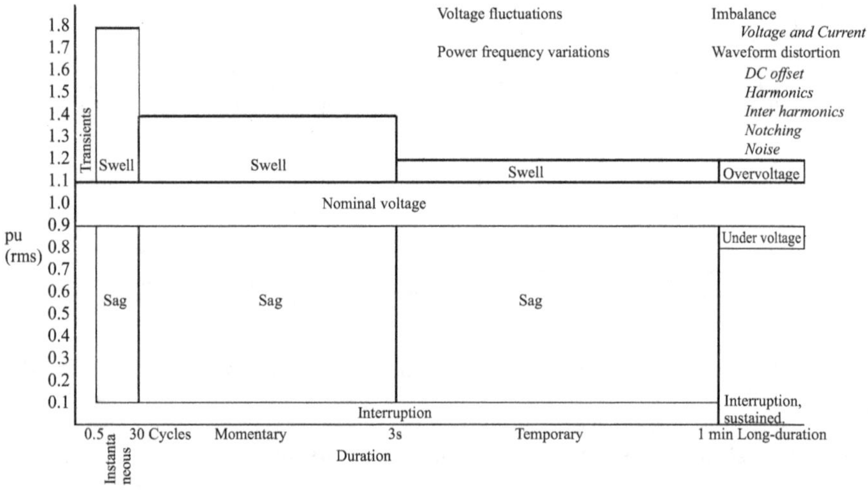

Figure 2.23 Summary of power quality issues with respect to their duration

2.3.7 Summary of power quality terms

It may be noted that power quality issues discussed in the above sections can be broadly categorized based on their duration and the magnitude of voltage. Accordingly, it can be grouped into four categories:

(a) Transients up to about 50 m/s
(b) Short-duration voltage sag, swell or interruption lasting up to about 1 min
(c) Long-duration over/undervoltage or interruption lasting more than 1 min
(d) Steady state – waveform distortion (DC offset, harmonics, inter-harmonics, notch, noise), imbalance, fluctuating voltage or frequency variation are of this category.

It may be seen that waveform deviation of undervoltage, overvoltage, imbalance and interruptions can occur regardless of linear or non-linear loads connected to the system. Whereas large amounts of non-linear loads, mostly of power electronics and emerging distributed energy resources such as solar PV, contribute to waveform distortion, i.e., harmonics, inter-harmonics, notches, noise, etc. Summary of these power quality issues is presented in Figure 2.23.

2.4 Standards and limits

Power quality has attracted enormous attention from both industries and researchers in recent times due to its adverse effects on the system. Hence, they jointly formulated standards and limits to cope with the issues. Table 2.8 indicates limits as per a few standards.

Table 2.8 Summary of power quality issues, limits and standards

Categories	**Typical spectral content**	**Duration**	**Voltage magnitude**	**Reference standards**
Momentary or transient disturbances	5 ns rise to 5 MHz	<50 ns to 50 ms	0 to 8 pu	IEEE 1159 IEC 61000 Part 4-4
Voltage sag/ voltage dips	Power frequency	Half cycle to several seconds.	0.9–0.1 pu	IEEE 1159 IEC 61000 Part 4-11
Undervoltage	Power frequency	Time > 1 min.	0.8–0.9 pu	IEEE Std 1159
Interruption (sustained)	Power frequency	Time > 1 min.	<0.1 pu	IEEE Std 1159
Voltage Swell	Power frequency	0.5 cycle–1.0 min	1.1–1.8 pu.	IEEE Std 1159
Overvoltage		>1 min	1.1–1.2 pu	IEEE Std 1159
Frequency deviation	Power frequency	<10 s	+0.1 Hz 49.90–50.05 Hz	IEEE Std 1159–2019 Indian Electricity Grid Code
Voltage unbalance/ imbalance	Power frequency	Steady state	3% for 33 kV and above	IEEE 519 Central Electricity Authority (CEA) - Technical Standard for Grid Connectivity
Voltage fluctuations	<25Hz	Intermittent Short-term flicker severity $P_{st} < 1.0$ Long-term flicker severity $P_{lt} < 0.65$	0.1%–7% During distributed generation synchronization within +5%	IEEE Std 1159 IEC 61000-3-3 CEA-Technical Stand. For connectivity for distributed generation
DC offset	DC	Steady state	Less than 0.5% of rated output	IEEE Std 1547 CEA - Technical Stand. For connectivity for distributed generation
Harmonics voltage	Integer multiple of power frequency	Steady state	THD – 5% with single harmonic content not exceeding 3%	IEEE 519 CEA- Technical Standard for Grid Connectivity

(Continues)

Table 2.8 (Continued)

Categories	Typical spectral content	Duration	Voltage magnitude	Reference standards
Harmonics current	Integer multiple of power frequency	Steady state	Individual and total as per IEEE 519	IEC 61000 Part 3-2 and Part 3-4 CEA-Technical Stand. For connectivity for distributed generation
Inter-harmonics For 60 Hz system	Non-integer multiple of power frequency	Steady state	0%–5% Based on short-term flicker severity $P_{st} < 1.0$	IEEE 519 IEC 61000-4-7
Notches		<0.5 cycles	Less than complete loss of voltage up to 0.5 cycles	IEEE 519
Noise	Below 200 kHz	Cyclic /series of random voltage change	About 1% of voltage	IEEE 1159

2.5 Problems and solutions

Problem 2.5.1

A 440 V (line-line), 50 Hz, 3-ϕ source is connected to a three-phase RL load with $R = 1\ \Omega$ and $L = 1$ mH. Write an expression for the instantaneous value of the line current and calculate the neutral current. Verify it using MATLAB® simulation.

Solution:

$$\mathrm{V_{L-L}} = 440\ \mathrm{V} \quad V_{ph} = \frac{440}{\sqrt{3}} = 254.03\ \mathrm{V} \quad V_m = \sqrt{2}\ \ V_{ph} = 359.253\ \mathrm{V}$$

$$\mathrm{Z} = 1.048\ \Omega \quad X_L = 0.314\ \Omega \quad \phi = 17.44^\circ$$

$$I_m = \frac{V_m}{Z} = 342.799\ \mathrm{A}$$

$$v_a = V_m\ Sin\ \omega t = 359.253\ Sin\ \omega t\ \mathrm{V}$$

$$v_b = V_m\ Sin\ (\omega t - 120^\circ) = 359.253\ Sin\ (\omega t - 120^\circ)\ \mathrm{V}$$

$$v_c = V_m\ Sin\ (\omega t - 240^\circ) = 359.253\ Sin\ (\omega t - 240^\circ)\ \mathrm{V}$$

$$i_a = Im\ Sin\ (\omega t - \phi) = 342.799\ Sin\ (\omega t - 17.44^\circ)\ \mathrm{A}$$

$$ib = Im\ Sin\ (\omega t - \phi - 120) = 342.799\ Sin\ (\omega t - 137.44)\ \mathrm{A}$$

$$ic = Im\ Sin\ (\omega t - \phi - 240) = 342.799\ Sin\ (\omega t - 257.44)\ \text{A}$$

$$\text{Neutral current} = ia + ib + ic = 0$$

The MATLAB model of the system is shown in Figure 2.24. The output voltage and current waveforms are shown in Figure 2.25.

From Figure 2.25, current lags voltage as an inductive load is connected to the system, and voltage and current are balanced in all three phases. The waveform for neutral current is given in Figure 2.26 It is shown that the neutral current is zero under balanced conditions.

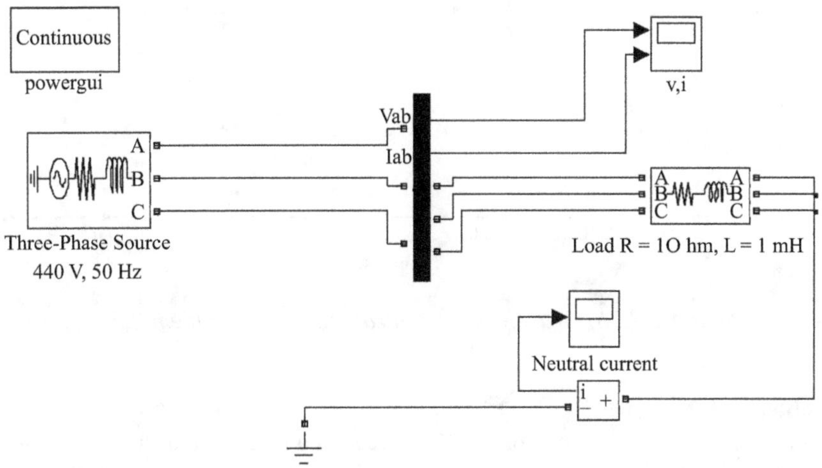

Figure 2.24 MATLAB model of three-phase system with R–L load – Problem 2.5.1

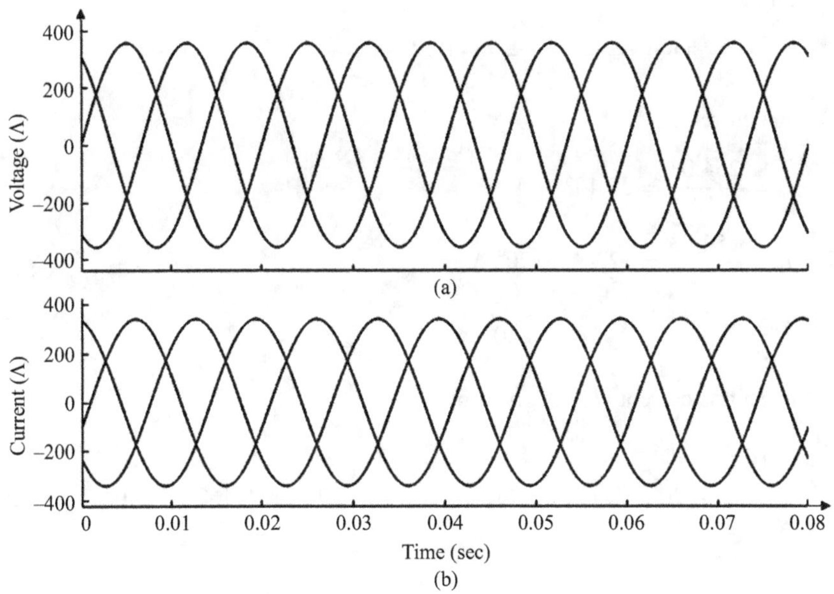

Figure 2.25 Waveforms of line voltage and current – Problem 2.5.1

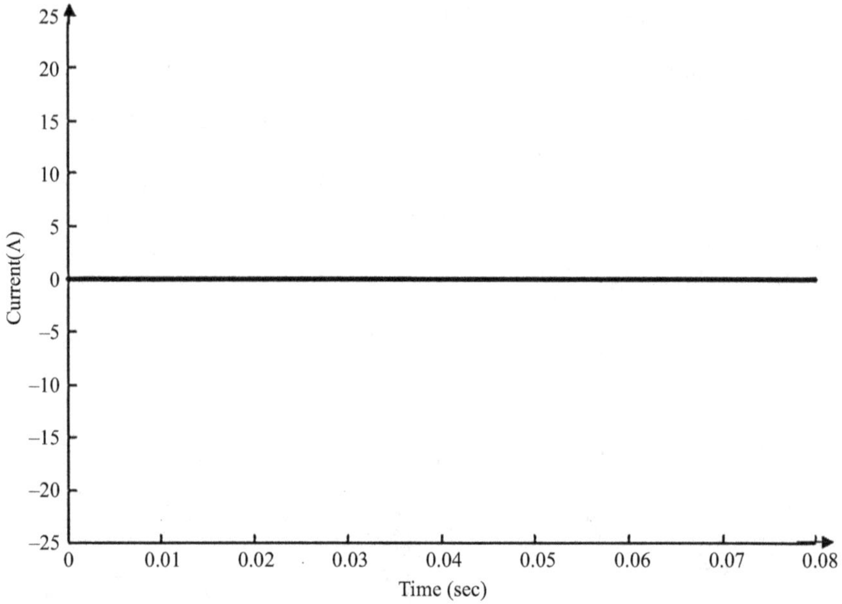

Figure 2.26 Waveform of neutral current – Problem 2.5.1

Problem 2.5.2.
A 440 V, 50 Hz, three-phase balanced source is connected to a three-phase unbalanced load with Ra = 10 Ω, Rb = 20 Ω and Rc = 30 Ω. Calculate neutral current. Verify it using MATLAB simulation.

Solution:

$$\text{Phase voltage } V_{ph} = \frac{440}{\sqrt{3}} = 254.03 \text{ V}$$

Line currents:

$$I_a = \frac{254.03}{10} = 25.403\angle 0 \text{ A}$$

$$I_b = \frac{254.03}{20} = 12.7\angle -120^\circ \text{ A}$$

$$I_c = \frac{254.03}{30} = 8.468\angle -240^\circ \text{ A}$$

As neutral current (I_N) is given by:

$$I_N = I_a + I_b + I_c$$

$$I_n = I_a + I_b + I_c = 25.403\angle 0 + 12.7\angle -120^\circ + 8.468\angle -240^\circ = 15.2 \text{ A}$$

MATLAB simulation results: The waveform of supply voltage and line current is given in Figure 2.27, and the neutral current is given in Figure 2.28.

It is seen from Figure 2.28 that the neutral current is sinusoidal and symmetrical, and the peak value is 15.2*$\sqrt{2}$= 21.4 A.

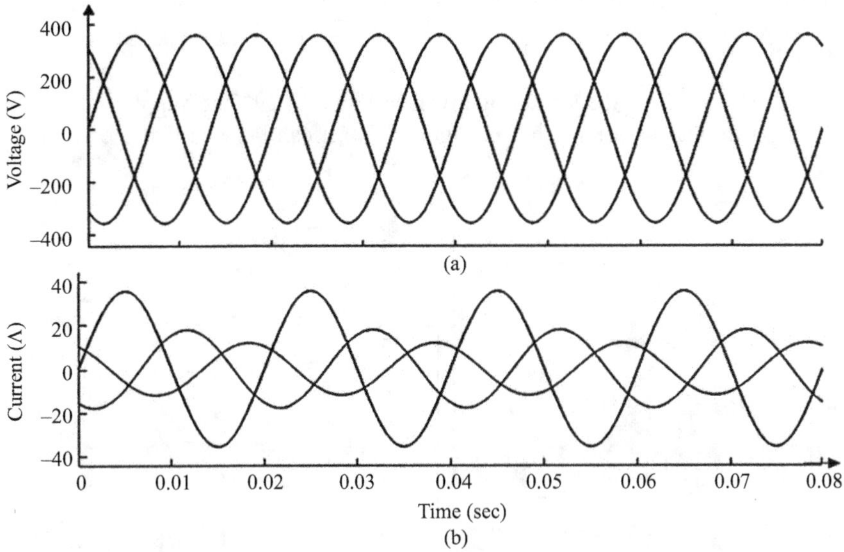

Figure 2.27 Waveform of voltage and current – Problem 2.5.2

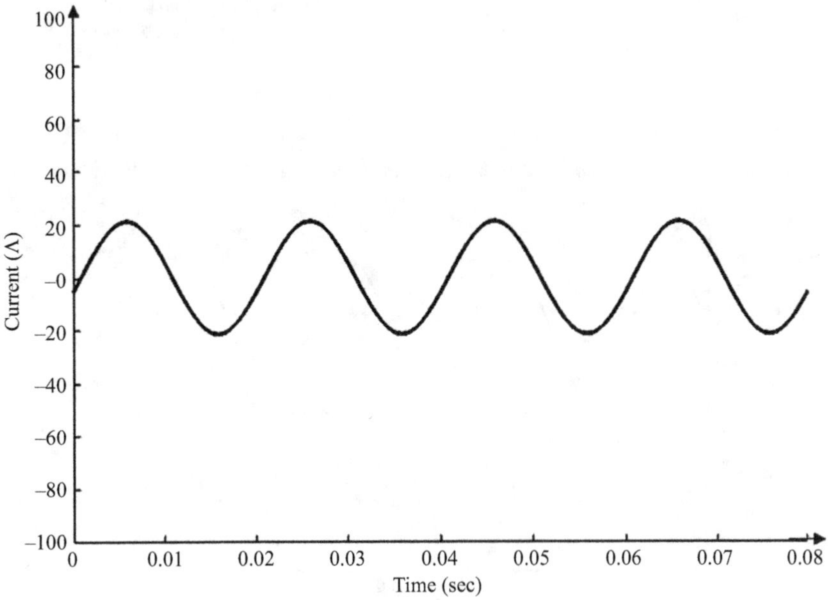

Figure 2.28 Waveform of neutral current – Problem 2.5.2

Problem 2.5.3

Consider a 440 V, 50 Hz three-phase system with a single-phase diode rectifier connected to each phase. Assume a resistive load of 10 Ω and a capacitor of 8000 μF connected at the output of each rectifier. (i) Plot input current in any one phase and its harmonic spectrum. Comment on the harmonics present. (ii) Plot neutral current and its harmonic spectrum and comment on the results.

Solution:

(i) MALAB model of the system is given in Figure 2.29. The supply voltage and current in phase-a are plotted in Figure 2.30. It can be seen that the voltage is

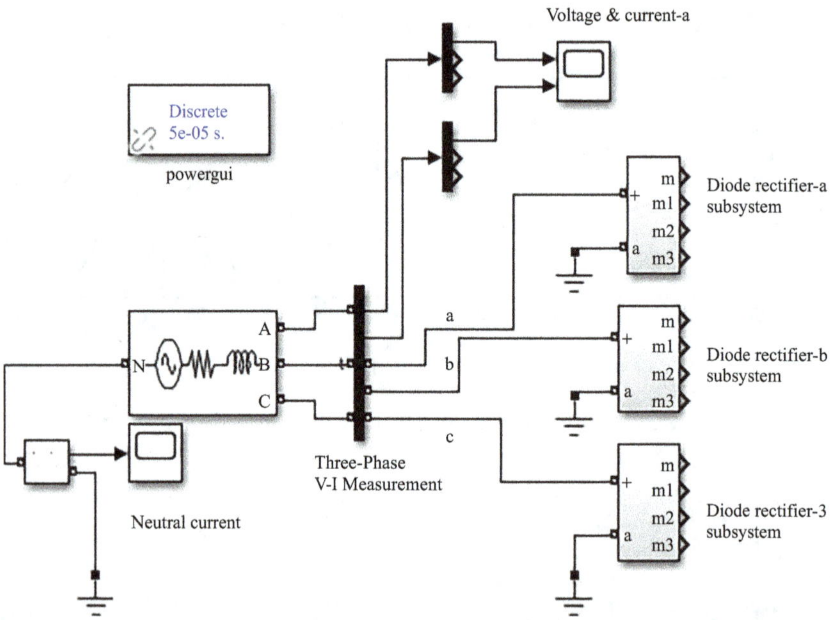

Figure 2.29 MATLAB model with rectifier load in each phase – Problem 2.5.3

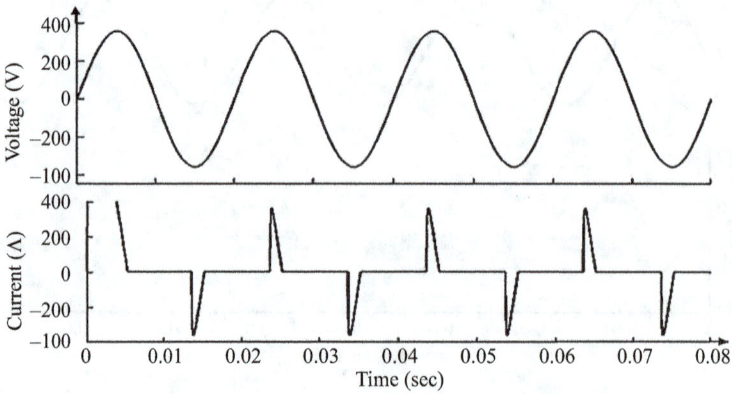

Figure 2.30 The waveform of supply voltage and line current – Problem 2.5.3

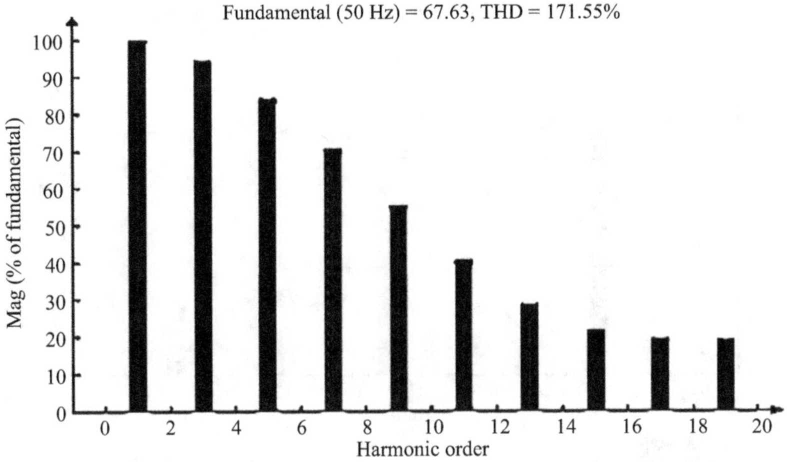

Figure 2.31 Harmonic spectrum of line current – Problem 2.5.3

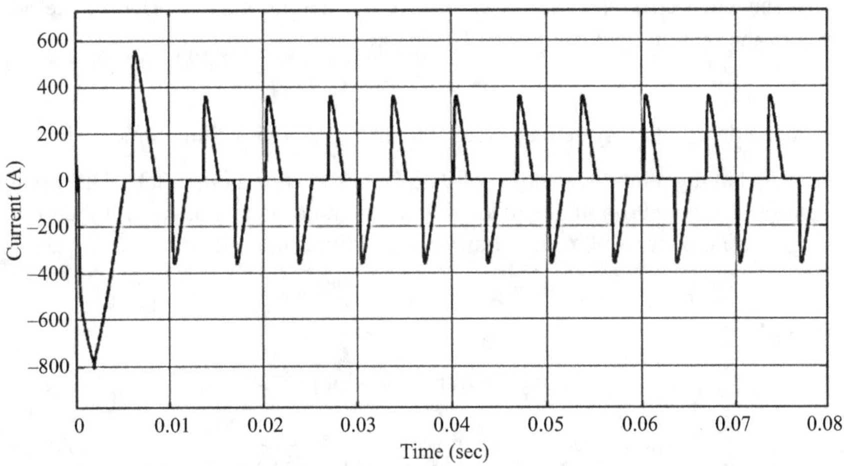

Figure 2.32 Neutral current – Problem 2.5.3

sinusoidal while the current is non-sinusoidal. The harmonic spectrum of line current is plotted in Figure 2.31. As a single-phase rectifier is connected in each phase, the characteristic harmonics are of the order 1, 3, 5, 7, 9, 11, etc., i.e., odd harmonics are present in the line current. The fundamental component of line current is 67.47 A (max).

(ii) Plot neutral current and its harmonic spectrum and comment on the results

The neutral current and its harmonic spectrum are plotted in Figures 2.32 and 2.33, respectively. It is found that the 50 Hz component is negligible in neutral

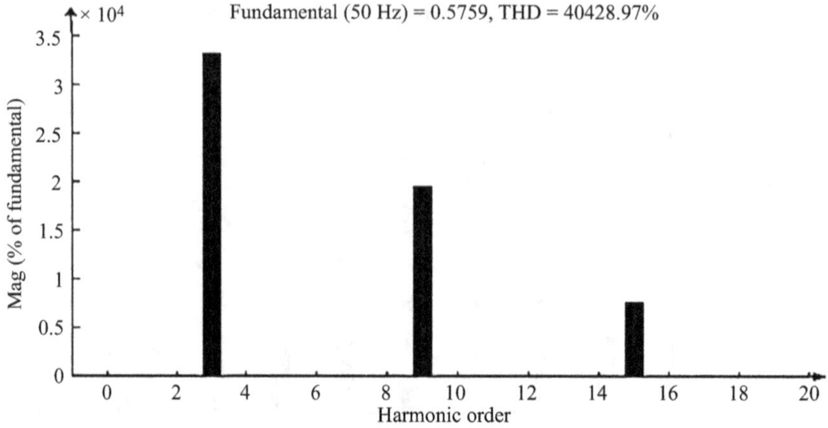

Figure 2.33 Harmonic spectrum of neutral current – Problem 2.5.3

current, and only the triple-order harmonic component, such as 3, 9, 15, etc., having a zero sequence characteristic, is present in the neutral current. Hence, the adequacy of the neutral conductor shall be checked.

Problem 2.5.4

Consider a distribution system connected to an 11 kV/415 V transformer using a $3\frac{1}{2}$ core cable having a total resistance of 0.25 Ω as shown in Figure 2.34. Discuss the effect of starting and running a 30 kW induction motor with a 0.8 power factor and having an efficiency of 90% that draws about 5 times the full current using a DOL starter.

Solution:

$$\text{Rating of Induction motor} = 30 \text{ kW}$$

$$\text{Full load current of the motor} = \frac{30 \times 1000}{\sqrt{3} \times 415 \times 0.8 \times 0.9} = 58 \text{ A;}$$

assumed pf as 0.8 and efficiency as 90% Starting current = 5 × 58 = 290 A

From the phasor diagram shown in Figure 2.35, the line-to-line voltage drop of the 3-phase line,

$$\Delta V = \sqrt{3} \times I \times r \cos \phi$$

where I is the load current, r resistance, x reactance and ϕ phase angle between voltage and current. Assuming a starting power factor of 0.5,

Line-to-line cable voltage drop during starting,

$$\Delta V = \sqrt{3} \times 290 \times 0.25 \times 0.5 = 62.79 \text{ V} = \frac{62.79}{415} = 15.13\%$$

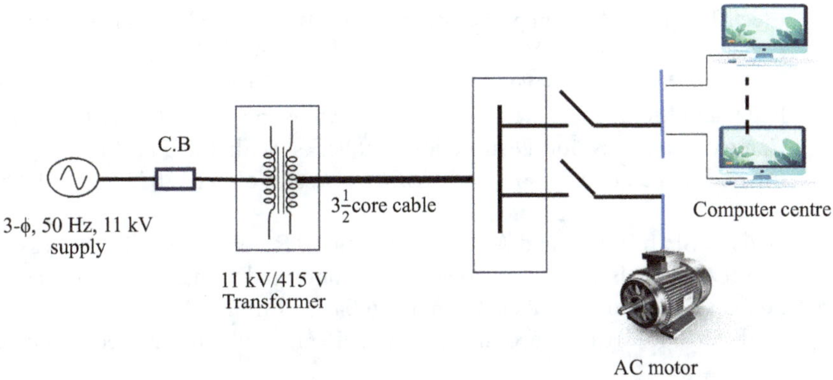

Figure 2.34 Distribution system – Problem 2.5.4

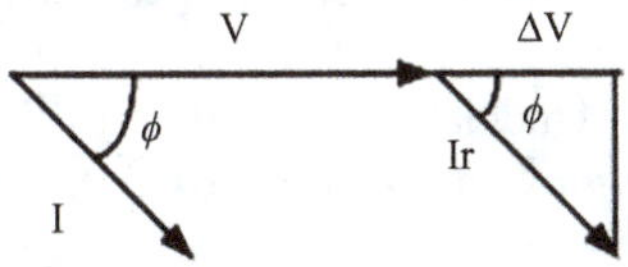

Figure 2.35 Phasor diagram – Problem 2.5.4

Voltage at distribution board = 415 – 62.79 = 352.21 V $= \dfrac{352.21}{415} = 84.87\%$

Voltage at terminals of computer = 352.21 or 84.87% of rated voltage
Line-to-line cable voltage drop ΔV during running $= \sqrt{3} \times 58 \times 0.25 \times 0.8 =$ 20.09 V $= \dfrac{20.09}{415} = 4.85\%$

Voltage at distribution board = 415 – 20.09 = 394.91 V $= \dfrac{394.91}{415} = 95.15\%$

Voltage at terminals of computer = 394.91 V or 95.15% of rated voltage

The impedance of the power cable to the distribution board for the motor causes a voltage drop of 62.79 V, or 15.13%, during starting and 20.09 V, or 4.85%, during running conditions. Because of the high voltage drop of 15.13% or under voltage conditions [refer Table 2.2] for the duration of starting, computer malfunction or need to provide with an alternative reliable power supply. Alternatively, reduce the voltage drop by limiting the starting using an electronic drive or a larger-sized cable that offers reduced resistance or providing a dedicated feeder for the computer facility.

References

[1] Sayed Mohammad Mousavi Gazafrudi, Adel Tabakhpour Langerudy, Ewald F. Fuchs, and Kamal Al-Haddad, "Power Quality Issues in Railway

Electrification: A Comprehensive Perspective", *IEEE Transactions on Industrial Electronics*, Vol. 62, No. 5, 2015, pp. 3081–3090.

[2] Darioush Razmi, Tianguang Lu, Behnaz Papari, Ehsan Akbari, Gholamreza Fathi, and Mojtaba Ghadamyari, "An Overview on Power Quality Issues and Control Strategies for Distribution Networks With the Presence of Distributed Generation Resources", *IEEE Access*, Vol. 11, 2023, pp. 10308–10325.

[3] Ishtiyaq Shafi Rafiqi, and Abdul Hamid Bhat, "Role of UPQC In Addressing Power Quality Issues in the Power Grid Connected with Renewable Energy Sources", held in SRINAGAR, India on 04–06 July 2022.

[4] IEEE Std 1159 IEEE Recommended Practice for Monitoring Electric Power Quality.

[5] 7th CEER-ECRB benchmarking report on the quality of electricity and gas supply 2022, Council of European Energy regulators (CEER).

[6] Report on Power Quality of Electricity Supply to the Consumers August 2018 – Forum of Regulators, Central Electricity Regulatory Commission, India.

[7] Ned Mohan, Tore M Undeland, and William P Robbins, *Power Electronics Converters, Applications, and Design*, 1995, Second Edition, John Wiley & Sons.

[8] IEEE Std 1100 IEEE Recommended Practice for Powering and Grounding Electronic Equipment.

[9] IEC 60044-4 Surge Arrester, Part 4 Metal Oxide Arresters without air gaps for AC system.

[10] IEC 60364-4-44 Low-voltage electrical installations – Protection for safety - Protection against voltage disturbances and electromagnetic disturbances.

[11] IEC 61000-3-3 Electromagnetic compatibility (EMC) – Part 3-3: Limits – Limitation of voltage changes, voltage fluctuations and flicker in public low-voltage supply systems, for equipment with rated current ≤ 16 A per phase and not subject to conditional connection.

[12] Reactive Power Management and Voltage Control Ancillary Services (VCAS) in India, March 2021, Power System Operation Corporation of India. https://posoco.in › wp-content › uploads › 2021/08

[13] Central Electricity Authority (Technical Standards for Connectivity to the Grid) Regulations, 2019.

[14] IEC 60034 Rotating electrical machines Part 26: Effects of unbalanced voltages on the performance of three-phase induction motors http://www.forumofregulators.gov.in/Study_Reports

[15] IEEE Std 519 IEEE Recommended Practices and Requirements for Harmonic Control in Electric Power Systems.

Chapter 3

Harmonics in power supply systems: sources, effects and regulation

Harmonics in power supply frequency are integer multiples of the fundamental power frequency of current or voltage

Goals of the chapter

After going through the chapter, the reader shall get a good idea about the most important power quality issue, 'harmonics', in today's power system, where power electronic-based equipment is increasingly used in transmission, distribution and utilization areas along with emerging distributed energy generation and microgrids. The chapter would also be intended to have a better awareness about the sources of current and voltage harmonics, the impact on equipment and sectors susceptible to disturbances under harmonic environment. To enhance the concepts, simulations using MATLAB® are additionally included. Finally, the reader shall get familiarised with the various international standards such as IEEE and IEC on harmonics.

3.1 Introduction

In the previous chapter, we discussed power quality issues, their various terms that define the deviations in the pure sinusoidal nature of power supply, along with its causes and effects. With the advancement of power electronic equipment in recent times, there is a growing challenge to power quality due to harmonics. Thus, in recent times, harmonics have become an important topic for industry, research and development. As detailed, harmonics are produced by non-linear equipment. These types of equipment draw current not in proportion to the applied voltage waveform, as in the case of linear loads. Most of the power electronic-based equipment, viz. motor drives, electric traction, battery charging for various applications, including for electric vehicles (EV), switch mode power supplies (SMPS), light emitting diodes (LED), etc., generate harmonics in the power system. Transmission line performance enhancement equipment, such as STATCOMs, SVCs, and HVDC connections, also belongs to this category. Additionally, grid-connected renewable energy sources like solar photovoltaic (SPV), wind turbine generators (WTG) also inject harmonics into the system. The current harmonics thus injected result in distorted voltage waveforms that affect the entire power system and equipment connected in the system. A block diagram showing classification of conventional and power electronic equipment used in various areas of power systems such as generation, transmission, distribution and utilization are depicted in Figure 3.1.

3.2 Sources of voltage and current harmonics

As shown in Figure 3.1, generation of current and voltage harmonics in power system can be grouped into various sectors of electrical power, viz. (i) power generation, (ii) transmission and distribution and (iii) utilization. The following sections detail these sources for analysis.

3.2.1 Sources of harmonics in power generation

Conventional power generating stations such as hydro, thermal and nuclear have supplied electrical energy to homes and industries for decades, driving global economic growth. Though it is highly efficient for large-scale energy production, it depends on geographical resources like abundant water and fossil fuels like coal, lignite, petroleum gas, diesel, uranium, etc. These resources are localized and non-renewable. This has resulted in a global movement towards more reliable, sustainable and renewable alternatives. Also, the power plants that depend on fossil fuels like coal and natural gas are a major contributor to greenhouse gas emissions, significantly accelerating global warming. To reduce greenhouse gas emissions and offer sustainable alternatives, modern power generation increasingly integrates renewable energy sources like solar, wind, etc. This has resulted in distributed power generation into the existing power system. Conventional power generation generally produces fewer harmonics, whereas distributed power generation often injects considerable harmonics into the grid. The following subsection details the generation of harmonics from these generating sources.

3.2.1.1 Harmonics in conventional power generation

Over the past half century or so, the focus has primarily been on the design, construction and operation of large-capacity generating stations and associated transmission and distribution systems to supply electricity to the consumers. This emphasis on large-capacity generation is centred around conventional plants viz hydro (depends on the availability of water at sufficient head), fossil fuel-fired (both oil and coal) and nuclear power plants.

The generators in hydro stations are of relatively low speed (n) with a higher number of poles (p), to match the grid frequency $\left(f = \frac{pn}{120}\right)$ compared with that in fossil-fired or nuclear power plants, where high-speed (3000 rpm for 50 Hz or 3600 rpm for 60 Hz) generators are employed. While the MVA ratings of installed hydro generators are relatively governed by waterfall, fossil generators are moderate up to about 660 MW and nuclear power plants are high up to about 1000–1200 MW. What is common in these generators is that all are synchronous

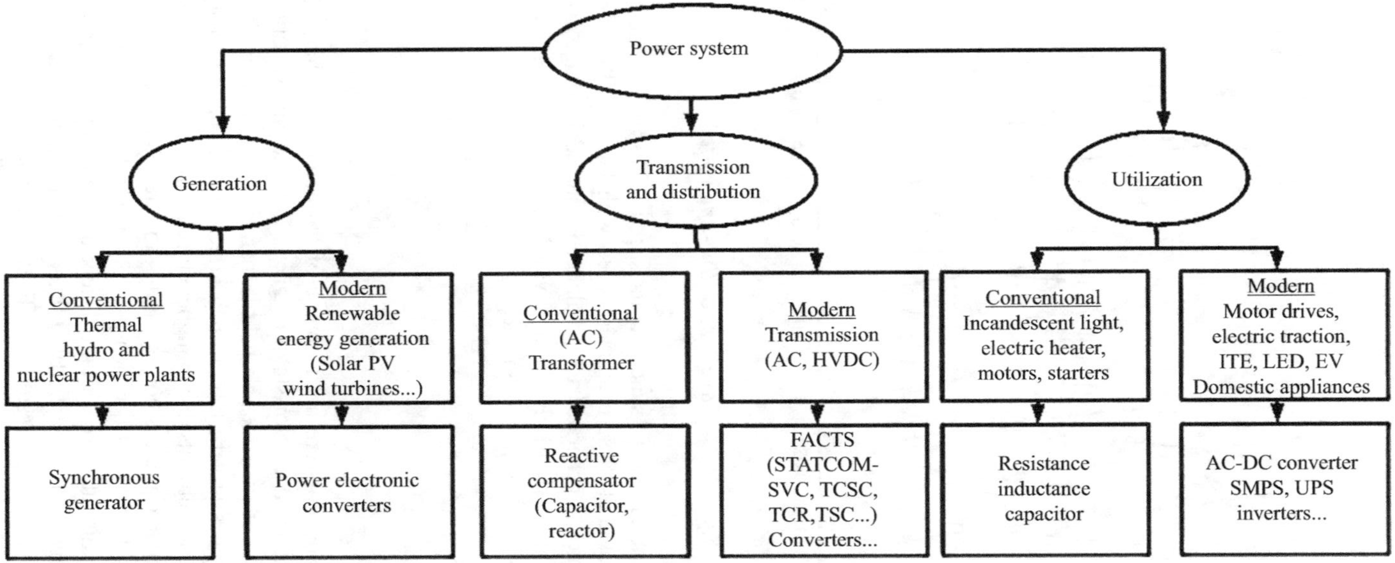

Figure 3.1 Block diagram of conventional and modern power generation, transmission, and utilization

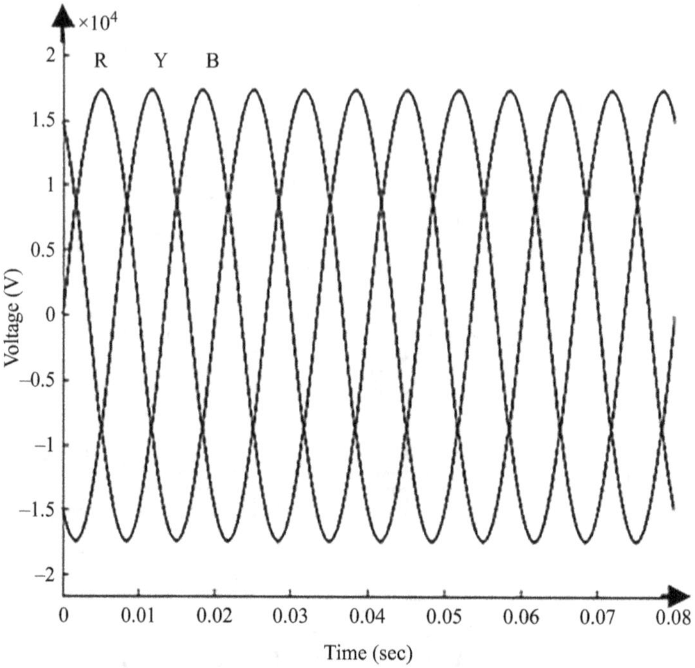

Figure 3.2 Waveform of line-line voltage

machines, rotating at the same synchronous speed with grid frequency (50 or 60Hz). While the rotor of these machines is fed with DC and 3-ϕ windings are distributed in the stator, it generates a 3-ϕ balanced sinusoidal voltage displaced by 120°. A typical generated sinusoidal voltage waveform is shown in Figure 3.2. The sinusoidal voltage is expressed mathematically as:

$$\begin{aligned} v_R &= V_m \sin \omega t \\ v_Y &= V_m \sin(\omega t - 120^\circ) \\ v_B &= V_m \sin(\omega t - 240^\circ) \end{aligned} \tag{3.1}$$

Synchronous generators are not a source of harmonics in voltage. However, due to asymmetry involved in rotor-stator magnetic flux path and variation in stator to rotor air gap, etc., it generates very little harmonics in voltage. Though the magnitude of harmonics is very low, it may affect interference to adjacent communication or protection circuits due to higher current and associated magnetic flux linking with it. Harmonic indices corresponding to interference are termed as telephonic interference factor (TIF), which is explained in the next chapter. With a view to minimising this interference on adjacent data communication circuits due to this stator harmonics in voltage, IEC 60034-1 standard for rotating machine, specify maximum limit of the resulting TIF as 1.5% for synchronous machines above 5 MW, considering measurement up to 5kHz. The standard also gives limiting values for other ranges of machines and weighting factors for different components of voltage harmonics [1].

3.2.1.2 Harmonics in renewable energy power generation

In recent years, there has been a phenomenal increase in the use of renewable energy (RE) that is abundantly available in the atmosphere for power generation. In line with the world over, India has also been integrating renewables, particularly wind and solar energy, with an exponential rise in line with the country's energy security and commitment towards greener energy targets. For synchronization of different systems, the magnitude, frequency and phase sequence of the incoming power system should match the corresponding AC grid parameters. All these systems generally use power electronics coupling as an intermediatory to match the synchronizing and other requirements, for example metering, harmonics, DC injection, reactive power exchange with the grid, etc. [2,3]. Grid connection of a renewable energy generator to the AC power grid is shown in Figure 3.3.

These power converters use switching semiconductor devices that operate at a few kHz for power conversion. These processes produce harmonics in the system, which distort the currents and voltages in the system. We shall discuss this in more detail in subsequent sections.

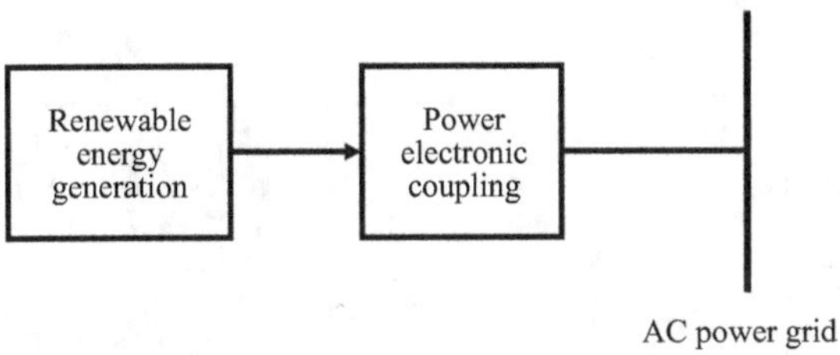

Figure 3.3 Block diagram of a renewable energy generator with an AC grid

3.2.1.2.1 Harmonics in solar photovoltaic generation

India lies in the equatorial sun belt of the earth, thereby receiving abundant solar energy. Most parts of India experience clear sunny weather for nearly 250–300 days a year. Solar energy is harnessed mainly in two ways: (a) Using the thermal content of solar radiation to generate steam and using turbines and generators to produce electricity and (b) converting the solar irradiation into electricity using solar photovoltaic (SPV) cells.

Solar photovoltaic cells are generally made up of either crystalline, polycrystalline, amorphous silicon semiconductor or compound semiconductors such as Cadmium Telluride and Copper Indium Gallium Selenide. When such cells are exposed to sunlight, they absorb photons and release free electrons. Thus, it converts sunlight to DC power (typically about 1.5 W/cell). As the single PV cell does not produce enough power, several cells are electrically connected in series and parallel to form an SPV module of higher power capacity (e.g., about 110 W at 12 V).

These modules are further connected in series (string) to add up the voltage and in parallel (array) to add up the current to realize a higher voltage, higher power source.

Electric power from SPV units is converted to alternating power by a DC–AC converter (inverter). Additionally, a DC–DC converter is often used to maximise the power from the SPV module. SPV inverters are either 1-ϕ or 3-ϕ, with the majority being 3-ϕ systems. Depending upon the application, SPV is termed as a standalone (isolated from the grid), grid-connected (grid-tied) or hybrid (connected to the grid and battery as backup source) SPV system. In case of grid-connected system, power generated from the SPV system can be utilized by the users for their own consumption or can be exported to the utility grid. If the power generated by SPV is more than the load connected within their own premises, the surplus power is delivered to the grid. In general, these are fully controlled, IGBT-based systems using pulse width modulation (PWM) technique to transfer the power to the grid. However, the switching operation of the IGBT injects harmonics into to the grid, which distorts the grid voltage. Figure 3.4 shows a simple grid-connected SPV system connected through a DC–DC converter, IGBT-based PWM inverter.

Figure 3.5(a) indicates a typical SPV voltage waveform, and Figure 3.5(b) shows its enlarged view. An enlarged view shows that the waveform contains higher frequency harmonics that depend on the switching technique used.

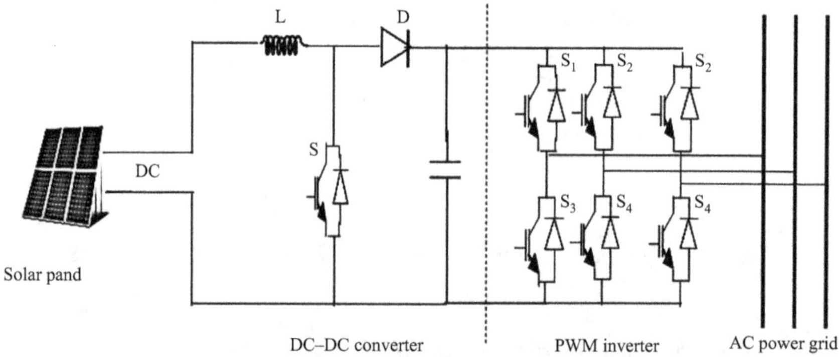

Figure 3.4 Scheme of a grid-connected SPV system

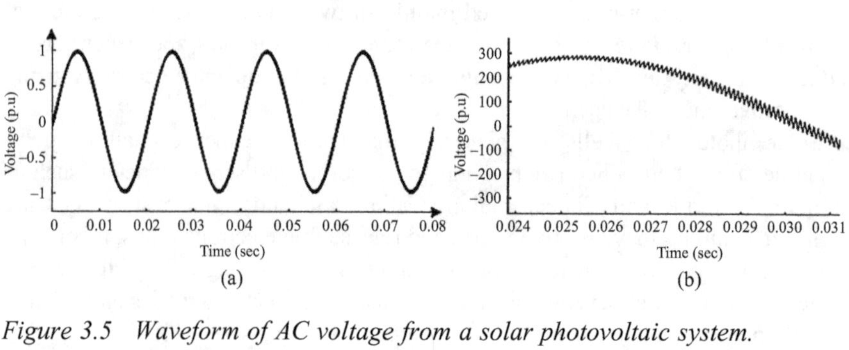

Figure 3.5 Waveform of AC voltage from a solar photovoltaic system. (a) Waveform of output voltage SPV system. (b) Enlarged view of the output voltage of SPV system.

3.2.1.2.2 Harmonics in wind energy power generation

In wind power plants, the kinetic energy of moving air is converted into mechanical energy by directing it on the rotor blade of the wind energy turbine (WET). The rotational energy of WET is then converted to electrical energy by the generator. After matching the synchronization requirement, it is connected to the grid for transmission of electrical power.

Wind energy is site-specific. Accordingly, during the last few decades, various schemes of wind turbine generator (WTG) systems have been built to harness wind energy. According to their construction, WTGs are grouped into vertical-axis and horizontal-axis wind turbines, with the majority (more than 98%) of installations forming in later group. Also based on installation as onshore and offshore, with the later forming WTG of higher capacity due to the availability of steady wind flow. Both onshore and offshore WTGs of larger capacity are interconnected to the main grid at a higher voltage level. However, offshore plants are interconnected with submarine cables and often with HVDC to the main grid. WTGs of the recent scheme use more power electronic equipment for sinusoidal power generation, synchronization and efficient control. Various types of wind energy systems are briefly explained in the following subsections [4].

Wind turbine generator (WTG): Type 1. The Type 1 wind turbine generator system is the oldest design among wind turbines, and its typical layout is shown in Figure 3.6. In general, a gear box is used to couple the turbine shaft with the generator. It may be noted that these systems do not use any power electronics and are fixed speed systems.

It employs a squirrel-cage induction generator for mechanical to electrical power conversion. Since it is not self-exciting, it needs an excitation source of grid frequency to generate the rotating magnetic field. It starts to produce power at a threshold wind speed (called the cut-in wind speed) of about 4 to 5 m/s. At very high wind speeds (typically beyond 25 m/s or so), the system is stalled to prevent damage to the turbine blades. This is called the cut-out speed of the wind. When the turbine rotates faster than the electrical grid frequency, the squirrel cage induction machine acts as a generator and produces real power. The turbine rotates close to electrical grid frequency and hence is virtually a constant speed machine. It requires reactive power support from the grid which is achieved by connecting switched

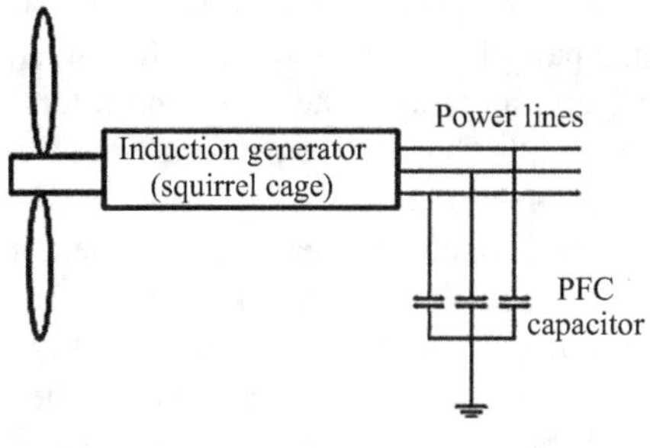

Figure 3.6 Scheme of a Type 1 wind turbine generator system

capacitor banks in the power line as shown in Figure 3.6. The power is controlled by passive stall control, active stall control and pitch control. These systems require a soft starter to reduce the impact of inrush current from the grid. These systems are cost-effective and reliable, however their response time is slow.

Wind turbine generator (WTG) with partial rated power converters: Type 2. A Type 2 WTG employs a wound rotor induction machine. The stator circuit is identical to that of the Type 1 WTG, with the addition of a resistor (through power electronics – typically a low voltage high current partial rating power converter) either internally within the rotor or externally mounted through slip rings. The variable resistor enables control of rotor current, which results in a variation of electromagnetic torque and the speed at which it is delivered. Through this resistor, a variation in speed above the synchronous speed in the range 0–10% can be achieved [5]. A simple layout of a Type 2 grid-connected wind energy system is shown in Figure 3.7. This scheme can tolerate higher wind speed while maintaining a fixed output power and these systems come under the category of variable speed wind turbines. Note that a Type 2 system also employs both a soft starter as well as a reactive power compensator in the form of a switched capacitor bank.

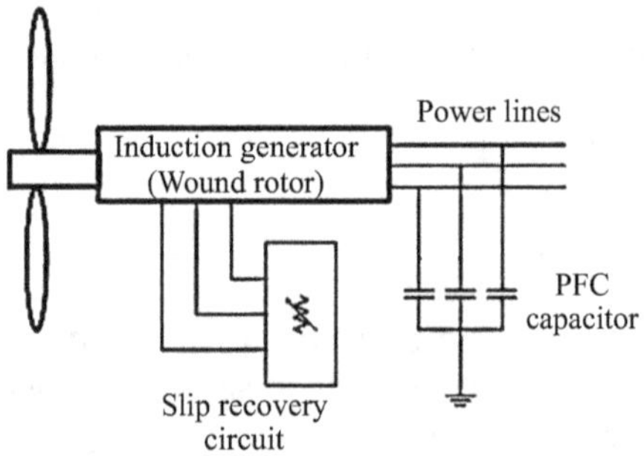

Figure 3.7 Scheme of a Type 2 wind turbine generator system

Wind turbine generator (WTG) with partial rated power converters: Type-3. In this type of WTG, a doubly fed induction generator or a doubly fed asynchronous generator is used. Instead of the power electronics supported slip recovery circuit or variable resistor used in type 2 WTG, a variable frequency ac excitation is applied to the rotor by inserting a pair of AC–DC and DC–AC converters between the rotor and the grid in a Type-3 configuration. The stator and rotor, are both connected to the grid as shown in Figure 3.8. Thus, active power always flows out from the stator and is independent of the operating state. The rotor, on the other hand, absorbs power when operating as a motor and delivers it when operating as a generator. The rating of the converters is about 30% of the machine rating and hence the name partial rated power electronics. By means of a Type 3 configuration, it is possible to accommodate up to 30% speed variation above or below the synchronous speed. It is also possible to control the reactive power or import and export reactive power.

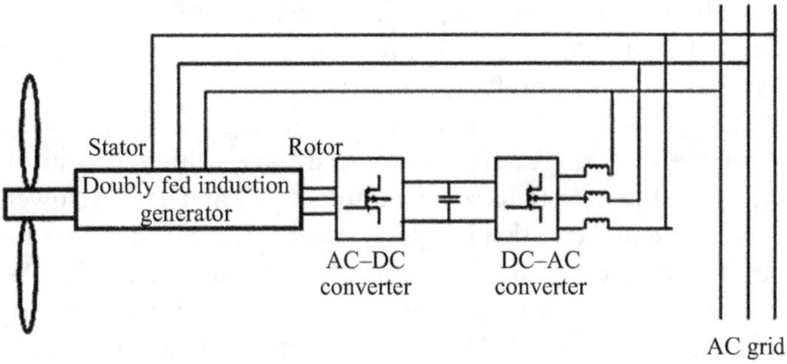

Figure 3.8 Scheme of a Type 3 wind turbine generator system

When Type 3 WTG systems operate in super synchronous mode, both the stator and the rotor support power flow, while in sub-synchronous mode, the electrical power flows into the rotor from the grid. Moreover, in this type of system, it is possible to independently control the active and reactive power by regulating the rotor flux by adjusting the rotor currents i_{rq}, i_{rd} respectively, in the d-q reference frame.

Variable speed turbine with full power rating converter: Type-4. The block diagram of the wind energy turbine system with a fully rated power converter scheme is shown in Figure 3.9. As the generator speed depends on the speed of wind, the frequency of the voltage at the terminals of the generator varies. Therefore, the generator output terminals cannot be directly connected to the grid. Instead, a back-to-back converter is introduced between WTG and the grid to match the grid requirement. In this case, the converter is of full capacity. Generally, an induction/synchronous generator or a multi-pole synchronous or PM synchronous generator is used in Type 4 systems. Neither a soft-starter nor a reactive power compensator is required for Type 4 systems. The power converter arrangement that connects stator with the grid can quickly control the active and reactive power besides holding the DC link voltage constant.

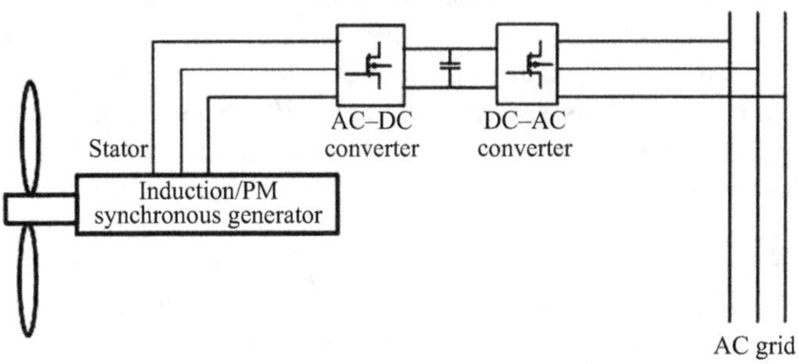

Figure 3.9 Scheme of a Type 4 wind turbine generator system

In terms of WTG performance, Type 1 and Type 2 WTGs are relatively earlier designs of lower capacity and require capacitors for reactive power compensation. However, they produce negligible harmonics as they are directly coupled to the grid without an additional switching interface. These systems cannot control the flow of reactive power and have a limited control over the active power. Also, these systems require a soft starter and have a slower response time when compared to Type 3 and Type 4 systems. Hence, islanding and standby operation are not possible with these kinds of systems. Generally, Type 3 and Type 4 WTGs are more popular for higher capacities. These variable-speed wind turbines are equipped with PWM controlled inverters schemes and are very efficient in controlling the active and reactive power flow to the grid; however, they spill harmonics into the grid. Type 4 WTGs are also popular among offshore generation systems. These systems have faster response times and lower investment but incur higher maintenance costs. These systems have standby and islanding operation capabilities.

Clearly, a WTG system also has many challenges like controlling harmonics, power factor, voltage regulation, etc., and needs to follow regulations as stipulated by the grid standards.

3.2.2 Sources of harmonics in power transmission and distribution system

The transmission and distribution system is mainly composed of (i) power transformers, (ii) transmission system power enhancement equipment (iii) utilization system load. The following sections detail these sources for analysis.

3.2.2.1 Harmonics due to power transformers

The main power transferring equipment in the transmission and distribution system is the transformer. At the generating station, it steps up the generated voltage to the transmission line voltage level at one or more stages and at the receiving side, steps it down to sub-transmission or distribution voltage level. It has windings on primary, secondary or tertiary depending on application and these are wound on cold-rolled-grain-oriented (CRGO) silicon steel. Due to the non-linearity of the magnetic characteristic of the silicon steel, i.e., flux density (B)–magnetisation force (H) curve, the excitation current has a significant amount of 3rd and 5th harmonic components. A typical magnetizing current, along with the predominant 3rd harmonic component, is shown in Figure 3.10. Since the no-load current is about 0.2%–2% of full load current, these harmonics have an insignificant effect except under core saturation, which is avoided by various protection schemes of the transformer. Still, measures like inter-linking of magnetic circuits, delta-connected winding primary or secondary windings are adopted to reduce harmonics. In some applications, an additional delta-connected tertiary winding, called stabilising winding, is provided for the third harmonic magnetizing currents to flow in a closed delta circuit, making induced voltages almost sinusoidal.

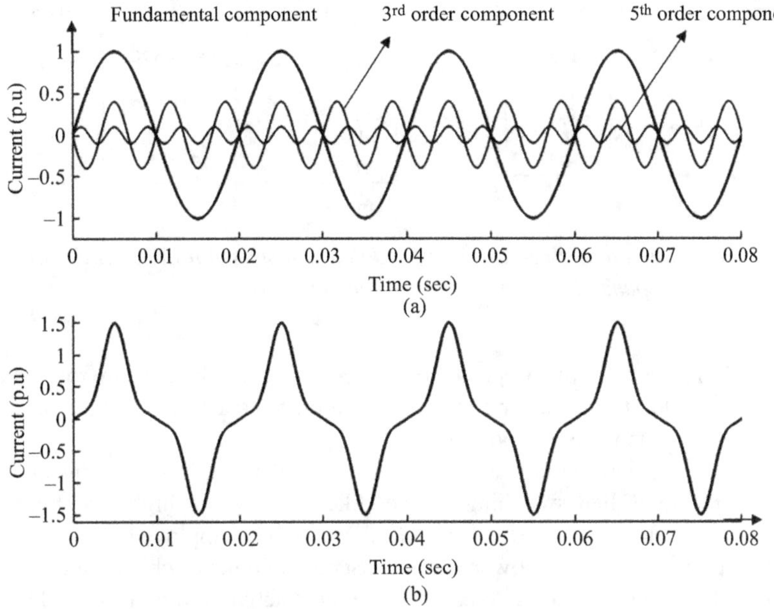

Figure 3.10 Typical waveform of magnetizing current of transformer. (a) Fundamental, 3rd and 5th order harmonic component and (b) total magnetizing current.

3.2.2.2 Harmonics due to transmission enhancement equipment

To make power transmission more efficient, flexible alternating current transmission system (FACTS) and HVDC systems are introduced. They have a lot of advantages in controlling various parameters such as voltage and power. However, it injects harmonics into the system [6]. We shall analyse these equipment and systems in the following subsections.

3.2.2.2.1 Flexible alternating current transmission system

As discussed in Chapter 1, power transmission lines are subjected to voltage variations due to changes in real and reactive power from generation to load, both under steady state and dynamic conditions. The voltage variations also depend on the impedance at the point of common coupling from the supply end, which is high at the downstream end of the transmission line. Accordingly, various types of compensation devices or systems have been developed and installed in the transmission lines that directly or indirectly control the reactance of the transmission line. Conventional methods of series and shunt compensation are by introducing passive elements like a capacitor or inductor using mechanical switches to draw leading or lagging compensating current. Such compensation, which can be represented as an independent current source in case of shunt compensation and an independent voltage source in the case of series compensation, as shown in

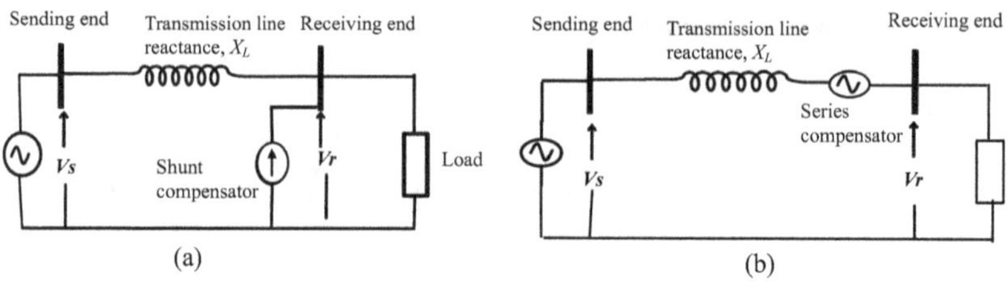

Figure 3.11 Equivalent circuit of compensated transmission line. (a) Shunt compensation and (b) series compensation.

Figure 3.11. These types of compensation will not inject harmonics into the system. However, they don't provide smooth compensation for reactive power, or in other words, dynamic control is not possible.

However, modern compensation techniques that replace the mechanical switches with controlled switching devices like thyristors or high switching frequency devices such as MOSFET or IGBT are widely employed for control of voltage or power in today's power systems. Such equipment, which are classified under FACTS devices, are used to compensate the reactive power consumed by the transmission line, which is the main culprit for voltage at various parts and the power transfer capability of power system. Thyristor-controlled compensation techniques are very effective for achieving dynamic control of various parameters; however, switching of thyristor introduces harmonics into the system. For example, under no load or lightly loaded conditions, reactive power (var) compensation can be achieved by connecting a thyristor controlled reactor (TCR) in parallel with the loads as shown in Figure 3.12(a). TCR consists of antiparallel thyristor switches in which one device conducts during the positive half cycle and the other during the negative half cycle. A transformer is used to interface a variable reactor X_L in parallel with the grid. As the load condition changes, the value of the shunt reactor must also change. This is done by controlling the firing angle (α) of the thyristor. This will vary the magnitude of the current through the shunt reactor, which indirectly controls X_L. At $\alpha = 0°$, the current through TCR is sinusoidal, and it becomes non-sinusoidal as α varies from 0° to 90°. Though it is non-sinusoidal, its fundamental component will always lag by 90° with respect to voltage. A TCR with inductor value 5 mH with $\alpha = 30°$ is supplied with a 230 V (peak) supply is simulated in MATLAB® and the waveform of TCR voltage and current is plotted in Figure 3.12(b). It is seen that this current is non-sinusoidal, and its fundamental component lags the voltage by 90°. The harmonic spectrum of TCR current is given in Figure 3.12(c). Peak value of fundamental component of current is 54.08 A and THD is 38.21%.

The current through TCR is given by:

$$i_L(t) = \frac{1}{L}\int_{\propto}^{\omega t} v(t)dt = \frac{V}{\omega L}(\sin \omega t - \sin \alpha) \tag{3.2}$$

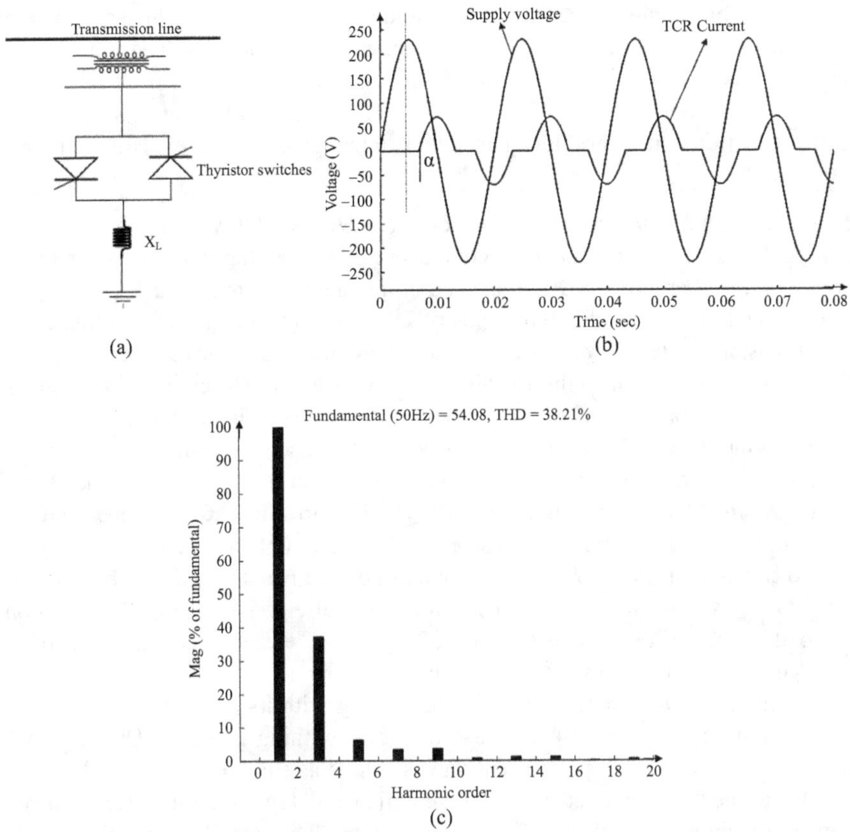

Figure 3.12 Thyristor-controlled shunt reactor. (a) Connection to a transmission line, (b) waveform voltage across TCR and its current and (c) harmonic spectrum.

where α is measured from the peak of the voltage and varies between 0 and $\frac{\pi}{2}$

$$v(t) = V_m \cos \omega t, \; L \text{ is inductance of reactor}$$

When it is non-sinusoidal, in addition to the fundamental component of current, odd harmonic components 3, 5, 7... etc. are also present in the waveform. In other words, TCR injects an odd harmonic component of current into the system. If the firing angle is different in each cycle, TCR injects non-characteristic harmonics too. Similarly, all other thyristor-controlled compensators such as thyristor switched capacitor (TSC), thyristor-controlled series compensator (TCSC) etc., used for better performance of the power system, may inject harmonics into the system. These harmonics may affect various other equipment connected to the system. IEEE 519-1992 indicates that TCR generates current harmonics of 3rd order as high as 13.78%, of 5th order 5.05%, of 7th order 2.58% of fundament component, etc. Thus, odd harmonics are injected into the system whose magnitude reduces as the order of harmonics increases. If the fundamental component of the current is higher, the magnitude of harmonics also increases proportionally. These harmonics can be reduced using high switching frequency devices such as MOSFET, IGBT, etc. Such equipment are SVC, STATCOM, SSSC, active filters, etc., which do the similar function with very few harmonics injections, and is explained in later chapters.

3.2.2.2.2 High-voltage direct current (HVDC) systems

As described in Chapter 1, the core component of HVDC transmission is the power converter. It converts AC to DC at one end of the line and converts back DC to AC at the other end. There are different types of HVDC systems converter configurations using thyristors or IGBTs for power conversion. A simple scheme of sending end side of an HVDC system using a thyristor-based six pulse AC to DC converter is shown in Figure 3.13. A similar arrangement will be there in the receiving end to convert DC to AC and its interface with the AC grid. The source reactance in the figure represents the transformer and transmission line reactance till the point of installation of the HVDC system. A smoothing reactor is used at the DC side to filter AC component present during the conversion process so that a steady DC is obtained at DC side [6]. As detailed earlier, converters produce harmonics of order given by $pn \pm 1$, where p is the number of pulses and $n = 2, 3 \ldots$. Thus, for a six-pulse converter, input current contains odd harmonic components such as 5, 7, 11, 13… etc., and the magnitude of 'h' order harmonic components is $\frac{1}{h}$ of fundamental component.

A simulated waveform of input current along with its converter terminal voltage is shown in Figure 3.14. Because of the smoothing inductor, DC current is constant, and hence the input current waveshape is as shown in Figure 3.14(a).

Due to the source inductance L_s, there will be a delay for the transfer of current from one switch to another during commutation. This results in notches in the terminal voltage of the converter as shown in Figure 3.14(b). The harmonics present in the input current can be reduced using 12 pulse converters, where the order of harmonics is 11, 13, 23, 25 … Based on CIGRÉ surveys on AC harmonic filters for numerous HVDC schemes, for harmonic elimination, the limit for individual harmonics is from 0.5% to 1.5% and total harmonic distortion from 1% to 4% [6].

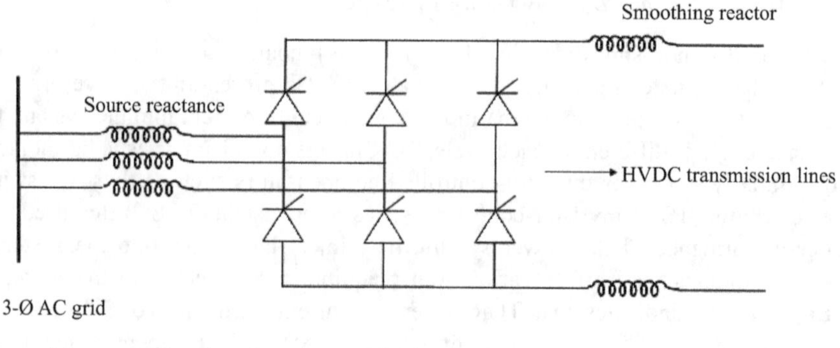

Figure 3.13 Simple scheme of AC to DC converter of HVDC system

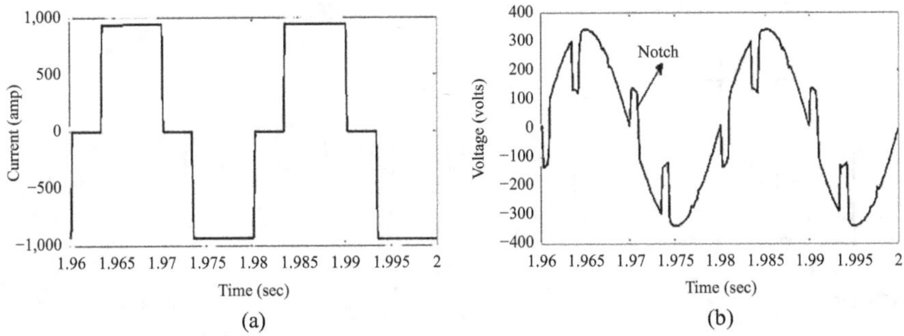

Figure 3.14 Simulated waveform of a six-pulse HVDC converter station. (a) Input current and (b) terminal voltage.

3.2.3 Source of harmonics at utilization

The utilization sector of electrical power is broadly classified as (i) residential and commercial, (ii) industrial and (iii) transportation. The power requirement for the residential equipment is low compared to the power consumed by an industry. However, all the equipment require different types of power supplies depending on the rating or capacity and reliability of the load. As the power supply forms the critical part of any of the loads, we shall start discussing, on various on types of power supply.

Power supplies

Power supplies for the recent electrical appliances use rectifiers of different types, such as 1-ϕ or 3-ϕ at their front end to convert AC to DC. Depending on its applications, it is further categorized as (i) line frequency uncontrolled AC–DC rectifiers, (ii) line frequency-controlled rectifiers and inverters and (iii) switch mode power supplies. In line frequency rectifiers, an AC power supply is available on the front end, whereas for inverters, it is DC on the front end [7]. In uncontrolled supplies, the conversion is done by a diode; however, in case of controlled supplies, it is by a switchable device like a thyristor and hence require additional gate control circuit. In both cases power circuits remain the same. We shall briefly discuss the performance of these most common power supplies.

Half-wave rectifier

Half-wave rectifier is one of the simplest forms of converter. It consists of one diode or thyristor to convert AC to DC. The rectifier conducts only during the positive half cycle, and hence the input current will be zero during the negative half cycle. Figure 3.15 shows a half-wave rectifier connected to a 1-ϕ, 50 Hz supply. The waveform of its input current, simulated using MATLAB along with its harmonic spectrum, is given in Figure 3.15. Harmonic spectrum shows the presence of DC and even harmonic components in its waveform. The DC current injection by the rectifier may lead to core saturation in case of transformers, depending on the flux level. Hence, it is rarely used in high-power applications.

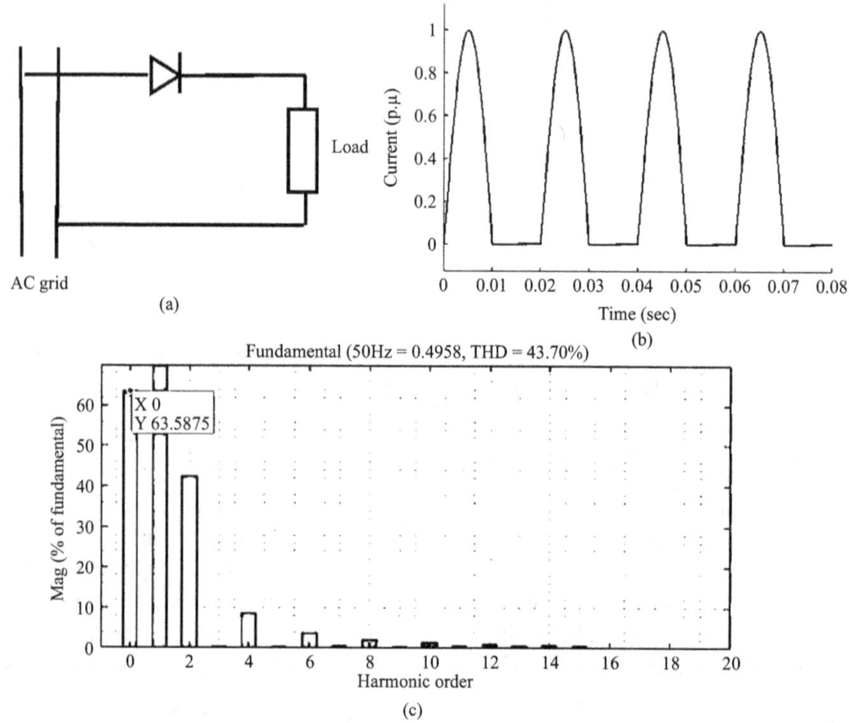

Figure 3.15 1-ϕ half-wave rectifier: (a) layout, (b) typical current and (c) harmonic spectrum

Single-phase bridge rectifier

A single-phase rectifier is the simplest and most common line frequency AC–DC rectifier. Figure 3.16(a) shows a scheme of a 1-ϕ uncontrolled rectifier, diode rectifier. For a controlled bride, the diode is replaced with a switchable device like a thyristor. During positive cycle, the supply voltage is greater than the capacitor voltage, capacitor charges to its peak voltage. On the other hand, during the negative cycle, when the supply voltage is less than the capacitor voltage causes the capacitor to discharge through the load. The capacitor discharges till the supply voltage becomes greater than the capacitor voltage. During this period, the current drawn from the system will be zero. The amount to which the capacitor discharges depend on the RC time constant of the circuit. For a high value of the filter capacitor, discharge will be less. During the negative cycle of the supply voltage, when the capacitor voltage becomes less than the supply voltage, the capacitor draws current from the system and charges to its peak voltage again. As the value of the capacitor increases shape of the current becomes peakier. Even if the value of the capacitor increases so high to give a DC output voltage, it has ripples and injects harmonics. As the value of the capacitor increases output voltage of the rectifier becomes closer to ideal; however, input current results into a peaky current. A 1-ϕ bridge rectifier is supplied with a 230 V (rms), 50 Hz

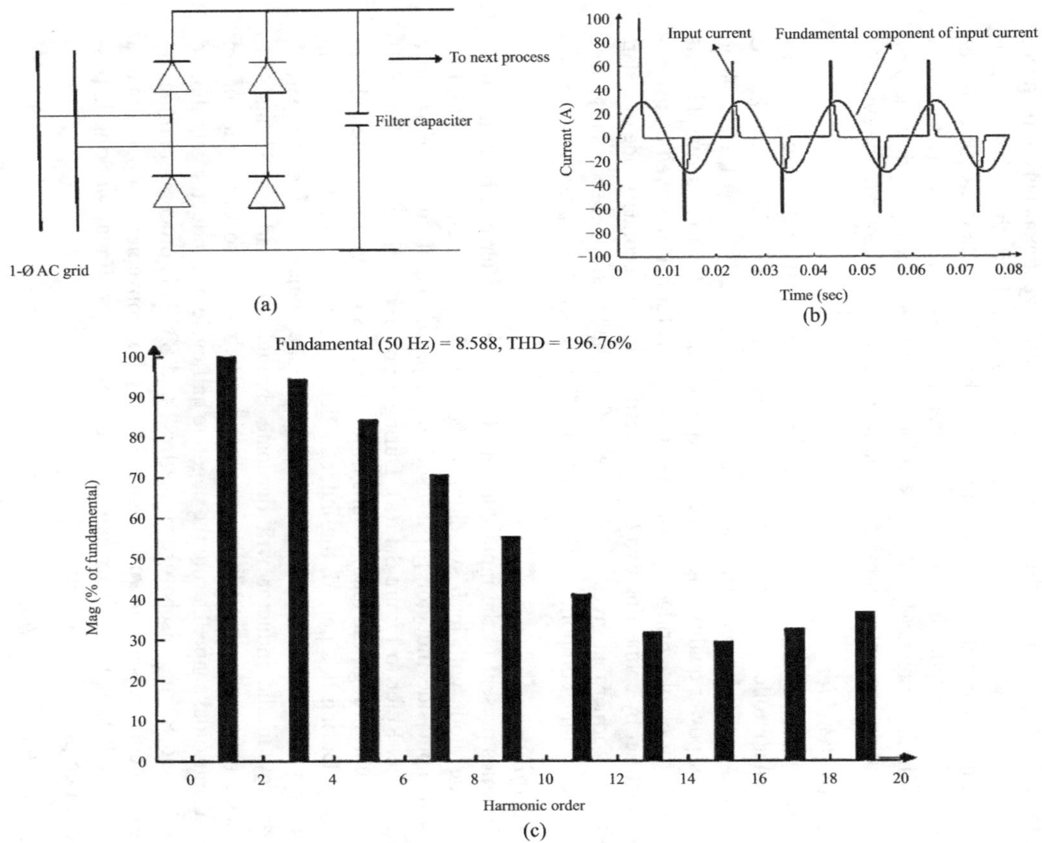

Figure 3.16 MATLAB simulated 1-ϕ rectifier with R = 20 Ω and filter capacitor of 3000 μF (a) scheme of single phase rectifier (b) waveform of input current (c) harmonic spectrum of input current

supply, and is simulated in MATLAB with $R = 20\ \Omega$ and a filter capacitor of 3000 μF. The input current drawn by the AC to DC converter becomes non-sinusoidal as shown in Figure 3.16(b), and its harmonic spectrum is given in Figure 3.16(c).

Such input current consists of 3, 5, 7 . . . etc. order of harmonics, which affects the performance of various components connected to the power system. The fundamental component of current is the only useful component that contributes to the transfer of power from input to output of the rectifier. Other harmonic components are produced by the rectifier; however, these are unwanted components as far as the power system is concerned. Frequency analysis of the input current shows THD as high as 196%. As the value of the filter capacitor increases, THD increases. These harmonics result in a low power factor in the system. Also, the peak of input current may be higher than the peak of fundamental component of current, which increases the crest factor of the system.

Three phase rectifier

Three-phase rectifiers are like a 1-ϕ rectifier discussed earlier. It also has line-frequency uncontrolled and controlled schemes. As detailed earlier on HVDC systems, a 3-ϕ, 6 pulse rectifier generates a harmonic of 5, 7, 11, 13, etc., which results in poor power quality. Frequency analysis of the input current shows THD as high as 31% for 6 pulse bridge rectifiers [9,10]. THD is further improved by using a 12-pulse rectifier by connecting two 6-pulse rectifiers on the secondary side of a star/star/delta transformer that gives a 30° phase shift on the transformer secondary. This eliminates harmonic components of the 5th and 7th orders in the current waveform.

PWM rectifier

The harmonics present in the input current of a diode rectifier can be reduced by replacing the diode with high switching frequency devices such as MOSFET or IGBT. An input side inductor [8] is used to shape the input current close to a sinusoidal one. Refer to Figure 3.17(a). During a positive cycle, the diodes of S1 and S4 conduct through the load. At this time, turn on S3 by giving a gate pulse to its terminal. During this period, the inductor gets shorted through S3 and S4 and stores energy. Then turn off S3 by removing its gate signal. During this period, stored energy in the inductor will dissipate through the load via S1 and S4. Continue turning on and off during the positive half cycle so the input current follows a sinusoidal shape. During the negative half cycle, the inductor gets shorted through S1 and S2 and discharged through S2 and S3. This process continues so that the input current is as shown in Figure 3.17(b). For convenience, operation of a single phase PWM rectifier is explained here, however, a similar working is possible for three phase PWM rectifier too.

Uninterruptible power supplies

Uninterruptible power supplies (UPS) are equipment meant to improve power quality. But it produces harmonics in currents. All these devices consist of a 1-ϕ or 3-ϕ rectifier at the front end of the. The rectifier can be an uncontrolled rectifier using diodes or a controlled rectifier using thyristors. And, in both cases, harmonics will be injected as discussed in the previous sections. However, the order and magnitude of each harmonic component depend on the configuration used [11].

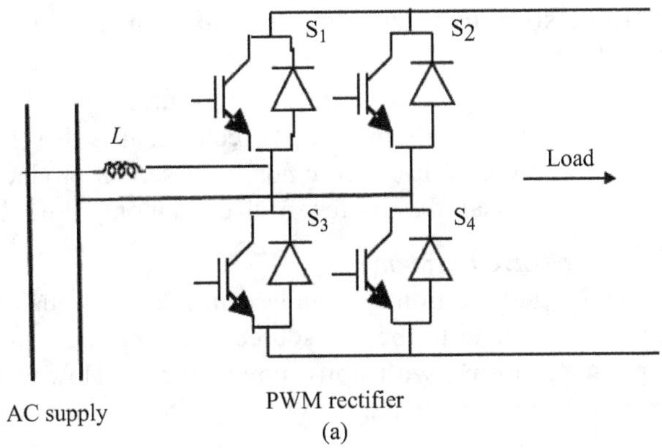

(a)

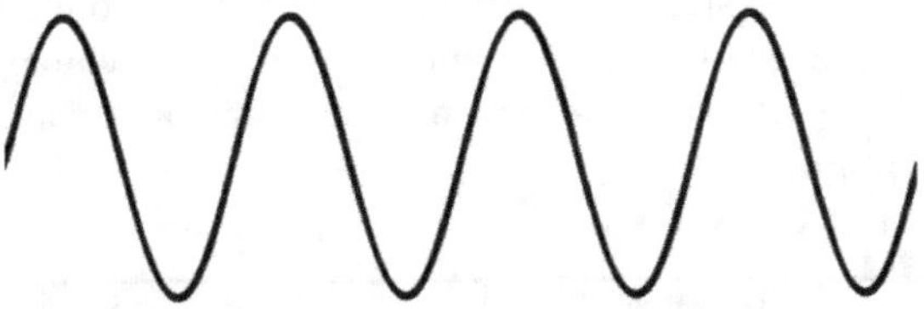

(b)

Figure 3.17 1-ϕ PWM rectifier. (a) Layout of PWM rectifier and (b) waveform of input current

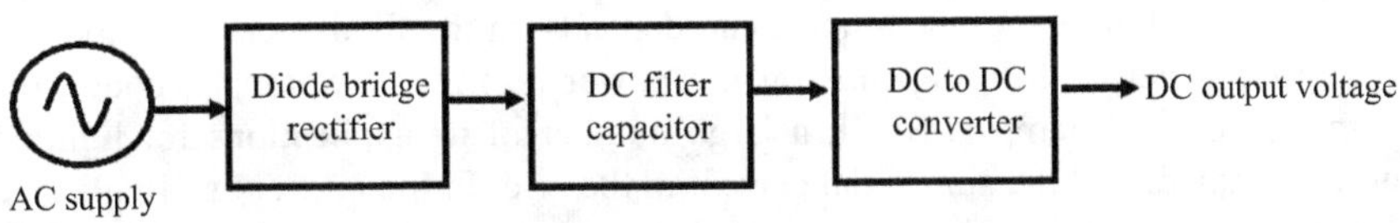

Figure 3.18 Block diagram of switch-mode power supply

Switch-mode power supply

Similarly, in an office large number of computers are used with switch mode power supply (SMPS). A typical block diagram of an SMPS is shown in Figure 3.18.

In this case too, as a 1-ϕ diode rectifier is at the front end of the circuit, which draws a similar peaky current as shown in Figure 3.16(b), and the order of harmonics present is 3, 5, 7.... As per IEEE 519-1992, the magnitude of current drawn by SMPS is with 3rd order as high as 81%, 5th order as 60.6% and 7th order as 37% of the fundamental component, and this magnitude goes on decreasing as the order of harmonics increases [9]. Although the power ratings of each of such equipment are low, simultaneous usage of many such residential or office appliances and their long hours of operation cumulatively cause significant distortion in the current waveform.

3.2.3.1 Harmonic sources in residential and commercial application

In most of the electrical appliances in recent times, utility power supply is first converted to a DC voltage, usually by a diode rectifier. It is then further utilized depending on its applications. What is common to these is it produce current harmonics to the system. We shall discuss few of the commonly used items.

Light emitting diode (LED) system

The current drawn by the traditional incandescent lamp is sinusoidal in nature. As it presents a resistive load to the AC source, the input current will be sinusoidal and in phase, resulting with unity power factor. However, it has low energy efficiency and has been replaced by more efficient gas discharge lamps such as fluorescent, compact fluorescent, sodium vapour lamps, etc. In recent times, these discharge lamps are also being replaced by the more efficient Light Emitting Diode (LED) system. The driver (ballast) unit converts the electric power of a low-voltage AC into the appropriate direct voltage and current required for the LED luminaire. A typical block diagram of an LED lighting scheme is shown in Figure 3.19.

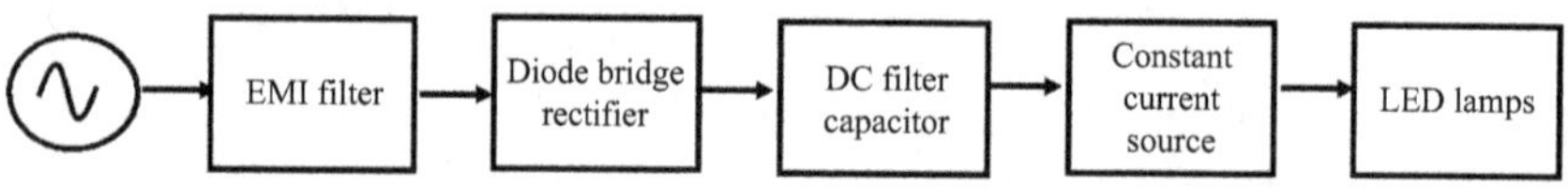

Figure 3.19 Block diagram of LED light system

An LED lighting fixture uses a 1-ϕ rectifier with a large value of filter capacitor for its operation and thus injects harmonics of order 3, 5, 7... etc. into the system. The magnitude of each harmonic depends on the fundamental component of current, which in turn depends on the power rating of the lamp. Though the magnitude of each lamp is small, a large number of its applications for lighting purposes may distort the sinusoidal current in the grid. In India, every self-ballasted LED lamp for general lighting services that works on a 1-ϕ supply shall comply with the requirement of harmonics [12] as per Indian Standard 14700 (part 3/sec 2)/ IEC 61000-3-2: 2018 [13,14].

Light dimmer and fan regulator

A light dimmer circuit or fan regulator is used to control the intensity of light, which is another source of harmonics in the power system. A circuit diagram of a TRIAC light dimmer circuit is shown in Figure 3.20(a). The pot resistor and the DIAC act as the triggering circuit to control the firing angle (α) of the TRIAC. For a fan regulator, the current is nearly the same. A light dimmer circuit supplied with 230 V, 50 Hz is simulated in MATLAB with $\alpha = 60°$ and a load resistor of 100 Ω, and it is seen that the input current of such a circuit is non-sinusoidal, as shown in Figure 3.20(b) and injects harmonics into the system. The order of harmonics in the input current, along with its magnitude, is shown in Figure 3.20(c). As the firing angle (α) is increased, the voltage across the lamp is decreased. Hence, the intensity of light decreases. As the firing angle increases, distortion in the waveform increases.

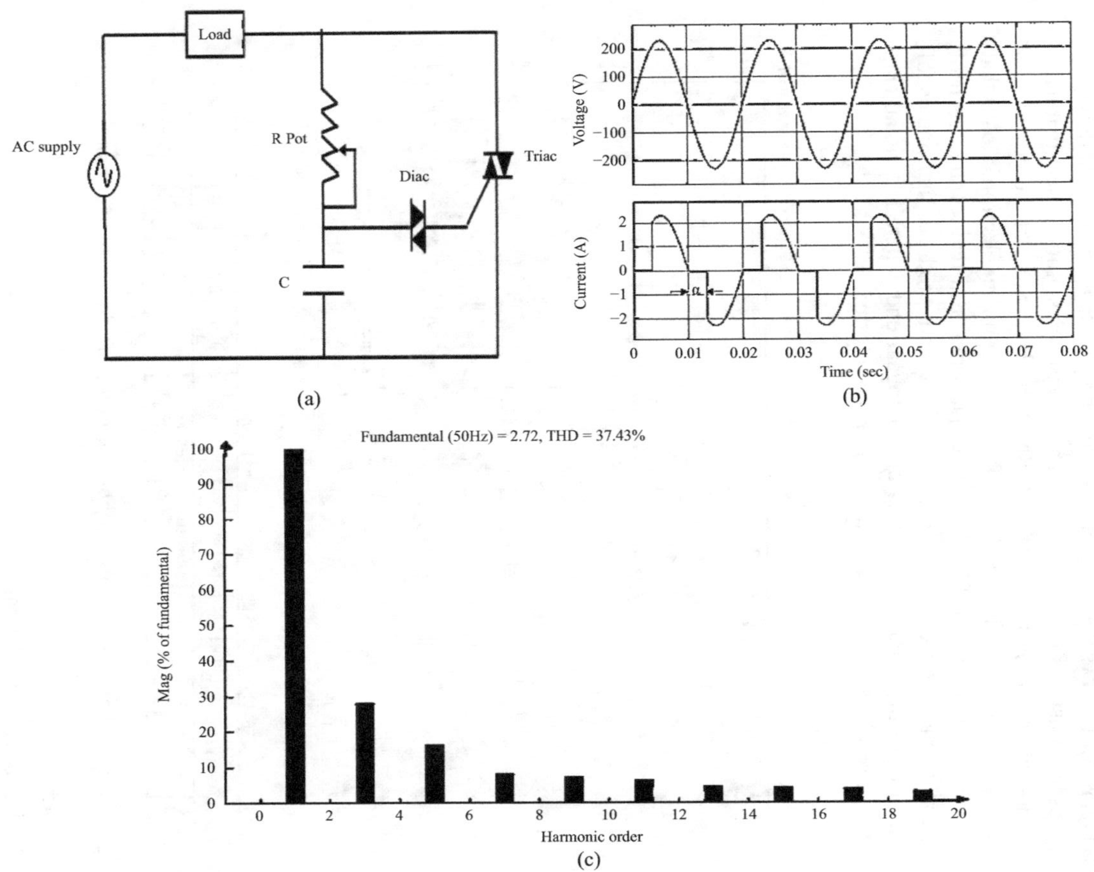

(a) (b) (c)

Figure 3.20 Light dimmer circuit. (a) Scheme of light dimmer circuit, (b) waveforms of supply voltage and load current and (c) harmonic spectrum of input current.

Information Technology and datacentre

Information technology equipment (ITE) is connected to AC power supply. The power supply unit of the ITE converts the AC supply to a lower voltage by an AC–DC rectifier or switch-mode power supply unit (SMPS). 1-ϕ ITE often uses SMPS that produce very high triplen harmonic component of current, as detailed earlier. Figure 3.21 shows the waveform of the supply voltage and current drawn by a laptop observed using a power analyser. These harmonics produce heat that has a determinantal effect on performance and life of the ITE. Most of the power supplies, viz. 1-ϕ and 3-ϕ rectifiers, SMPS, UPS and other power electronic loads that contain switching devices are found in large quantities at a data centre. Each of them operates under different loading conditions and produces harmonics of varying nature. At the common coupling or switchgear end, where individual current is added, results a random distorted current waveform.

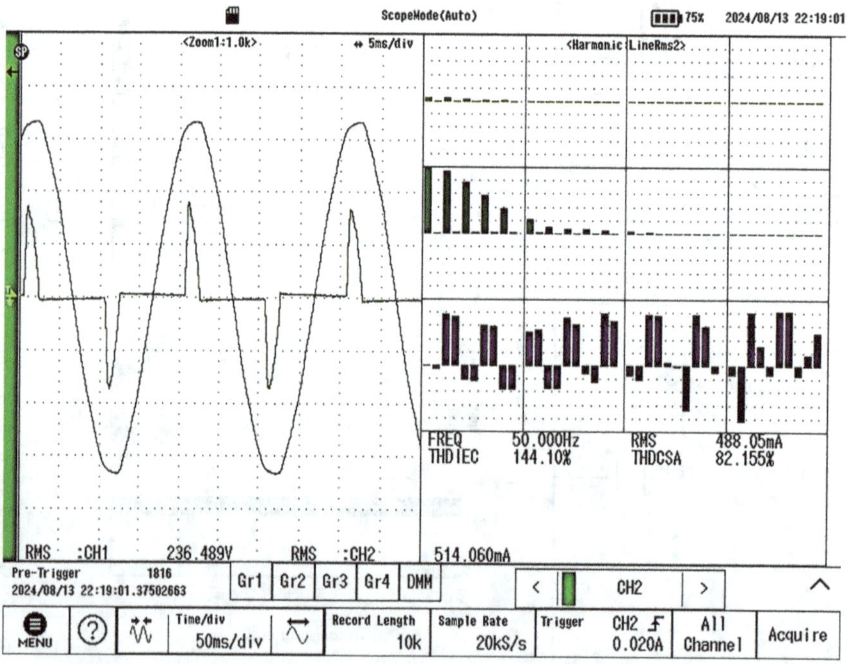

Figure 3.21 Wave form power supply and current drawn by a computer

3.2.3.2 Industrial consumers

The most common source of harmonics at the industrial unit is the AC to DC power converters. It is used in applications like motor drives, arc furnaces, uninterruptible power supplies (UPS), etc. Most of these system uses 3-ϕ or 1-ϕ AC to DC converter at its front end, as shown in Figure 3.22. These converters may use a diode, thyristor, IGBT or MOSFET as switching devices depending on their application.

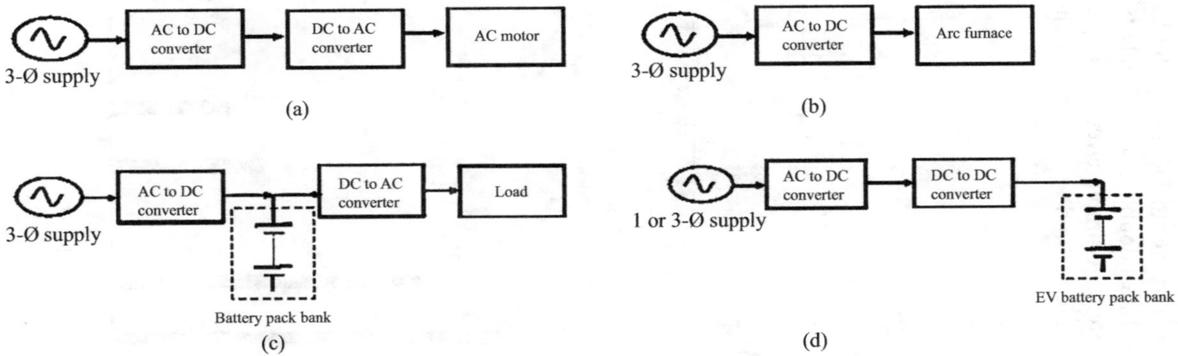

Figure 3.22 Block diagram of converter application. (a) AC drives, (b) arc furnace, (c) uninterrupted power supply and (d) EV charging.

Variable frequency drives

As discussed in Chapter 1, variable frequency drives (VFD) for motors are increasingly used to drive large pumps, blowers, heating, air conditioning, elevators, etc., because of their numerous advantages. These drives use rectifiers at their front end to convert AC to DC using diodes or other switching devices such as thyristor. It is like a double conversion UPS without a battery or intermediate energy storage item. A simple circuit diagram of AC drives using an uncontrolled six-pulse converter and an inverter is shown in Figure 3.23.

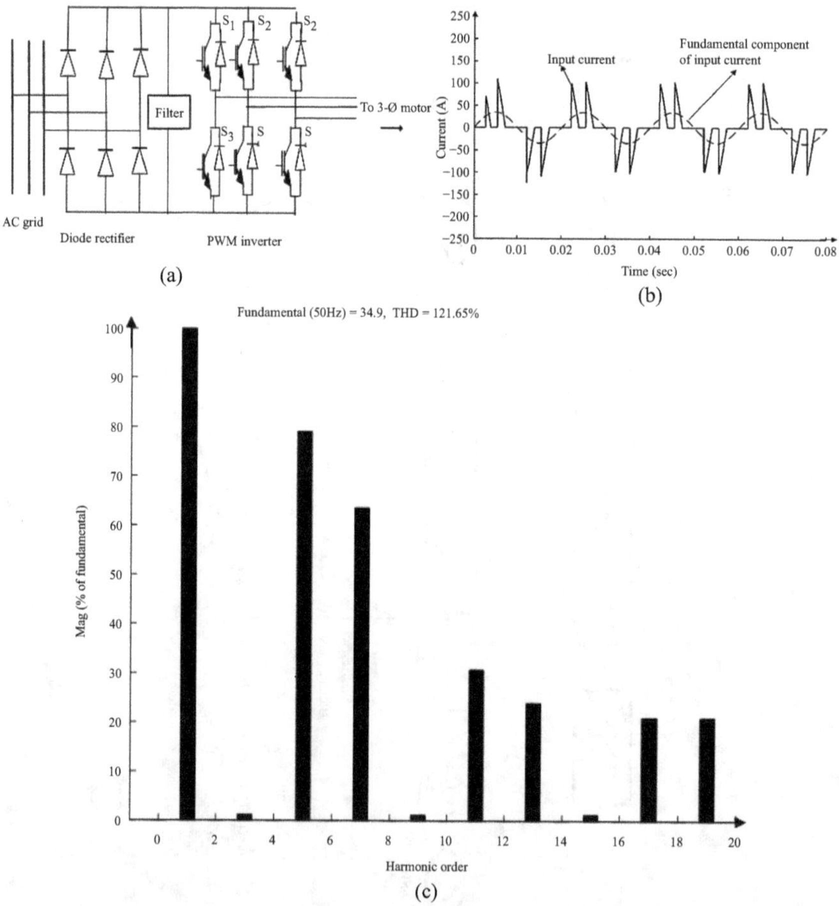

Figure 3.23 AC motor drive circuit diagram and waveform. (a) Circuit of AC drives, (b) typical waveform of input current and (c) harmonic spectrum.

A 440 V, 50 Hz three-phase diode rectifier with R = 20 Ω and filter capacitor of 1000 μF is simulated in MATLAB and input current along with its harmonic spectrum is shown in Figure 3.23(b) and (c), respectively. It is seen that the order of harmonics present in input current is 3, 5, 7 . . . . (odd harmonics) with THD of 121.65% while fundamental component of current is in phase with the voltage.

By replacing diode with thyristor, a phase-controlled rectifier is accomplished where DC bus voltage can be controlled unlike diode rectifier by varying firing angles of thyristors. By controlling the firing angle over 90°, DC-bus voltage can be made negative, which means bipolar DC voltage is possible. This allows a regenerative flow of power from the DC-bus back to the power supply or in other words power reversal is possible. However, the generation of harmonics at the front end of the rectification process is inevitable.

Electric arc furnaces

One of the major harmonics-producing industrial loads is electric arc furnaces (EAF). Because the process inside the furnace is irregular, the requirement of active and reactive power is fluctuating. Also, the arc current is non-periodic and contains unpredictable harmonic frequencies of both integer and non-integer orders. Harmonic content also varies with smelting and refining the cycle of ore in the furnace. It is shown that arc furnaces draw even harmonic components of current from the system, which results in an unsymmetrical current waveform. IEEE 519-1992 indicates harmonics of order 2, 3, 4, 5, 7, etc. are produced during the melting of steel. For the production of steel, EAF requires about 300–600kWh/Ton and hence it is connected to the grid at high voltage through a transformer. Figure 3.24 shows the scheme of an electric arc Furnace [16].

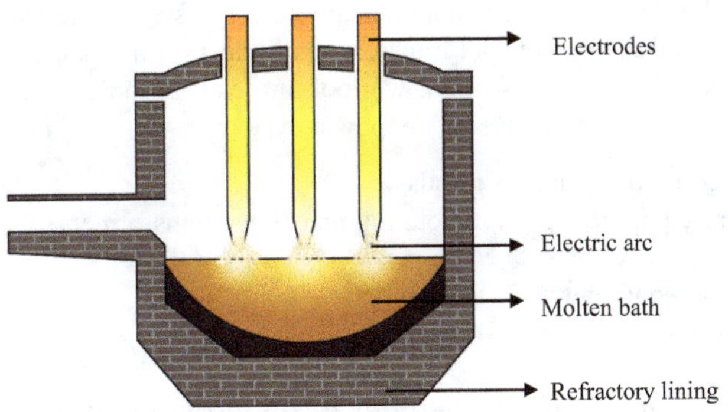

Figure 3.24 Electric arc furnace (https://commons.wikimedia.org/wiki/File: Electric_Arc_Furnace.png#/media/File:Electric_Arc_Furnace.svg)

3.2.3.3 Transportation

Electric traction and electric vehicle are the two most advanced mode of land transportation used in recent times. Following sections briefly discuss their effect on power quality.

Electrical traction

Now a days Indian railways use 66/25 kV to 220/25 kV transformers for supplying 1-ϕ power to the 25 kV overhead lines for the locomotives [15]. The AC electric/AC–DC electric locos are fitted with a 1-ϕ rectifier bridge for feeding the DC supply to traction motors. These rectifiers introduce harmonics in the current on the 25 kV AC power supply, apart from unbalance and load fluctuations, and due to 1-ϕ loading on the traction overhead line.

Charging stations for electric vehicles

We have witnessed a tremendous increase in the electric mobility sector in recent times. As the electric vehicles (EV) battery needs to be frequently charged, the number of charging stations is coming up over roadsides and is more concentrated in urban areas. In this case too, the front end of the charging system is an AC to DC converter, which takes non-sinusoidal current as explained in the previous section. Research is going on to replace AC–DC converters with a PWM rectifier, where charging and discharging the battery from and to the grid is possible with good power quality. Inductance (L) at the input side, along with PWM rectifier, helps the EV system to improve power quality by drawing a sinusoidal current closer to an ideal one from the grid.

3.3 Effects of harmonics

In the previous section, we have seen some of the non-linear equipment connected at various parts of the power system that distort the voltage and current in the system and thus introduce harmonics in the power system. It reduces the life of various electrical equipment like transformers, induction motors, etc., and affects their efficiency. In the next section, we look in detail at the effects of these harmonies on the various elements of the power system.

(i) Point of common coupling or bus
(ii) Electrical equipment such as rotating machines, transformers
(iii) Power capacitor
(iv) Electronic equipment
(v) Power factor

3.3.1 *Effect of voltage harmonics at point of common coupling*

In the earlier sections, we have seen the generation of harmonics and distortion in the transmission line current due to various power electronic equipment, which acts as a non-linear load to the power system. Consider a four-bus system as shown in Figure 3.25, where load 1 is linear, and loads 2 and 3 are non-linear in nature. The

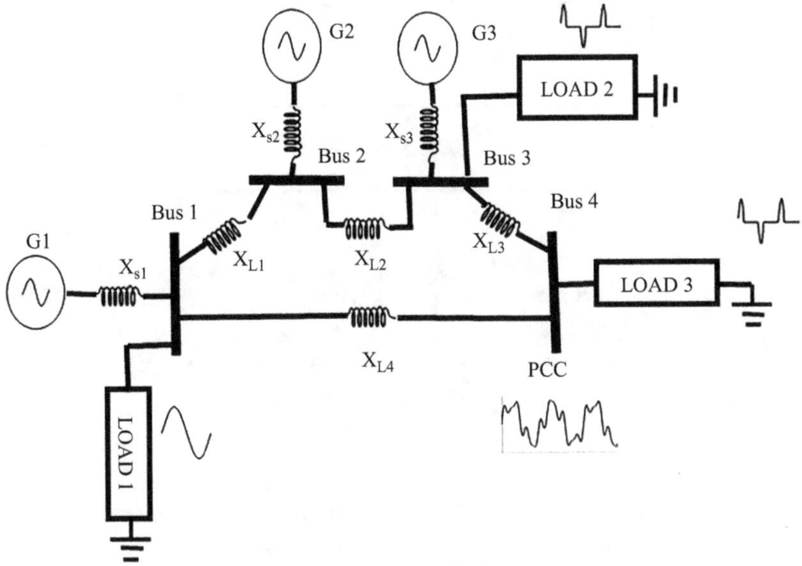

Figure 3.25 Generation of non-sinusoidal voltage at the point of common coupling

non-sinusoidal current drawn by the non-linear loads causes non-sinusoidal voltage across the transmission line reactance (X_L). Consider Bus 4. The voltage at Bus 4 is the difference between the sending end voltage and the voltage drop across the transmission line reactance to Bus 4 from the sending end. As the voltage drop across the line is non-sinusoidal, the voltage at Bus 4 is also non-sinusoidal. Thus, it may affect other loads connected to the system.

The effect of current harmonics depends upon the system impedance. In a large system, short circuit capacity is a measure of system impedance. Hence, the limit on current distortion depends on the short circuit ratio (SCR). SCR is given by:

$$SCR = \frac{I_{sc}}{I_L} \tag{3.3}$$

where I_{sc} is the short circuit current, and I_L is the load current.

I_{sc} depends on the positive sequence reactance (x_{PCC}) of the network up to PCC, which is the Thevenin equivalent reactance till PCC of the system. Consider a simple single-line diagram of a power system shown in Figure 3.26. The Thevenin equivalent reactance till PCC (x Pu) is given by:

$$x(pu) = X_{G1} \,||\, X_{G2}(pu) + X_T(pu) + X_{TL}(pu) \tag{3.4}$$

$$I_{sc} = \frac{V_{PCC}}{x_{PCC}} = \frac{1(pu)}{x(pu)} \tag{3.5}$$

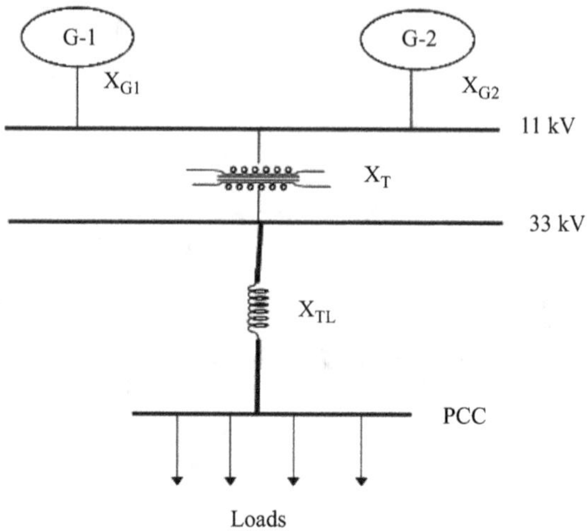

Figure 3.26 Single-line diagram of a simple system

Thus, when x_{pu} is small, *Isc* will be high. This generally occurs closer to the generator end, where equivalent reactance is small. As voltage is less affected by the changes in current at this point, the generator side exhibits higher stiffness. The voltage harmonic limit is decided based on the impact of current harmonics on the system. The limit for THD in current is set to ensure that it does not cause the voltage THD to exceed its permissible value. Thus, current harmonics exceed their limit, voltage distortion may cross their threshold value and impact overall power quality. When the system stiffness is high, the limit on allowable harmonics is also higher. This indicates that it can accommodate larger harmonics in currents without causing significant voltage distortion. Conversely, in a weaker system with low SCR, even small harmonics in currents can cause considerable voltage distortion.

IEEE 519-2014 indicated voltage and current harmonic limit from low voltage through 161kV. It shows that as voltage level increases individual voltage harmonic limit reduces from 5 to 1% along with THD from 8 to 1.5%. This is because of magnitude of individual harmonic is high at higher voltage level, so less percentage limit is permitted. Similarly, the standard also shows that as $\frac{I_{SC}}{I_L}$ increases, limit of individual harmonic increases along with TDD.

3.3.2 *Effects of harmonics on capacitor*

Capacitors are extensively used in a transmission, distribution and utilization system for reactive power support or power factor control. When it is under sinusoidal voltage supply condition, it draw sinusoidal current of fundamental frequency. However, when it is installed under harmonic environment, it may cause parallel or series resonance depending on the circuit conditions. Parallel resonance offers high impedance path to flow of current while series resonance results in low impedance path to flow of current. At a particular frequency when reactance of capacitor bank is equal to reactance of transmission line, parallel resonance may occur which cause high voltage distortion and current amplification in the system. As an example, consider single line diagram of a sample power system with linear loads along with a capacitor bank (X_c) as shown in Figure 3.27. Assume that the reactance of transmission line including the transformer reactance till the point of installation of capacitor bank is X_s. If a non-linear load is connected at this point, at a particular frequency when $X_c = X_s$, a tank circuit is created as shown in the Figure 3.28, the harmonics in current (*Ih*) generated by the load will oscillate through the tank circuit. [9,17].

At a load centre, transformer is connected to a capacitor bank at its secondary side and harmonic load at primary side. When $X_c = X_t$ series resonance occur in such circuit as shown in Figure 3.29 and results in a high voltage distortion.

Thus, there may be parallel or series resonance in the system. In case of series resonance, overheating may takes place whereas under parallel resonance, over voltage may result. This may lead to failure of both the capacitor and the other

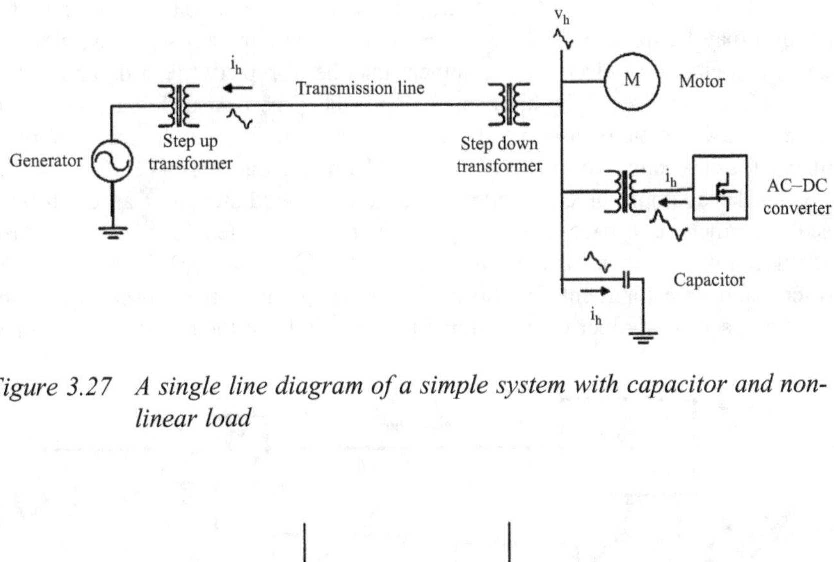

Figure 3.27 A single line diagram of a simple system with capacitor and non-linear load

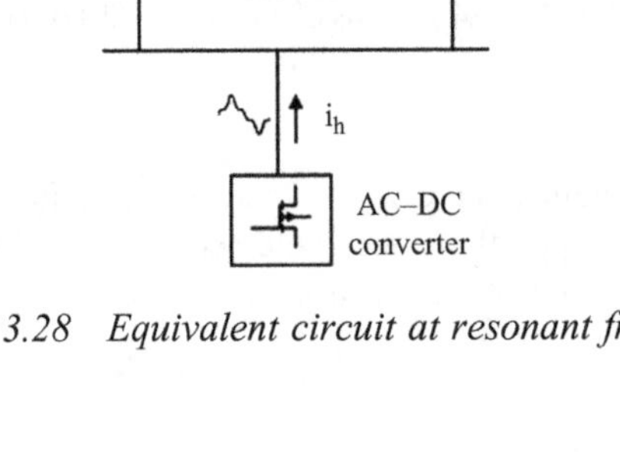

Figure 3.28 Equivalent circuit at resonant frequency

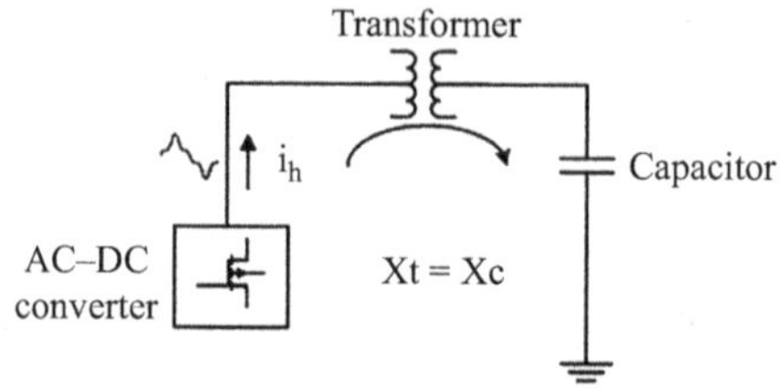

Figure 3.29 Equivalent circuit under series resonance

equipment connected in the system [18]. So, it is necessary to design the capacitor compensator properly by considering the harmonics present in the system [19].

3.3.3 Effects of harmonics on electronic equipment

The operation of most of the electronic equipment depends upon zero crossing of the waveform, while converter power supplies use the peak of the voltage waveform to maintain the filter capacitors at full charge. Depending on the voltage distortion, the waveform may be flattened, although the rms value may remain same. With severe distortion, operation of electronic equipment may be disrupted. Metering and instrumentation are also affected by harmonic components. The flow of harmonics in current in a power system also results in telephone interference. Hence, electronic equipment is susceptible to malfunction due to harmonic distortion [17].

Consider a simple block diagram where a controlled switch of an electronic-based equipment is operated by a pulse which is generated based on the zero-crossing point of voltage or current of power line as shown in Figure 3.30. The zero-crossing detector identifies the zero-crossing point of the sensed voltage or current and sends a proper control signal to the switch for the desired outcome.

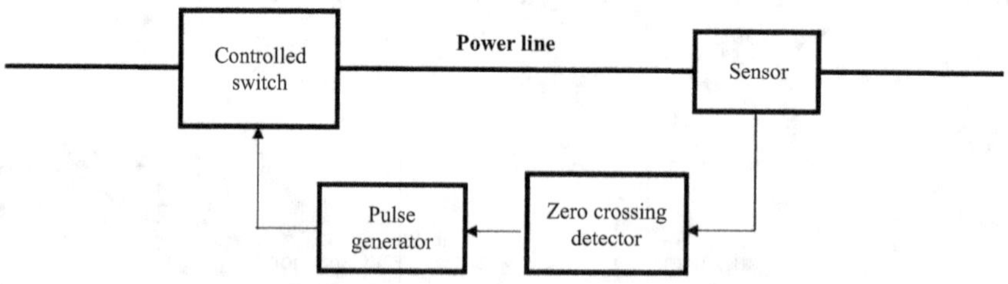

Figure 3.30 Block diagram of a power electronic switch using zero voltage detector

Assume that the voltage at PCC is distorted as shown in Figure 3.31, which has more than three crossing points in one cycle. Zero crossing detector senses these points and send wrong control signal to switches, resulting mal operation of the equipment concerned.

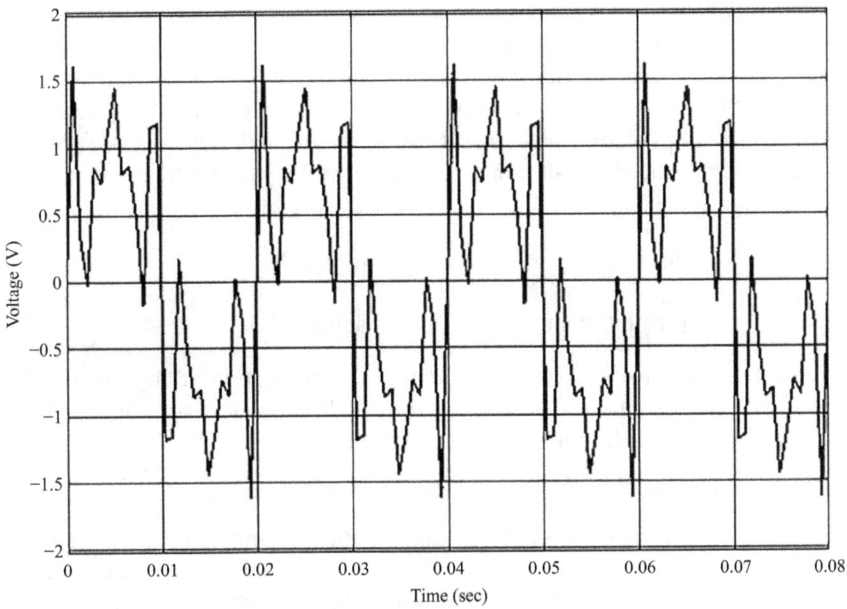

Figure 3.31 Distorted sinusoidal voltage waveform

3.3.4 Effect of harmonics on power system neutral

Generally, a neutral conductor is present in distribution side where 1-ϕ loads are connected. Under balanced harmonic free conditions, the neutral current is zero; however, under a harmonic environment, the triplon (multiple of 3rd) order harmonic (3rd, 6th, 9th, ...) component of current with zero sequence characteristic has a severe impact on both the power distribution system and the devices connected to it. Unlike positive and negative sequence harmonic currents, for example, triple frequency harmonics in currents $(I_{h3a}, I_{h3b}, I_{h3c})$ are in phase with each other and add arithmetically instead of cancelling at the neutral point. Where, I_{h3a}, I_{h3b}, I_{h3c} are 3rd order harmonic components of current in a, b and c phases,with zero sequence characteristic.

$$I_n = 3I_{h3a} \tag{3.6}$$

Thus, for a 3-ϕ, 4-wire, balanced non-linear system with triplon harmonic I_{h3}, neutral current (I_n) is given by,

$$I_n = 3\sqrt{\sum_{h=3(2k-1)}^{\infty} I_h^2}, \quad k = 1, 2, 3, \ldots \tag{3.7}$$

Thus, neutral current can be as high as 3 times the triple frequency phase current.

In simple terms, if 2 A fundamental frequency current is flowing in each phase of a 3-ϕ, 4-wire, balanced linear load, then neutral current (I_n) is 0 A. whereas if 2 A, 3rd order harmonics in current is flowing in each phase of the system due to non-linear load, then neutral current (I_n) is 6 A. This will overheat the neutral conductor, may lead to fire or failure. However, a system without fundamental is rarely seen.

For an unbalanced non-linear system with triplon in neutral current, (I_n) is given by

$$I_n = \sum (\text{unbalanced portion}) + (\text{non} - \text{triplen}) + (\text{triplen harmonic portion of fundamental}) \tag{3.8}$$

And for a 3-ϕ balanced system, the unbalanced portion of fundamental frequency is zero and hence the neutral current is reduced by the same amount. Thus I_n can be as much as $\sqrt{3}$ times the phase current.

Thus, neutral current (I_n) can vary between, $3I_{h3}$ to $\sqrt{3}\, I_{ph}$

This led to overloading of the neutral conductors and calls for special consideration while designing the system. IEEE 1100 recommend neutral conductor rated at 200% of the phase conductor where the load is primarily made up of switch-mode power supplies. This is very much relevant to information technology equipment (ITE). Thus, a neutral conductor needs to be doubled the rated normal conductor cross-sectional area.

3.3.5 *Effects of harmonics on power factor*

Power factor (*PF*) is defined as the ratio of active power (*P*) to apparent voltampere (*S*) in power system [9].

$$PF = \frac{P}{S} \tag{3.9}$$

Power factor in the system is maintained close to unity to draw optimum current from the utility and to keep the voltage variation at various parts of the system under limit. Under sinusoidal conditions, the apparent voltampere is the vector sum of real and reactive voltampere consumed, and *S* is given by:

$$S = \sqrt{P^2 + Q^2} \tag{3.10}$$

If *PF* is low, the system has to supply more current (or *S*) to meet the same power demand. Ideally, *P* should be equal to *S* and hence $PF = 1$ for an efficient system. However, most of the loads are inductive and non-linear in nature, will go below unity. Under the environment, *PF* will be further affected by harmonics as apparent voltampere is given by:

$$S = \sqrt{P^2 + Q^2 + D^2} \tag{3.11}$$

where *D* is known as the distortion voltampere due to harmonics.

So, when distortions in voltage and current increase, D increases and results in a low power factor. At low power factor, the system requires a higher voltampere rating of generator, transformer, transmission, and distribution line equipment. Thus, the system's real power capacity is reduced with a reduction in power factor. In other words, operating at high power factor enables serving more consumers. A detailed discussion on power factor under sinusoidal and non-sinusoidal conditions is presented in Chapter 5.

3.3.6 Effects of harmonics on electrical equipment

The effect of voltage and current harmonics on major electrical equipment, such as rotating machines and transformers is explained below.

3.3.6.1 Effects of harmonics on motors

Motors are the largest loads in the power system. When an induction or synchronous motor is connected to a sinusoidal power supply, it will draw sinusoidal current. However, when it is fed from a non-sinusoidal supply voltage, it will draw a current similar in shape to the voltage as shown in Figure 3.32.

As seen from Chapter 2 Table 2.1, the characteristic of each harmonic component depends on its order. The direction of rotation of the magnetic field or torque produced in motor depends on the characteristics of each harmonic component. Harmonics of order 4, 7, etc. have positive sequence characteristics, rotate in the same direction as that of fundamental component, and of order 2, 5, etc. have negative sequence characteristics, rotate in the opposite direction. Thus, the harmonics produce a pulsating torque in the motor that results in increased vibration and reduced bearing life. Harmonic pairs, such as 5th and 7th order harmonics, yield intermediate harmonics in current (6th) in the rotor, which increases rotor heating. Also, non-oriented core materials that are generally used in rotating machines produce higher magnetic loss under non-sinusoidal voltage conditions.

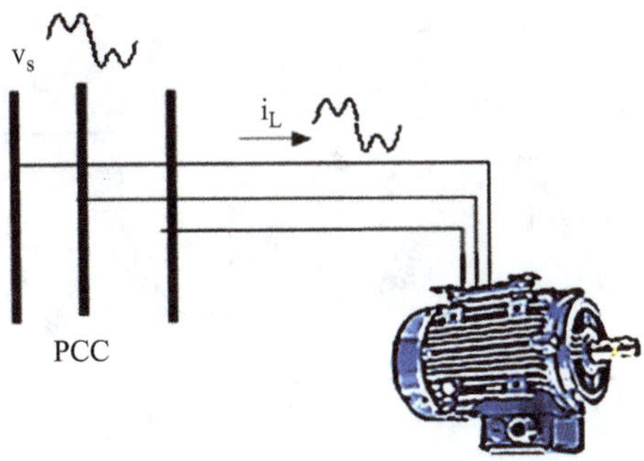

Figure 3.32 Induction motor connected with a non-sinusoidal supply voltage

Additionally, it causes increased cogging (refusal to start) or crawling (high slip). Overall, harmonics cause increased heating, vibration, reduction in efficiency and equipment's life.

3.3.6.2 Effects of harmonics on transformers

The transmission and distribution system consists of a large number of transformers of different classes of voltage, type and capacity. Additionally, various utilization equipment also contains transformers. In fact, a transformer is a major capital equipment in the power system. Whatever be its cooling medium (air, oil, cast resin, etc.) and its circulation (natural, forced, etc.), transformer losses are categorized as

(a) No load loss – It is the sum of no load, I_n^2R loss in winding and core magnetisation loss.
(b) Load losses – It is the sum of load, I_l^2R loss in winding, and due to stray magnetic flux linked in the winding and other parts such as core, clamps, transformer tank and structural material. Thus, cause induced voltage resulting eddy current in winding and other parts.

We shall analyse it by considering non-sinusoidal voltage at the bus level where a transformer is connected with non-linear loads, as shown in Figure 3.33.

No load loss under harmonics in voltage

No load losses mainly consist of hysteresis and eddy current loss. These losses are related to the design transformer magnetic core and its material, i.e., cold rolled grain-oriented steel (CRGO). Under non-sinusoidal voltage conditions, CRGO core material produces a square-shaped hysteresis loop with a larger area, implying increased magnetic losses or core loss, which is the addition of eddy current (W_E) and hysteresis loss (W_H).

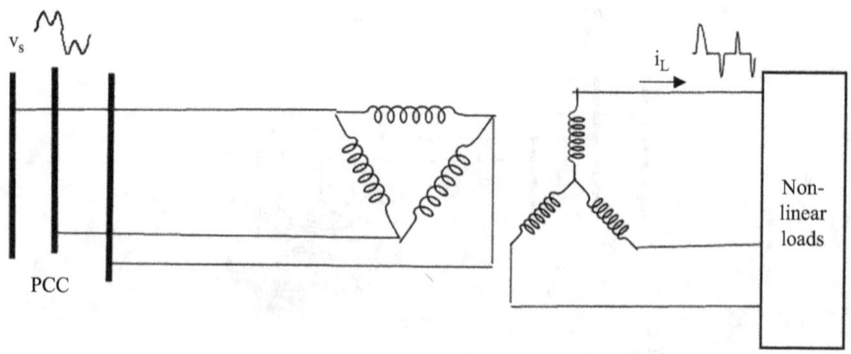

Figure 3.33 A Δ-Y transformer connected to non-linear loads

Thus, a transformer with a power supply containing up to 'h' order of harmonic components of voltage. The core losses (P_{coreL}), can be represented by:

$$P_{coreL} = W_E + W_H = k_E \sum_{n=1}^{h} B^2_{\max-n}(nf)^2 + k_H \sum_{n=1}^{h} B^{\sigma}_{\max-n}(nf) \tag{3.12}$$

where k_E and k_H are constants corresponding to eddy current and hysteresis loss, respectively.

$B_{max\text{-}n}$ is the maximum flux density corresponding to harmonic order n,
σ is Steinmetz coefficient, and f is the fundamental frequency.
Thus, as the order of harmonics increases, W_E and W_H will increase.

Load loss due to harmonics in current

Load loss is composed of winding I^2R loss and stray loss. Copper loss (P_{CL}) under non-sinusoidal current conditions is given by:

$$P_{CL} = \sum_{n=1}^{h} I_{ln}{}^2 (R_p + R_s^{'}) \tag{3.13}$$

where,

I_{ln} is the rms value of the nth order harmonic component of the load current
R_p is the primary winding resistance and
$R_s{}'$ is the equivalent secondary winding resistance with respect to the primary.

As the order of harmonic in current increases copper loss increases due to increase in rms value of current.

Winding eddy current loss due to harmonics in current

Winding eddy current is due to the current flowing through the windings of a transformer. As the magnetic flux has a bigger inclination to fringe towards low reluctance path of the core, inner winding has a greater eddy current loss than the outer winding. Also, highest eddy current loss occurs in the end conductors of the inner winding due to higher radial flux density in this region and as the radial flux passes through the width of the rectangular winding.

Winding eddy current loss (P_{EC}) is a function of square of the load current and frequency. However, it depends on cross section of conductor and frequency. Thus, it can be approximately written as:

$$P_{EC} \approx P_{EC-R} \sum_{n=1}^{h} \left[\frac{I_{\ln}}{I_R}\right]^2 n^2 \approx P_{EC-R} K \tag{3.14}$$

where, P_{EC} is the winding eddy current loss

$P_{EC\text{-}R}$ is the winding eddy current loss under rated conditions
I_R is the rated rms fundamental frequency current
K is the multiplication factor, called harmonic loss factor, and is like K-factor.[1]

Thus, winding eddy current loss also increases with harmonic component of load current.

Stray loss due to harmonics in current

Other stray losses (P_{OSL}) in the core, clamps and structural parts also increase, because of harmonics in current and frequency. IEEE Std C57.110 [20], indicate it increases by a harmonic frequency exponent factor of 0.8. For dry-type transformers, as the losses are dissipated to air, the temperature rise in these areas will be less dangerous than in the windings or oil-immersed transformers.

$$P_{OSL} = P_{OSL-R} \sum_{n=1}^{h} \left[\frac{I_{\ln}}{I_R}\right]^2 n^{0.8} \tag{3.15}$$

Total loss

As detailed above, the total loss under non-sinusoidal conditions is obtained by adding all the losses detailed in the above section [20,21]. Thus, the total loss P_{TL} is given by:

$$P_{TL} = P_{coreL} + P_{CL} + P_{EC} + P_{OSL} \tag{3.16}$$

Hence, it can be concluded that the main effect of waveform distortion on transformer are the additional loss due to harmonics, resulting in higher operating temperature of the transformer. Other problems include possible resonance due to harmonics and core vibrations. Thus, for transformers under harmonics condition, the following measures are to be adopted.

For existing installation:

(a) Install harmonic filters of adequate capacity so that harmonics are bypassed (shunt) or blocked (series) by the filter circuit [22].
(b) De-rate the transformers based on the measured harmonics so that the losses and temperature rise are within the design limit.

For new transformer installation:

(a) Transformers shall match with harmonic requirement: Harmonics produce increased stray losses in windings, core material and structural parts. To limit losses, stranded conductor, higher-sized neutral conductor, reduced flux density in core, more spacing between core and tank and magnetic shielding of transformer tank is often considered in design.
(b) Consider delta (Δ) connected winding for circulation of triplen harmonics in current in a transformer.
(c) Consider a transformer with increased cooling for the additional losses due to harmonics in voltage and current.

[1]UL 1561

(d) K-rated transformers with the anticipated harmonic conditions to handle increased heat due to non-linear loads.
Even order harmonic components in current may result in saturation of the transformer core. So, a more stringent limit is imposed on even harmonic components of the current. Indian states have mandated these industries to limit individual current harmonics and total demand distortion (TDD) as per IEEE 519-2014.

Normally, transformers are designed to operate in a less harmonic environment (THD <5% as per IEC 60076). However, with the change in nature of load, its performance is greatly affected by current and voltage harmonics.

3.3.7 Summary of effects of harmonics on power system

To conclude, harmonics adversely affect various electrical equipment and systems of both utility and customer. Thus, it impairs the overall performance of the system. It can be summarized as follows:

(i) Reduction in power factor and thus system capacity
(ii) Increased neutral current necessitating higher cross sectional area of neutral conductor
(iii) Additional losses in transmission and distribution transformers, motors, capacitors, conductors, switchgears, etc.
(iv) Reduction in efficiency of power generation, transmission, distribution and utilization equipment
(v) Possibility of overloading, overheating and accelerated ageing, thus shortening the life of components of the power system
(vi) Higher probability of resonance
(vii) Higher order harmonics increase the skin effect and thus increase effective conductor resistance, thereby resulting in higher losses and temperature
(viii) High frequency components can cause interference on nearby metallic communication circuits, thus data loss or data corruption
(ix) Spurious operation of protective system and electronic components.
(x) Increased probability of power supply interruption
(xi) Increased cost of power generation, transmission, distribution and utility bills

3.4 Standards and regulation

Having realized the consequences of individual and total harmonic distortion, IEEE and IEC have developed appropriate standards to restrict the spilling of harmonics into the power system. The philosophy adopted in these standards is to limit the current harmonic injection from individual equipment and consumers so that it does not cause unacceptable overall voltage distortion in the power system. In India, electricity regulatory commissions have formulated harmonic limits on distribution and transmission systems. Table 3.1 briefly summarizes harmonic sources, order of harmonics and limits as per the standard.

Table 3.1 Summary of harmonic sources, order of harmonics and limits as per standard

Application	Source of harmonics	Order of harmonics generation	Limits	Reference standard
Power Generation				
Conventional power generation	Synchronous machines	Negligible harmonics	More than 5MW, TIF – 1.5%	IEC 60034 rotating machines [1]
RE generation Solar photovoltaic (SPV) Wind Turbine Generator (WTG Type 3&4)	DC to AC converter DC–AC converter	Odd order Depends on switching frequency and topologies used	IEEE 519 THD $< 5\%$ IEEE 1547	CEA (Technical std for connectivity of distributed generation) MNRE std for PV grid tie inverter [2,3]
Transmission and distribution				
AC sinusoidal environment	Transformers	3,5 . . .	5% THD	IEC 60076 [23]
AC non-sinusoidal	Transformers	Depends on load	Capacity rating	IEEE C57.110 [19] UL 1561 [24]
Flexible AC transmission SATCOM, SVC	Thyristor IGBT	Odd order	IEEE 519	IEEE 519
Thyristor-controlled shunt reactor (TCR)	Thyristor	Odd order 3,7,11,15, 19,23	IEEE 519	IEEE 519
Thyristor switched capacitor (TSC)	Thyristor	Odd order	IEEE 519	IEEE 519
Thyristor-controlled series compensator (TCSC)	Thyristor	Odd order	IEEE 519	IEEE 519
HVDC	AC to DC DC to AC converters	(12 pulse) odd order such as 11,13,23, 25 . . .	Individual harmonic 0.5–1.5% and THD from 1% to 4%	IEC/TR 62001: 2003
Utilization				

(Continues)

Power supplies	1-ϕ Diode half-wave rectifier	DC component, 2,4,6 . . .	IEEE 519	IEC 61000-3
	AC to DC converters 1-ϕ rectifier	Odd order 3,5,7 . . .		
	AC to DC 3-ϕ, 6 pulses rectifier	Odd order such as 5,7, 11,13,17,19.		
	AC to DC 3-ϕ, 12 pulses rectifier	Odd order such as 11,13,23,25. . .		
	Switch mode power supplies (SMPS)	Odd order 3,5,7,9,11, 13,15		
	3-ϕ PWM rectifier	Negligible harmonics		
	Uninterruptible power supplies (UPS)	Odd harmonics	IEC 62040	IEC 62040 IEC 61000-3
Residential and commercial application				
Low-power electronic or electrical equipment	Power supplies	Odd order 3,5,7,9,11,13, 15,17,19 . . .	IEC 61000-3-6	IEC 61000-3-6
Light emitting diodes (LED)	AC to DC converters 1-ϕ rectifier	Odd order	IEC 61000-3-2	IEC 61000-3-2
Information Technology Equipment	Switch mode power supply (SMPS).	Odd order	–	–
Industrial application				
DC power supply	AC–DC Rectifiers	Odd order	IEC 60146	IEC 60146
	Motor drives	Odd order such as 5, 7, 11, 13, 1 . . .	IEC 61800-3	IEC 61000-3-2, 3-4,3-12 [14,15,25]
	Arc furnace	Even harmonics	IEEE 519	IEEE 519
Transportation				
Electric traction	AC–DC Rectifiers	3,5,7,9,11	THDv 5%, Individual 3% THDi 8%	RDSO, Indian Railways [19]
Charging stations for electric vehicles (EV)	AC–DC Rectifiers	Odd harmonics such as 5,7, . . .	IEEE 519	CEA Regulations (Standard for connectivity of DG Resources) [2]

3.5 Conclusion

The chapter detailed various sources of voltage and current harmonics in the power generation, transmission and utilization sector in recent times. It has effects on various components of the power system. It is seen that it adversely affects the performance of the other equipment in the system as well as harmonic-producing equipment, both at the consumer and utility end. However, the extent to which harmonics can be tolerated depends on the susceptibility of the equipment or component. The least vulnerable ones are those whose main function is heating, and the most vulnerable ones are those whose design assumes a perfect sinusoidal wave shape.

3.6 Problems and solutions

Problem 3.1 Consider a system with 230 V, 50 Hz, 1-ϕ diode rectifier and a filter capacitor of $C = 1000$ μF connected to a load of $R = 10\ \Omega$ (Figure 3.34). Carryout a simulation study with MATLAB and show the current waveform and harmonic spectrum. Also, analyse the result.

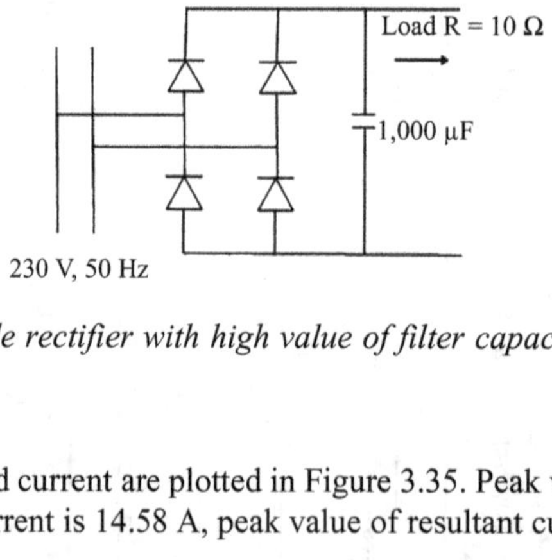

Figure 3.34 Diode rectifier with high value of filter capacitor – Problem 3.1

Solution:
The input voltage and current are plotted in Figure 3.35. Peak value of fundamental component of the current is 14.58 A, peak value of resultant current is around 42 A and THD is 97.52%.

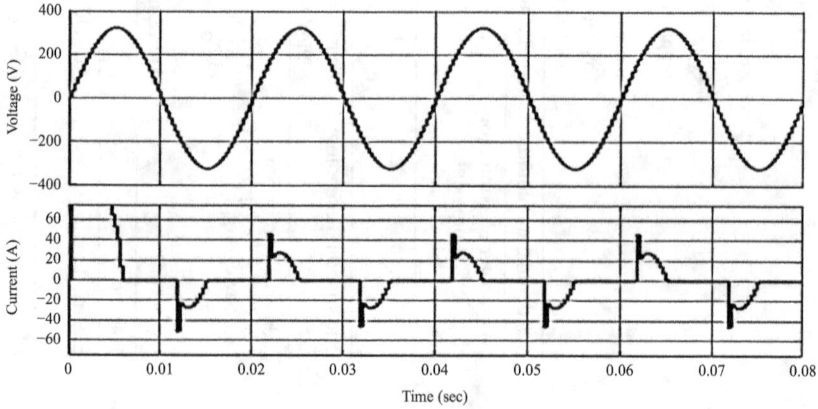

Figure 3.35 Waveforms of supply voltage and current of 1-ϕ diode rectifier

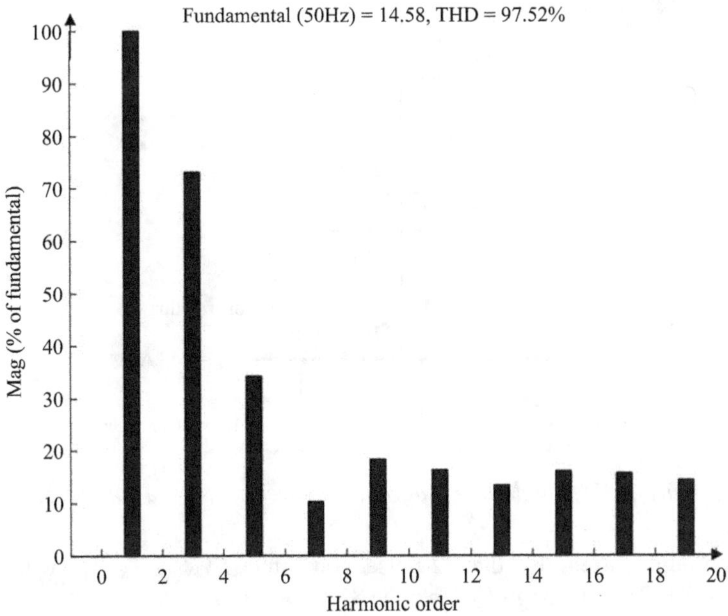

Figure 3.36 Waveform of current and its harmonic spectrum of a diode rectifier

The harmonic spectrum of the supply current is shown in Figure 3.36. The order of harmonics present in the current is odd harmonics such as 3rd, 5th, 7th, etc.

Problem 3.2 A 12-pulse AC to DC converter with a rating of 400 kVA is connected to a 440 V bus along with a linear load of 500 kVA as shown in Figure 3.37. Assume short circuit current at 440 V bus as 15 kA and the maximum value of the fundamental component of converter current is 625 A. Calculate (a) whether the installation meets the compliance requirement of IEEE-519. (b) If it is not meeting the compliance, suggest a proper solution. Neglect the harmonic component above 15.

Solution:
Given data:

$$440 \text{ V system, } I_{sc} = 15\ kA$$

Rating of 12 pulse converter = 400 kVA, Fundament component of converter current = 625 A

Rating of linear loads = 500 kVA

$$\frac{\text{Liner load}}{Non-\text{linear load}} = \frac{500}{400} = 1.25$$

Step 1. Calculation of the short circuit ratio of the system

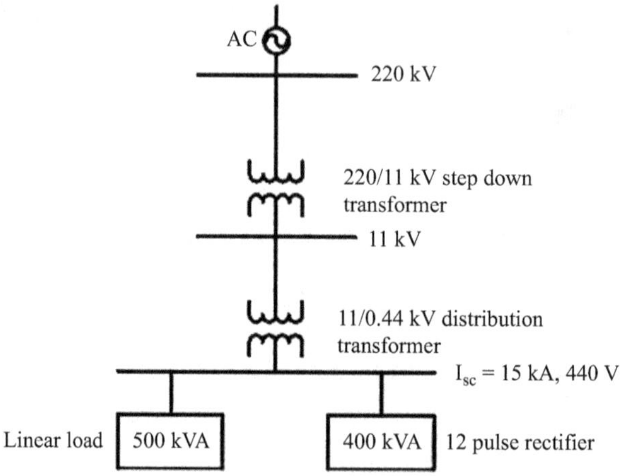

Figure 3.37 Single line diagram of power supply – Problem 3.2

Maximum current by the 12-pulse converter, $I_{CONVERTER-\max}$ with rating 400 kVA, 440 V, assuming efficiency as 83%

$$I_{CONVERTER-\max} = \frac{400 \times 1000}{\sqrt{3} \times 440 \times 0.83} = 632.37\text{ A}$$

Maximum current by linear loads, $I_{LINEAR-\max}$ with rating 500 kVA, 440 V is:

$$I_{LINEAR-\max} = \frac{500 \times 1000}{\sqrt{3} \times 440} = 656.08\text{ A}$$

$$\textit{Total maximum current } (I_{\max}) = 632.37 + 656.08 = 1288.45\text{ A}$$

$$\textit{Short circuit ratio} = \frac{I_{sc}}{I_{\max}} = \frac{15000}{1288.45} = 11.64$$

As short circuit ratio is less than 20, as per IEEE-519, the limitations are

$$\textit{TDD limit} \leq 5\%, 11 \textit{ to } 16 \textit{ order of harmonics limit is } 2\%$$

Step 2. Calculation of THD and TDD without compensation

TDD is calculated based on maximum demand of the system and magnitude of harmonic component of current.

$$\textit{Maximum demand} = 1288.45\text{ A}$$

For a 12 pulse converter, order of harmonics 11, 13, 23, 25... neglecting higher order harmonics,

$$I_{11\max} = \frac{I_{CONVERTER-1}}{11} = \frac{625}{11} = 56.82A;\ I_{13} = \frac{I_{CONVERTER-1}}{13} = \frac{625}{13}$$
$$= 48.08\text{ A}$$

$$I_h = \sqrt{I_{11}^2 + I_{13}^2} = \sqrt{56.82^2 + 48.08^2} = 74.43 \text{ A}$$

$$\%\text{THD} = \frac{I_h}{I_{CONVERTER-1}} \times 100 = \frac{74.43}{625} \times 100 = 11.9\%$$

The magnitude of the harmonic current added into the power system remains the same when a converter is added to linear loads in the system. However, TDD will vary with a linear to non-linear current ratio.

$$\%TDD = \frac{I_h}{I_{\max}} \times 100 = \frac{74.43}{1288.45} \times 100 = 5.78\%$$

TDD of individual orders of harmonics, TDD_{11} and TDD_{13} are given by:

$$\%TDD_{11} = \frac{I_{11\max}}{I_{\max}} \times 100 = \frac{56.82}{1288.45} \times 100 = 4.41\%$$

$$\%TDD_{13} = \frac{I_{13\max}}{I_{\max}} \times 100 = \frac{48.08}{1288.45} \times 100 = 3.73\%$$

TDD is 5.78% greater than 5%, and individual 11th and 13th order harmonics exceeding 2%, as per IEEE519, are not allowed. So, suitable harmonic filters are to be provided to limit the harmonics. As linear to non-linear load ratio decreases, TDD increases.

Step 3. TDD calculation with compensation

To maintain an 11th harmonic component less than 2% in the system, the maximum value of the magnitude of each component is:

$$I_{11\max} = I_{13\max} = \frac{2 \times 1288.45}{100} = 25.77 \text{ A}$$

$$I_h = \sqrt{I_{11}^2 + I_{13}^2} = \sqrt{25.77^2 + 25.77^2} = 36.44 \text{ A}$$

THD of the converter with harmonic filter is:

$$\%\text{THD}_{withfilter} = \frac{I_h}{I_1} \times 100 = \frac{36.44}{625} \times 100 = 5.83\%$$

With shunt harmonic filter, the percentage of total and individual TDD is calculated as:

$$\%TDD = \frac{I_h}{I_{\max}} \times 100 = \frac{36.44}{1288.45} \times 100 = 2.83\%$$

$$\%TDD_{11} = \frac{I_{11\max}}{I_{\max}} \times 100 = \frac{25.77}{1288.45} \times 100 = 2\%$$

$$\%TDD_{13} = \frac{I_{13\max}}{I_{\max}} \times 100 = \frac{25.77}{1288.45} \times 100 = 2\%$$

It is seen that both TDD and individual TDD are within the limit and acceptable as per IEEE-519. With the implementation of filter, the THD of the converter reduced from 11.9 to 5.83%.

Problem 3.3 A distribution system shown in Figure 3.38 supplies non-linear loads of 1-ϕ information technology equipment and an EV battery charging facility. Loads are fed through a $3\frac{1}{2}$ core, 35 mm^2 copper conductor PVC cable (current rating 110 A) through an 80 kVA, 11 kV/440 V, 50 Hz dry-type distribution transformer. Check the adequacy of the system and observation on the cable and transformer for feeding the loads.

Solution:

$$\text{Given: Balanced non-linear load current per phase} = 100 \text{ A}$$
$$\text{Cable type} = 3\frac{1}{2} \text{ core, } 35 \text{ mm}^2 \text{ copper conductor PVC cable}$$
$$\text{current rating} = 110 \text{ A per phase}$$
$$\text{Transformer (Dry type)} = 80 \text{ kVA, } 11 \text{ kV}/440 \text{ V}$$

For non-linear balanced load, neutral current (I_n) can vary between, $3I_{h3}$ *to* $\sqrt{3}I_{ph}$

Since the 3rd harmonic component is not exclusively given,

$$\text{Neutral current} = \sqrt{3}I_{ph} = \sqrt{3} \times 100 = 173 \text{ A}$$

As per IEEE 1100, recommend practice for non-linear load, neutral conductor to be rated at 200% of phase conductor, i.e., 200 A.

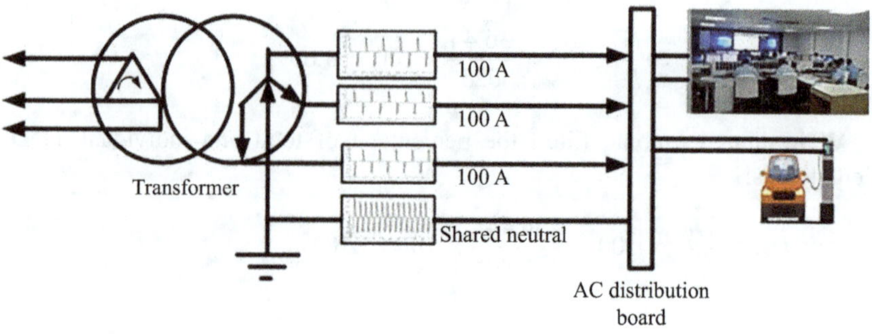

Figure 3.38 Distribution system non-linear load – Problem 3.3

Therefore, neutral conductor needs to be rated for twice the phase conductor cross-sectional area, i.e., 200 A. Thus, the neutral conductor is to be 70 mm^2 copper.

This is to avoid overloading of the neutral conductors due to harmonic current drawn by the non-linear loads.

$$\text{Transformer capacity for linear loads} = \sqrt{3} \times 440 \times 100 = 76.2 \text{ kVA}$$
$$\text{Given transformer} = 80 \text{ kVA}, \ \Delta/Y$$
$$\text{Margin available for non-linear loads} = 80 - 76.2 = 3.8 \text{ kVA}.$$

This may not be adequate for feeding non-linear loads. Hence transformers need to be k rated depending on harmonic spectrum of the load current.

Transformer primary winding Δ. This facilitates circulation of 3rd harmonic component within transformer primary winding.

Problem 3.4 Consider a system with 230 V, 50 Hz, 1-ϕ diode rectifier and a filter capacitor of C = 1000 μF shown in Figure 3.34, now connected to a load of $R = 10\ \Omega$ and $L = 100$ mH. Calculate the rms value of load voltage and current. Also, calculate the fundamental and rms current supplied by the source, along with the fundamental and actual power factor of the source.

Solution:

Given data: $V_s(t) = 230\sqrt{2}\ \sin\omega t$ V, $\omega = 2\pi f$ rad/s and $f = 50$ Hz, $Z_L = 10 + j31.4\ \Omega$ (RL load). The value of L is very high, hence the current drawn by the load will be of constant value.

Now, for calculating the rms load output voltage (V_d) for a single-phase diode bridge rectifier is given by:

$$V_d = \frac{2\sqrt{2}V_s}{\pi} = 0.9\ V_s = 0.9 \times 230 = 207 \text{ V}$$

Therefore, the RMS load current value is given by:

$$I_d = \frac{V_d}{Z_L} = \frac{207}{32.95} = 6.28 \text{ A}$$

Now, for calculating the value of the source current, the Fourier analysis of the square wave source current needs to be done.

$$i_s(t) = a_0 + \sum_{n=1}^{\infty} a_n \cos n\omega t + \sum_{n=1}^{\infty} b_n \sin n\omega t$$

$$a_0 = 0 \text{ and } a_n = 0$$

Due to half wave symmetry and odd wave symmetry and hence only sine terms will be present.

$$b_n = \frac{2}{\pi}\int_0^{\pi} I_d \sin n\omega t \cdot d\omega t$$

$$b_n = \frac{4I_d}{n\pi}$$

Thus, the expression for the instantaneous value of the source current is given by:

$$i_s(t) = \sum_{n=1}^{\infty} b_n \sin n\omega t = \sum_{n=1}^{\infty} \frac{4I_d}{n\pi} \sin n\omega t$$

The fundamental current component is given by:

$$i_1(t) = \frac{4I_d}{\pi} \sin \omega t$$

The peak value of the fundamental component of the source current is $i_{m1} = \frac{4I_d}{\pi}$ and the rms value of the fundamental component of the current is given by:

$$i_{rms1}(t) = \frac{2\sqrt{2}I_d}{\pi} = 0.9 \times 6.28 = 5.652 \text{ A}.$$

Actual RMS value of current = 6.28 A
Fundamental power factor = $\cos\phi_1 = 1$
Actual power factor = $\left(\frac{I_1}{I_d}\right)\cos\phi_1 = 0.9$
The average real power that the source supplies can be calculated as:

$$P = V_s I_{rms1} \cos\phi_1 = 230 \times 5.652 = 1300 \text{ W}$$

The apparent power supplied by the source:

$$S = V_s I_d = 230 \times 6.28 = 1444.4 \text{ VA}.$$

Power factor = P/S = 1300/1444.4 = 0.9

This shows that although the fundamental power factor is 1 and the reactive power drawn is zero, the circuit is still drawing harmonic power from the source, and hence actual power factor will be less than 1.

References

[1] IEC 60034-1: Rotating electrical machines – Part 1: Rating and performance.
[2] Central Electricity Authority (Technical Standards for Connectivity to the Grid) Regulations, https://cea.nic.in/regulations-category/connectivity-of-distributed-generation-resources.
[3] Ministry of New and Renewable Energy Draft Standard on "Technical requirements for Photovoltaic Grid Tie Inverters" to be connected to the Utility Grid in India, https://mnre.gov.in.
[4] IEC 61400-27-1 Wind energy generation systems – Part 27-1 Electrical Simulation models – Generic models.

[5] Technical Application Papers No.13 Wind power plants, ABB, https://library.e.abb.com/public/92faf0c1913f5651c1257937002f88e8/1SDC007112G0201.pdf.

[6] Indian standard IS 16071: 2013 High-voltage direct current (HVDC) Systems—Guidebook to the specification and design evaluation of A.C. Filters.

[7] IEC 60146-1-1 Semiconductor converters – General requirements and line commutated converters – Part 1-1: Specification of basic requirements.

[8] M. Pichan, A. A. Ahmad, A. Abrishamifar, and M. Fazeli, "Simple and Efficient Design and Control of the Single Phase PWM Rectifier for UPS Applications", *2017 8th Power Electronics, Drive Systems & Technologies Conference (PEDSTC)*, 14–16 February 2017, Mashhad, Iran.

[9] IEEE Std 519 IEEE Recommended Practice and Requirements for Harmonic Control in Electric Power Systems.

[10] N. Mohan, T. M. Undeland, and W. P. Robbins (eds.), *Power Electronics: Converters, Applications and Design*, 3rd Edition.

[11] IEC 62040 Uninterruptible power systems (UPS).

[12] Bureau of Energy Efficiency (Particulars and Manner of their Display on Labels of Self-ballasted LED Lamps) Regulations, 2017, Ministry of Power Notification New Delhi, The 27th December, 2017.

[13] Indian Standard 14700 (Part 3/Sec 2) Electromagnetic compatibility: Part 3 Limits, Section 2 Limits for harmonic current emissions.

[14] IEC 61000 3-2: Electromagnetic compatibility (EMC) – Part 3-2: Limits – Limits for harmonic current emissions (equipment input current ≤ 16 A per phase).

[15] IEEE Std C57.110 IEEE Recommended Practice for Establishing Liquid Immersed and Dry-Type Power and Distribution Transformer Capability when Supplying Nonsinusoidal Load Currents.

[16] https://commons.wikimedia.org/wiki/File:Electric_Arc_Furnace.png#/media/File:Electric_Arc_Furnace.svg.

[17] IEEE Std 1100, IEEE Recommended Practice for Powering and Grounding Electronic Equipment.

[18] High Power Static Uninterruptible Power Supplies for Nuclear Power Plants – Requirements and Operating Experience: *George Sebastian and others 20th Annual Conference of Indian Nuclear Society*, January 4–6, 2010, Chennai.

[19] S. George, and V. Agarwal, "Effect of Harmonics on Reliability of Electrical Equipment", *International Conference on Quality, Reliability and Control, ICQRC* 2001, pp. R68-1.

[20] G. Sebastian and S. George, "Power Quality: Harmonics, Issues Regulations and Compensating Techniques", *ICQRSE 2008, Macmillan Advance research series Advances in Performance and Safety of Complex System.*

[21] IEC 61000 3-4: EMC –Part 3-4: Limits – Limitation of emission of harmonic currents in low-voltage power supply systems for equipment with rated current > 16A.

[22] UL 1561: Dry-Type General Purpose and Power Transformers.

[23] Functional requirement of power quality compensating equipment (PQCE), Traction Installation Directorate, Research Designs and Standards Organization, https://rdso.indianrailways.gov.in/.
[24] IEC 61000-3-12: Electromagnetic compatibility (EMC) – Part 3-12: Limits – Limits for harmonic currents produced by equipment connected to public low-voltage systems with input current >16 A and ≤ 75 A per phase.
[25] IEC 60076: Power transformers.
[26] F. Blaabjerg, F. Iov, R. Teodorescu, and Z. Chen, "Power Electronics in Wind Turbine Systems", *Proceedings of the 12th International Power Electronics and Motion Control Conference (IPEMC)*, 2006.

Chapter 4

Power quality and reliability indices

Power quality and supply reliability are the two important aspects of the modern power supply system. While the former refers to deviation from a perfect sinusoidal wave, the reliability focuses on the interruption of power supply.

Goals of the chapter

Previous chapters detailed waveform distortion, particularly harmonics, on emerging power systems and their adverse effects on the electrical supply. After going through this chapter, the reader shall get a broad idea about:

- Power quality monitoring
- Indices and waveform analysis
- Power supply interruption and system reliability
- Standards and regulations

4.1 Introduction

Uninterrupted electrical power supply with sinusoidal waveform has become a strategic topic in recent times for both utilities and their consumers. This is because of our increased dependance of automation processes in all sectors of daily life. Be it in transportation, commercial, business, manufacturing or industrial sectors. In the past, only conventional electrical equipment, such as motors and incandescent lamps were available and the economic impact of power supply loss was limited. Also, this equipment is not very sensitive to the power supply wave shape. With the advancement of microelectronics, digital technology, computers, power electronics, etc., the utilization of electrical energy has spread over a wide area. As discussed in the previous chapters, these items are more sensitive to power quality, viz., wave-shape, frequency and its magnitude, including interruption, even for very short-time. For example, a minor deviation beyond the permissible supply voltage to information technology or data centre units can cause data loss, equipment maloperation or even system failure. Computers, automation systems, digital technology devices, etc., are now an indispensable part of day-to-day life. Any loss of power supply thus has a huge impact on our economy and daily life. Thus, availability of a power supply conforming to the utilization equipment is a mandatory requirement. Power supply is provided by synchronously connected generating stations through a network of transmission and distribution systems to utilization equipment. Largely, utilization equipment is responsible for impairing power quality, and distribution agencies are mandated to provide reliable power supply that matches the sinusoidal voltage waveform of specified magnitude and frequency to the utilization equipment. Thus, in recent times, power quality and reliability of the supply are becoming the two important topics that continue to grow as society becomes more reliant on digital technology and process automation.

Although power quality and reliability are different topics, the reliability of power depends on the quality of supply and not vice versa. Power quality refers to the degree of conformance to electrical power supply to the sinusoidal waveform of both magnitude of voltage and frequency. It is the measure of electrical supply that meets the requirements of the equipment connected to the system. Greater deviation or poor power quality adversely affects devices or equipment connected to the supply by their early failure and thus may cause supply interruptions. Reliability, in simple terms, is the availability of power supply provided by the utility when it is required by the customer. Thus, a highly reliable supply need not be of good power quality.

Generally, higher reliability in power supply is achieved by adopting improvements in system design, manufacturing of equipment, operation and maintenance. Higher margin in conductor size to reduce possible temperature rise, improved material in manufacturing, redundancy, failure mode and effect analysis (FMEA), etc., are among the few steps that are considered in design. While in production, quality control in material, process, environmental and product qualification improves equipment reliability. During operation, periodic checks should be conducted to ensure it operates as per its design intent, including the connected power supply. In maintenance, preventive, predictive and condition monitoring are some of the steps that improve the availability of the system.

In many countries, including India, users are classified as consumers, and the distribution companies need to provide an uninterrupted power supply to the consumers [1]. Hence, it is necessary to monitor and record power supply to consumers on real-time basis. Some countries have also imposed economic penalties on utilities in case of deviation from specified values. Thus, the topic of power quality and supply reliability is evolving into a greater area of interest among industrial and academic forums. The following sections detail various monitoring instruments, indices in power quality, especially on harmonics, supply interruptions and system reliability.

4.2 Power quality monitoring

Assessing the power quality level [2] of equipment, facilities, customers, regions, or networks of interconnected generators, transmission lines, and distribution systems spread over a vast area is a challenging task. Before going ahead, we must set our objectives of measurement very clearly. Broadly, it is to minimize the risk of damage to electrical equipment and system interruption, and thus it includes:

(a) Identifying the source of imperfection in the power supply waveform
(b) Accessing the prevailing power quality level
(c) Check the compatibility of the equipment before its installation
(d) Predict the performance of the equipment or component in the environment
(e) Devising mitigation techniques
(f) Check the effectiveness of the corrective action.

As detailed earlier, the performance of most of the grid-connected equipment is affected by the quality of power from its transient state to long-duration variation in the rated value of voltage, phase angle and frequency. Hence, the response and accuracy of measuring devices or instruments shall match the range of interest. Also, the distance of signal transmission should not impair the accuracy of measurement. Thus, the monitoring location and types of instruments used depend on the objective of monitoring.

If the monitoring objective is to check or test the equipment, then the monitor should be placed close to the equipment. For example, for monitoring the performance of sensitive equipment like a computer or assessing the waveform distortion of a VFD, a monitor shall be placed very close to the equipment. And if the objective is to investigate the overall quality level, especially transients and other supply variations that originate from outside the facility, then the monitor should be placed at the incoming power supply.

The following sections briefly discuss the instruments and steps involved in assessing the power quality level.

4.2.1 Instrument transformers

Primary electrical quantities required for assessing the waveform parameters are the supply voltage and current. Considering the operator's safety, monitoring instruments are rated for low voltage and low current levels. Thus, for LV systems, voltage is sensed directly. However, for higher voltage systems, potential transformers (PT) are employed to step down (typically to ≤ 110 V) to the low-voltage level. For measurement at extra high voltages, Capacitive Voltage Transformers (CVT) consist of a capacitive voltage divider with an electromagnetic transformer, which is used considering a smaller size and cost of the corresponding PT. The oldest and simplest method of measuring current is by using a shunt resistor. However, the resistor is affected by heating due to power (I^2R) loss and voltage (IZ) drops across the measuring shunt. Hence, it cannot be used to measure a large current. This is done by using a current transformer (CT) in the power circuit. One type of CT is a clamp-on or split-core type that can be placed around the conductor without interrupting the circuit. The other type is solid core or toroidal core, where the primary winding is connected in series with the power circuit and secondary winding (typical rating < 5 A or 1 A) to the measuring instruments. Clamp-on type CT is generally useful for short-term measurement. Transformers for stepping

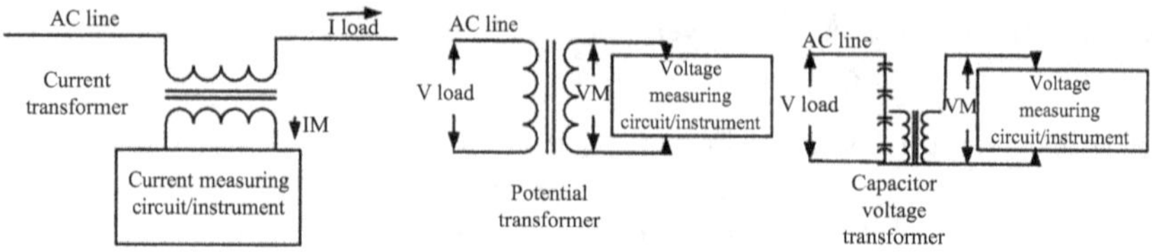

Figure 4.1 Line diagram of current and voltage measurement

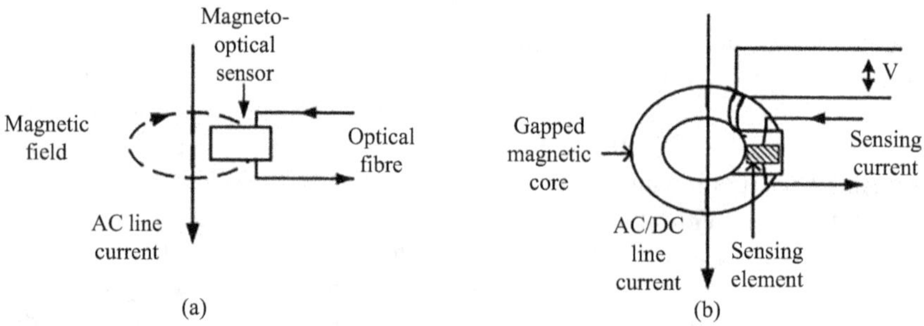

Figure 4.2 Line diagram of current and voltage measurement. (a) Based on optical sensing and (b) based on hall effect.

down the voltage or current for measuring purposes are grouped under instrument transformers. There are other new types of measuring devices using non-conventional instrument transformers – fibre optic and hall effect (where a semi-conducting wafer is placed in a magnetic concentrating ring). Once the voltage and current are within safe limits, Power Quality (PQ) parameters are measured using various types of analog and digital instruments. Line diagrams of conventional CT, PT and CVT, along with a block diagram of measuring instruments, are shown in Figure 4.1, and emerging current measuring transformers are shown in Figure 4.2. The emerging non-conventional CTs are either based on the interaction of the electromagnetic field produced by the line current on either an optical sensing medium or hall effect sensor. Compared to conventional measuring transformers, non-conventional transformers are smaller in size and lighter in weight. Additionally, electromagnetic interference on the measuring circuit is also reduced.

4.2.2 Monitoring instruments

As discussed, the quality of power supply is assessed by a set of parameters that vary from its amplitude, phase angle and frequency from its transient state to sustained long duration. Thus, we need different types of instruments ranging from simple portable instruments troubleshooting to very sophisticated hardware and software-based instruments for detailed power quality analysis.

The following types of instruments are generally used in assessing the power quality.

4.2.2.1 Flicker meters

As discussed in Chapter 2, a sudden change in supply voltage causes variations in the light output of the lighting sources. Based on the adverse effect of the fluctuation in voltage or the resulting severity of flicker on the human eye [3], acceptance norms are developed. IEC Standards 61000 part 3-3 [4] and part 3-15 details limits of voltage change, fluctuations and flicker in public low-voltage systems for equipment with rated current <16 A and <75 A per phase. Assessment of voltage fluctuations or resulting flicker on short-term and long-term basis based on the rms value of voltage change is detailed in IEC 61000-4-15 [5]. Voltage fluctuations are normally measured by portable handheld flicker meters or power quality analysers. It can also be analysed by simulation using software like ETAP, MATLAB®, etc.

4.2.2.2 Multi meters

It is the simplest and most popular instrument used for troubleshooting or checking and repairing electrical and electronic instruments and circuits. It is a very compact and portable instrument. It is available in analogue and digital versions and can measure multiple electrical parameters like DC and AC voltage and current. It is frequently used for measuring current, voltage, under voltage, overvoltage and imbalances.

4.2.2.3 Oscilloscopes

It is an instrument used for graphical or oscilloscopic display of variation of voltages, currents and phase angles, with its magnitude on the vertical (Y) axis and time on the horizontal (X) axis. The type of monitor depends on the data required to be analysed. For example, for measuring under or over voltage, a less sophisticated meter than that required for troubleshooting a VFD trip or for harmonic study. A high-resolution oscilloscope displays signal noise, transients, sag, swell, unbalance, distortion, frequency, etc. A portable oscilloscope is often used where monitoring is required on a temporary basis, like testing or assessing PQ on a short-time basis. It can store the data for the calculation of peak, average, rms, and other values of the supply signal.

4.2.2.4 Power quality analyser

PQ analyser is a digital instrument used for monitoring power quality. It is available for use in 1-ϕ or 3-ϕ systems and indicates the parameters either directly or facilitates downloading it to computers for analysis for troubleshooting. It is normally useful for measurement of phase or neutral current, voltage, dip, swell, interruption, waveshape deviation, transients, rapid voltage variation, frequency, etc., and allows calculation of unbalance, real and reactive power, power factor, harmonic analysis, THD, phase angle between voltage and current, etc. An advanced version of PQ analysers is provided with GPS time synchronization, module interfaces for USB, flash drives, Wi-Fi, Ethernet, PC connectivity, etc. Most of PQ monitors display information in time plots of current or voltage disturbance, a histogram of cumulative values, etc. Figure 4.3 shows a connection diagram of a power quality monitor connected on 3-ϕ system, and Figure 4.4 shows a typical waveform of the supply using a power analyser.

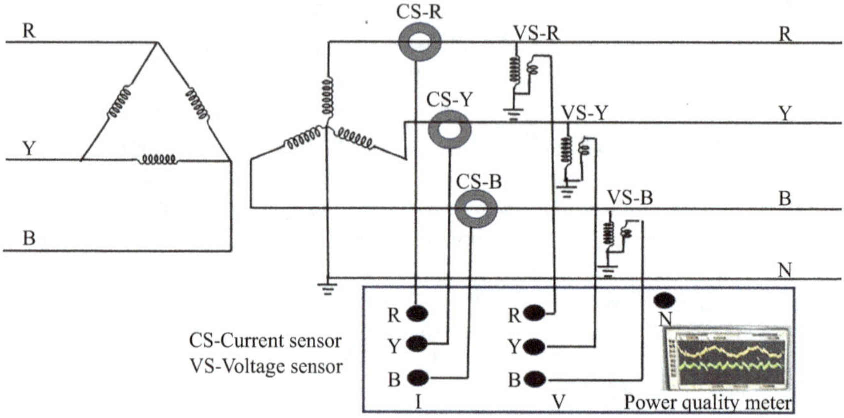

Figure 4.3 Connection diagram of typical power quality monitor

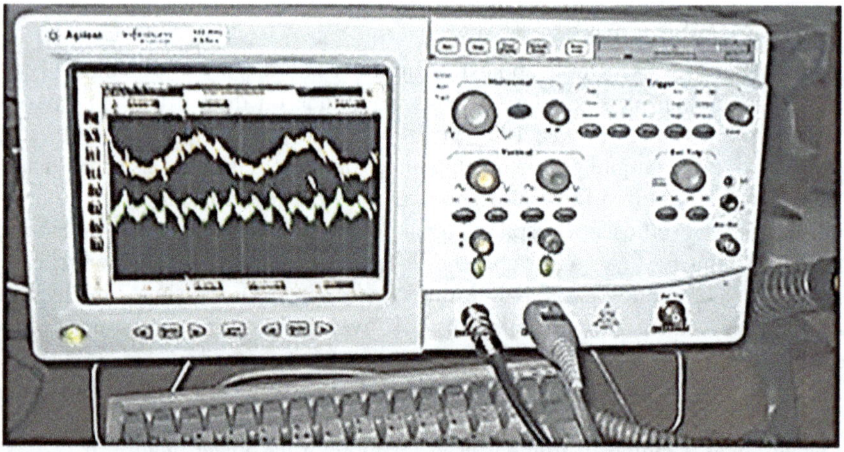

Figure 4.4 Typical waveform of supply using a power analyser

Power quality analysers are the most commonly effective and easy devices to take real-time readings and collect data. This data can be later transferred to computers for analysis.

While portable analysers are useful for troubleshooting or testing of equipment at a location, permanently installed analysers can monitor system disturbance since the system is unpredictable. Modern protection and control relays, also called Intelligent Electronic Device (IED), installed in the system serve to monitor PQ. This is detailed in the following section.

4.2.2.5 Intelligent electronic device

DSP-based protection relays have been widely used for over a decade in power generation, transmission and distribution systems for the protection of equipment and systems like generators, transformers, substations, overhead lines, cables, motors, etc. As detailed in Chapter 1, supply voltage and current are given as electrical input to the relay for sensing the fault condition. Based on the algorithm and type of required protection, the relay or IED calculates other circuit parameters such as rms values of voltage, current, frequency, active and reactive power, power factor, impedance, unbalance, etc. On sensing the fault condition, which is based on the calculated value and user-set trigger value, the relay trips the associated circuit breaker to isolate the fault in the system. IED also permits the user to define or add further logic called 'Programmable Scheme Logic' (PSL), in the relay. Normally, the processor in the relay has sufficient excess capacity to perform functions other than protection functions. Thus, with PSL, monitoring of other power quality parameters, such as fluctuations in voltage and frequency, harmonics, interruptions, etc., is done. Also, the relay registers pre-fault and post-fault voltage and current in its memory. The stored data are later transferred to computers for further analysis of waveforms and other parameters for troubleshooting of an event.

4.2.2.6 Power quality monitoring at central place

Power system is often subjected to various types of faults, resulting in supply interruptions. To analyse the event, pre-fault and post-fault data of various electrical parameters are required. Often, data stored in Disturbance Recorders (DR) or modern numerical protection relays installed in the associated system or transmission lines is helpful. Additionally, continuous monitoring, such as that implemented using Phasor Measurement Units (PMUs), is particularly desirable in critical areas or regions of the system. It is often done at a central control place in a power system. Figure 4.5 shows a block diagram of a typical central power quality monitoring system. The primary data are gathered by the instrument transformers or transducers, installed close to the source or point of measurement. For reducing the burdens and increasing the safety, it is then converted to numerical values by

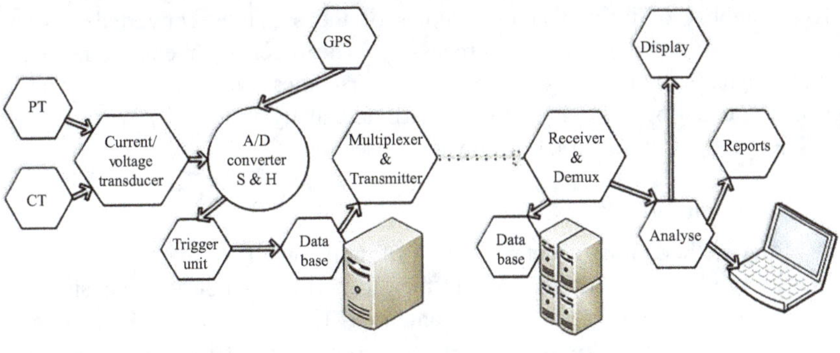

Figure 4.5 Block diagram of a typical power quality monitoring at a central place

the Analog to Digital (A/D) converters, including a Sample and Hold (S&H) unit to meet the data requirement of the processors that compute the required information based on the application. Since the occurrence of PQ disturbance is unpredictable, to save memory, it is often provided with a triggering unit to initiate the recording of wave forms whenever sensed data exceeds the setting or information based on Data Base (DB) stored. Often, the data captured is monitored at a remote control centre. Hence, the instrument needs to be time-synchronized. It is often based on timing information from Global Positioning Satellite system (GPS), Inter-Range Instrument Group (IRIG-B), etc. In the past, manufacturers often used their own proprietary data formats for the exchange of data among instruments for measuring, recording, and monitoring of PQ among utilities and customers. As the data format among different manufacturers is often different, it often encounters data communication problems when customers or utilities use instruments of different make and type. Communication protocols normally used include Ethernet, Transmission Control Protocol, and Internet Protocol (TCP/IP), Hypertext Markup Language (HTML), RS485, etc. With the adoption of standard data format as per IEC 61850 by the manufacturers, information exchange among different instruments is made easy. All the subsequent data processing is carried out by the relevant algorithm installed in PQ monitoring system.

4.3 Harmonic analysis

As we have discussed, one of the most emerging power quality issues is harmonics. It is inadvertently generated in a power system by the power electronic switching devices in addition to system-generated harmonics. Also, its magnitude, phase angle and order vary from ideal conditions during operation due to number of factors such as asymmetry in network impedances, unbalance in power supply, mismatches in firing angles among the switching devices, etc. These result in harmonics of integer and non-integer multiples of the fundamental [6], in the system, and are fluctuating or fast changing. Thus, it may require continuous monitoring of the signal. For assessing the harmonics, signals are brought to the power quality monitoring equipment or power quality analysers. With the microprocessor, a fixed number of instantaneous samples of the signal is converted to digital quantities and stored for further processing. The practice of examining a time domain signal's frequency component, or harmonics, especially those that are periodic or quasi-periodic, is known as harmonic analysis.

These are discussed in the following subsections.

4.3.1 Fourier analysis

One of the methods for the determination of harmonics is to mathematically construct the time domain waveform, and then to perform a Fourier analysis of this waveform to determine the harmonic content [7–10]. According to French mathematician Jean Baptiste Joseph Fourier, any continuous periodic function with a time period (T) can be shown as the addition of a constant term, a sinusoidal component corresponding to the fundamental frequency, and an infinite series of harmonic components at frequencies that are integer multiples of the base frequency.

Mathematically, the trigonometric form of the Fourier series of a periodic function $f(t)$ can be written as,

$$f(\omega t) = a_0 + \sum\nolimits_{n=1}^{\infty} (a_n \cos(n\omega t) + b_n \sin(n\omega t)) \tag{4.1}$$

where,

$$a_0 = \frac{1}{\pi}\int_0^{2\pi} f(\omega t) d\omega t \tag{4.2}$$

$$a_n = \frac{1}{\pi}\int_0^{2\pi} f(\omega t)\cos(n\omega t) d\omega t \tag{4.3}$$

$$b_n = \frac{1}{\pi}\int_0^{2\pi} f(\omega t)\sin(n\omega t) d\omega t \tag{4.4}$$

In terms of polar or cosine form, the frequency components of the signal can be written as:

$$c_n = \sqrt{a_n^2 + b_n^2}; \quad c_0 = a_0, \tag{4.5}$$

and

$$\theta_n = \tan^{-1}\left(\frac{-b_n}{a_n}\right) \tag{4.6}$$

$$x(t) = c_0 + \sum\nolimits_{n=1}^{\infty} c_n[\cos(n\omega t + \theta_n)] \tag{4.7}$$

Thus, a measured periodic continuous signal $x(t)$ in time domain is transformed into a series of discrete frequency components by Fourier Analysis. Many of the power quality monitoring devices, utilize Fourier analysis-based methods.

4.3.2 Discrete Fourier Transform (DFT)

To analyse harmonics in a discrete-time signal, we use the Discrete Fourier Transform (DFT) or the Fast Fourier Transform (FFT), which is considered an efficient algorithm to compute the DFT. Microprocessors and microcontrollers were introduced in early 1980 in various areas of the power system. The input analogue current or voltage signals are sampled and converted into a digital representation without loss of precise information. It is further processed according to the appropriate mathematical algorithm based on its function. These devices use DFT to implement their application algorithm. The mathematical definition of DFT is given as follows:

For a finite-length, discrete-time signal $x[n]$, where $n = 0, 1, 2, \ldots, N-1$, the DFT is defined as:

$$X[k] = \sum_{n=0}^{N-1} x[n] \cdot e^{-j\frac{2\pi}{N}kn},\ k = 0, 1, \ldots, N-1 \tag{4.8}$$

Here:

- $X[k]$: The complex spectrum (magnitude phase) at the kth frequency bin.
- N: Total number of samples.
- $e^{-j\frac{2\pi}{N}kn}$: Complex sinusoid representing the basis function at bin k.

bin: A bin refers to a discrete frequency component output from the DFT/FFT process. Each bin represents a specific frequency range in the signal's spectrum.

Now, to recover the time domain sequence from frequency domain coefficients, the Inverse DFT (IDFT) is used. The mathematical expression for the same is given as:

$$x[n] = \frac{1}{N}\sum_{k=0}^{N-1} X[k] \cdot e^{j\frac{2\pi}{N}kn} \tag{4.9}$$

This confirms that the transformation is invertible, allowing for full time-frequency representation. For doing harmonic analysis with DFT, a frequency mapping is done. Each bin k in DFT corresponds to a frequency:

$$f_k = \frac{k \cdot f_s}{N} Hz \tag{4.10}$$

where:

- f_k: Frequency represented by the kth bin.
- f_s: Sampling frequency (Hz).
- N: Number of samples.

Although DFT can give an accurate solution, but:

1. In general, to compute a one-point (one-sample) DFT, N multiplications and $N-1$ additions are required.
2. Now, to compute the N-point DFT, it requires N^2 multiplications and $N(N-1)$ additions.

Thus, DFT is useful for understanding the underlying theory, educational purposes, or when analysing very small signals where the computational load is negligible. However, microprocessors of 8 bits or 16 bits have limited memory and processing capacity. This restricts the number of samples of waveform that can be processed per cycle or the speed of operation of the relay, restricting itself mainly to the protection function. The output for both DFT and FFT is shown in Figure 4.6 for a signal containing a 50 Hz fundamental and 3rd and 5th harmonic components.

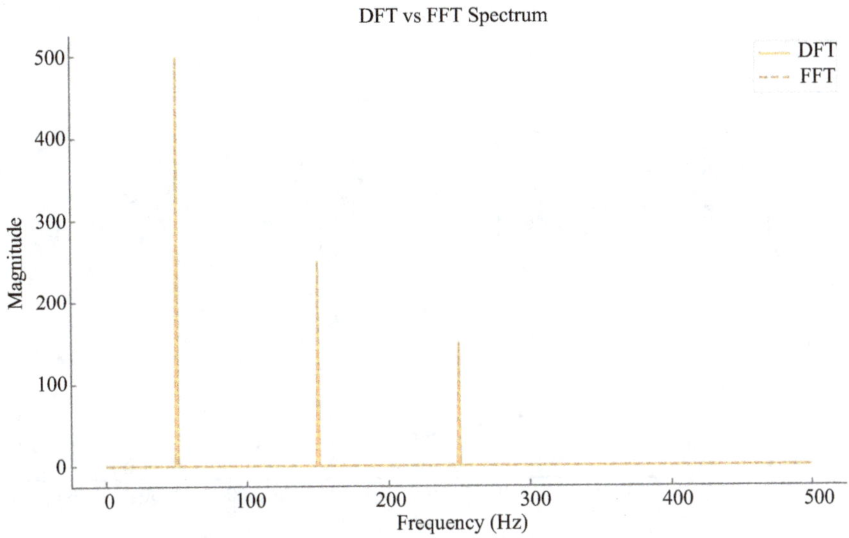

Figure 4.6 Frequency spectrum output for DFT and FFT

4.3.3 Fast Fourier transform

The FFT algorithm is extensively used to compute the DFT. The most common algorithm for computing the FFT is the Cooley–Tukey FFT algorithm, which reduces the computational complexity. By using FFT, the number of multiplications and the number of additions are reduced to $N\ \log_2 N$ additions and $\frac{N}{2}$ $\log_2 N$ multiplications. FFT to reduce the number of

1. Calculations significantly, it breaks the DFT into smaller DFTs recursively. Thus, FFT is preferred in most of the practical applications, especially where real-time or high-speed computation is needed (e.g., digital oscilloscopes, power quality analysers).
2. The signal length is a power of two (e.g., N = 1024, N = 2048, etc.).

Although the results shown for DFT and FFT are the same in Figure 4.6, from Figure 4.7, it can be inferred that computation time is far less for an FFT calculation. It is visible that there is a sharp dip at 1024 samples, while the time was rising earlier. This is because most high-performance FFT libraries perform faster when the number of samples N is a power of 2, like 256, 512, 1024, 2048, etc. These sizes are ideal for Cooley–Tukey FFT, which recursively splits the transform into halves-optimal when $N = 2^k$. Hence, at $N = 1024$ in Figure 4.7, the FFT performs much more efficiently, hence the dip is observed. For example, for $N = 1024$ input points, DFT will have to do 1 million operations, whereas for FFT it is just 10,240 operations.

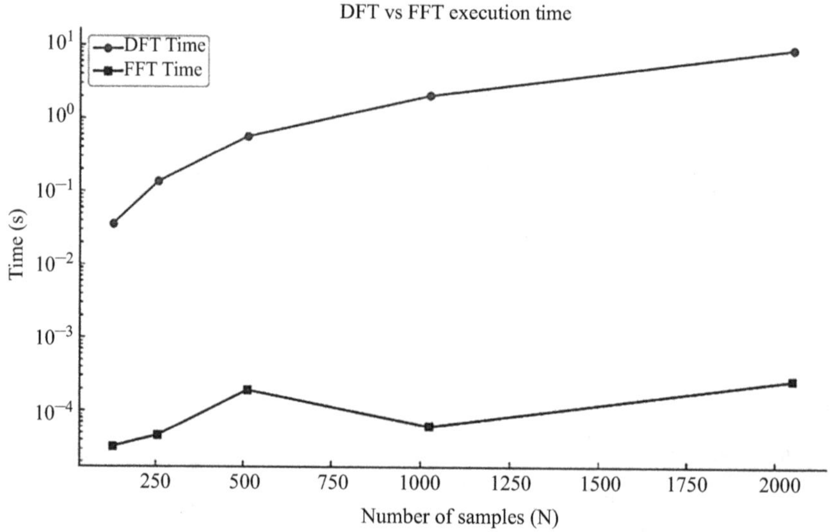

Figure 4.7 Run time comparison of FFT with DFT

4.3.4 Mathematical insight

The frequency resolution Δf in DFT/FFT is given by:

$$\Delta f = \frac{f_s}{N} \tag{4.11}$$

To achieve better resolutions, increase N and reduce f_s but ensure that the Nyquist criterion is satisfied: $f_s \geq 2f_{\max}$. Windowing is often used before the FFT to reduce spectral leakage (e.g., Hamming, Hanning windows). The zero-padding method is also used to improve visualization but doesn't increase actual frequency resolution. Figure 4.8 shows the effect of introducing a Hamming window in the FFT calculation to reduce the spectral leakage and gives accurate, sharp harmonic peaks.

The THD is calculated as:

$$\text{THD} = \frac{\sqrt{\sum_{k=2}^{\infty} |X[k]|^2}}{|X[1]|} \tag{4.12}$$

where $X[1]$ is the fundamental component and $X[k]$ are higher-order harmonics.

In conclusion:

1. DFT is the basis of frequency domain analysis, but computationally expensive.
2. FFT is the optimized algorithm for DFT and is the de facto standard for harmonic analysis.
3. FFT is preferred for performance, especially in real-time and high-sample-rate applications.

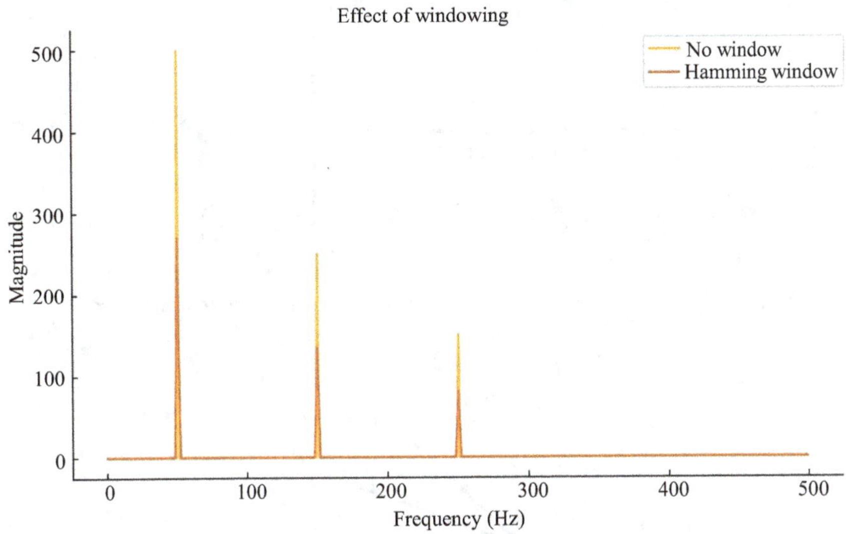

Figure 4.8 Effect of Hamming window on the FFT calculations

4.4 Harmonic indices

As described in previous chapters, the operation of a semiconductor rectifier produces characteristic harmonic components that depend on the number of pulses the rectifier produces during one cycle of fundamental frequency. Depending on the waveform's symmetry under steady state conditions, these characteristic harmonics [5] are grouped as odd when $f(-t) = -f(t)$, even when $f(-t) = f(t)$ or triplen, where frequency is a multiple of 3 times the fundamental waveform. Also, depending on rotational sequence, odd, even and triplen harmonics rotate in positive sequence, negative sequence and zero sequence with respect to the fundamental frequency. Table 2.6 in chapter 2 shows rotational sequences of harmonic frequencies up to 7th order harmonics. Additionally, it may also have non-characteristic harmonics due to unbalance in AC supply, asymmetry in converter transformer turns ratio and reactance, firing angle variation, etc. Further systems may have inter-harmonics [5,6] components that are not integer multiples of the fundamental frequency, often produced by arc furnaces, HVDC back-back systems, etc. All these harmonics adversely affect the performance of almost every electronic or electrical equipment connected to the systems. It also affects the performance of adjacent data or signal communication lines due to its electromagnetic coupling. However, the severity of impact depends on the magnitude of both the individual and total harmonic components. The following sections detail the most common indices [11–15] to quantified harmonics.

4.4.1 Individual harmonic distortion

Consider a transmission system with non-linear loads drawing current $i_L(t)$ as shown in Figure 4.9. Let the voltage at PCC be V_{PCC} and the transmission line current is $i_s(t)$.

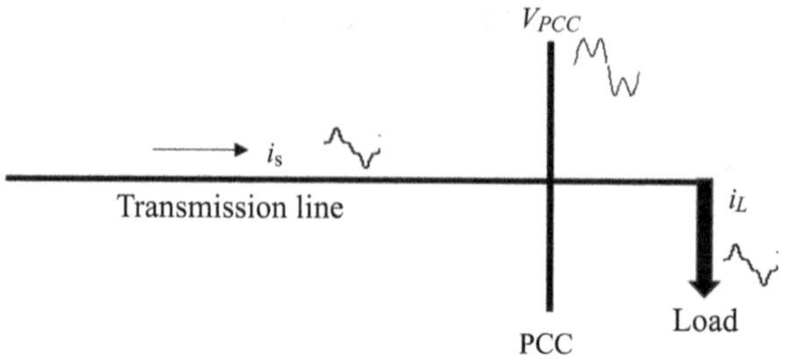

Figure 4.9 Transmission line connected to a non-linear load at PCC

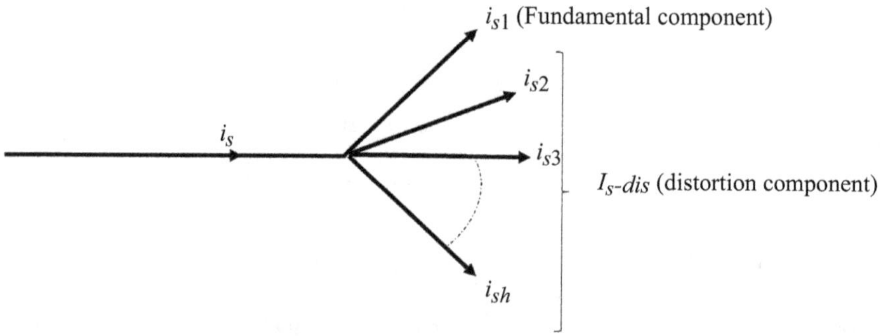

Figure 4.10 Representation of line current in terms of its fundamental and distortion components

The instantaneous value of line current $i_s(t)$ is a function of all individual harmonic components present.

$$i_s(t) = f(i_{s1}(t),\ i_{s2}(t),\ i_{s3}(t),\ldots,i_{sh}(t) = f(i_{L1}(t)\ \ i_{L2}(t),\ i_{L3}(t),\ldots,\ i_{Lh}(t) \tag{4.13}$$

Similarly, the rms value of line current (I_s) or load current (I_L) is also a function of the rms value of the individual harmonic component of respective current, as shown in Figure 4.10.

$$I_s = f(I_{s1}, I_{s2}, I_{s3}, \ldots, I_{sh}) = f(I_{s1}, I_{s2}, I_{s3}, \ldots, I_{sh}) \tag{4.14}$$

where h is the order of the harmonic component of current. Similarly, voltage at point of common coupling (PCC) is written as:

$$v_{PCC}(t) = f(v_{PCC1}(t), v_{PCC2}(t), v(t), \ldots, v_{PCCh}(t) \tag{4.15}$$

$$V_{PCC} = f(V_{PCC1}, V_{PCC2}, V_{PCC3}, \ldots, V_{PCCh}) \tag{4.16}$$

I_s, I_{s1}, $I_{s\text{-}dis}$ and $I_{s\text{-}peak}$ are the rms value of total, fundamental component, distortion component and peak value of source current, respectively.

Individual harmonic distortion [16] of hth in line current $(HD_{i_{s-h}})$

$$HD_{i_{s-h}} = \frac{I_{s-h}}{I_{s1}} \tag{4.17}$$

Similarly, individual voltage distortion $(HD_{v_{PCC-h}})$ is given by

$$HD_{v_{PCC-h}} = \frac{V_{PCC-h}}{V_{PCC1}} \tag{4.18}$$

4.4.2 Total Harmonic Distortion (THD)

Presence of harmonics is characterized by waveform distortion [6,17], and it is quantified by Total Harmonic Distortion (THD). $i_s(t)$ and I_s are given by:

$$i_s(t) = \sum\nolimits_{h=1}^{\infty} i_{sh}(t) \tag{4.19}$$

$$I_s = \sqrt{\sum\nolimits_{h=1}^{\infty} {I_{sh}}^2} \tag{4.20}$$

As distortion in the current is due to higher-order harmonic component of the current other than the fundamental component, the amount of distortion in the current $(i_{s-dis}(t))$ and I_{s-dis} are given by,

$$i_{s-dis}(t) = \sum\nolimits_{h=2}^{\infty} i_{sh}(t) \tag{4.21}$$

$$I_{s-dis} = \sqrt{\sum\nolimits_{h=2}^{\infty} {I_{sh}}^2} \tag{4.22}$$

THD is the ratio of the rms value of all harmonic content to the rms value of the fundamental quantity, expressed as a percentage of the fundamental component of voltage or current, respectively. Definitions in IEEE 519-2014 consider harmonic components up to the 50th order without inter-harmonics. THD in current (THD_i) is expressed as:

$$THD_i = \frac{\sqrt{\sum\nolimits_{h=2}^{\infty} {I_{sh}}^2}}{I_{s1}} \tag{4.23}$$

Similarly, THD in voltage at PCC (THD_{vPCC}) is given by:

$$THD_{vPCC} = \frac{\sqrt{\sum\nolimits_{h=2}^{\infty} V_{PCCh}^2}}{V_{PCC1}} \tag{4.24}$$

where, V_{PCCh} is the rms value of the h-order component of V_{PCC}.

Equation (4.6) can be written as:

$$I_s = \sqrt{I_{s1}^2 + \sum_{h=2}^{\infty} I_{sh}^2} = \sqrt{I_{s1}^2 + I_{s-dis}^2} \tag{4.25}$$

I_{s-dis} is given by:

$$I_{s-dis} = \sqrt{I_s^2 - I_{s1}^2} = \left[I_s^2 - I_{s1}^2\right]^{\frac{1}{2}} \tag{4.26}$$

Hence, THD_i can be written as:

$$THD_i = \frac{\sqrt{I_s^2 - I_{s1}^2}}{I_{s1}} = \sqrt{\left(\frac{I_s}{I_{s1}}\right)^2 - 1} \tag{4.27}$$

Resultant current in terms of THD is given by:

$$I_S = I_{S1}\sqrt{1 + THD_i^2} \tag{4.28}$$

If THD_i is zero, the rms value of current will be equal to the rms value of the fundamental, and the waveform will be a pure sine wave. As THD increases, distortion in the waveform increases. It can exceed 100% which indicates that the rms value of all higher-order harmonics present in the waveform is more than the fundamental value. Thus, THD gives information about how much the waveform is distorted from the sinusoidal waveform.

To show how THD increases with distortion, a diode rectifier with resistive load $R = 10\ \Omega$ supplied with 230 V, 50 Hz supply is simulated using the MATLAB platform. In this case, the input current is sinusoidal, as in Figure 4.11(a), with a peak value of 32.32 A. The output voltage has more ripples, as shown in Figure 4.11(b). The THD is close to zero, as given in Figure 4.11(c). To get ripple-free voltage, a capacitor filter with C = 3000 μf is used, and simulated input current and output voltage waveforms. When a filter is added to make the output voltage with less ripple, the input current deviates from sinusoidal and results in a distorted waveform. It is found input current is non-sinusoidal as in Figure 4.12(a) with 83.93A though the quality of output voltage is improved as shown in Figure 4.12(b). It is noted in Figure 4.12(c) that THD is 139.2% and odd harmonics 1,3,5,7…., etc., are present in the input current.

The THD gives a quick measure of distortion. However, it will not give detailed information on the full spectrum, and there is no rule for truncation point. It does not directly give information about the magnitude of the signal. The effect of distortion in current on voltage at PCC depends on both the magnitude and THD of its current. In other words, a high magnitude of current with low THD can have a bigger impact on the system voltage than a low magnitude of current with high THD. This ambiguity can be eliminated by using other harmonic indices known as total demand distortion, as explained in the next section.

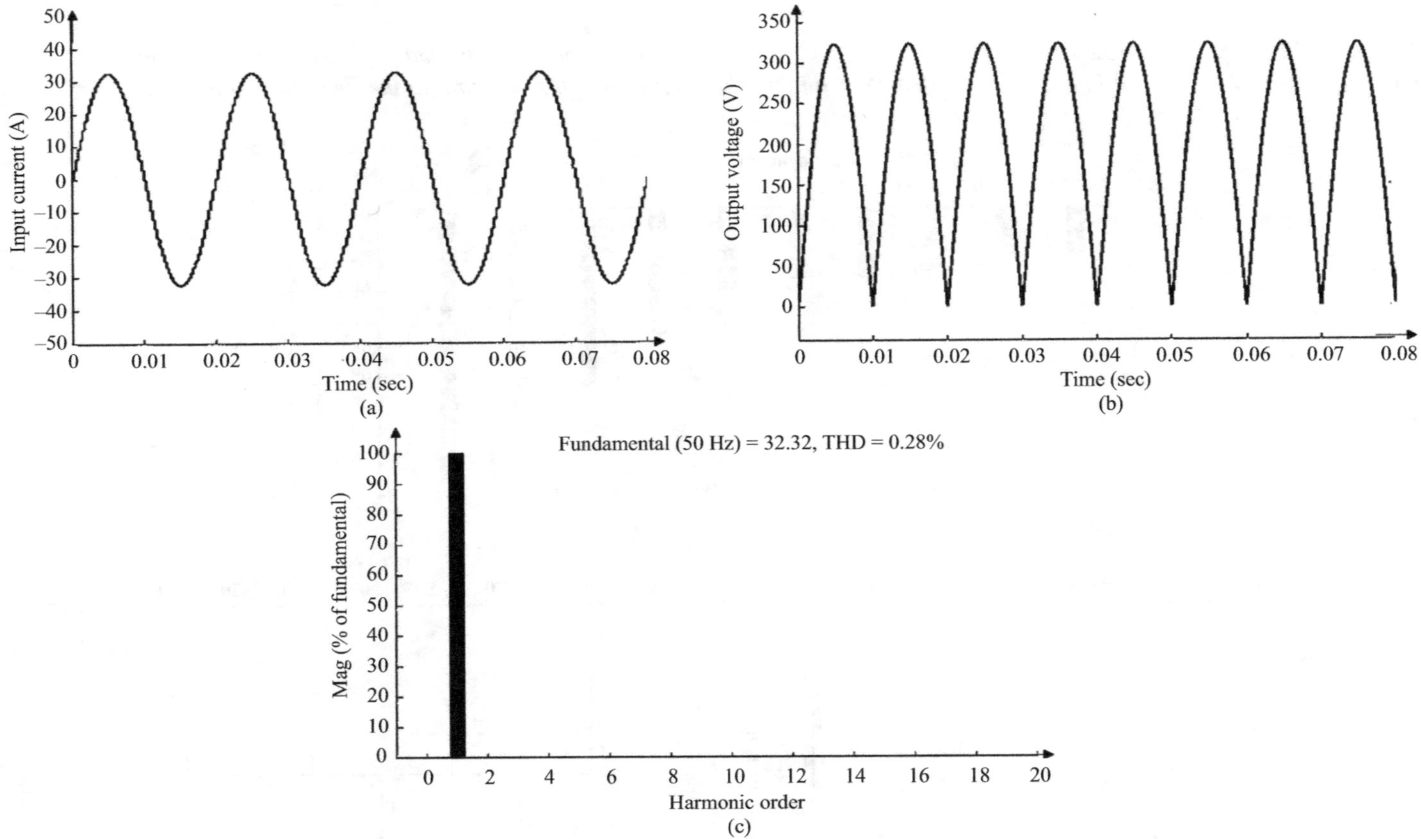

Figure 4.11 Waveform of input current of diode rectifier with $R = 10\ \Omega$. (a) Input current, (b) output voltage and (c) harmonic spectrum.

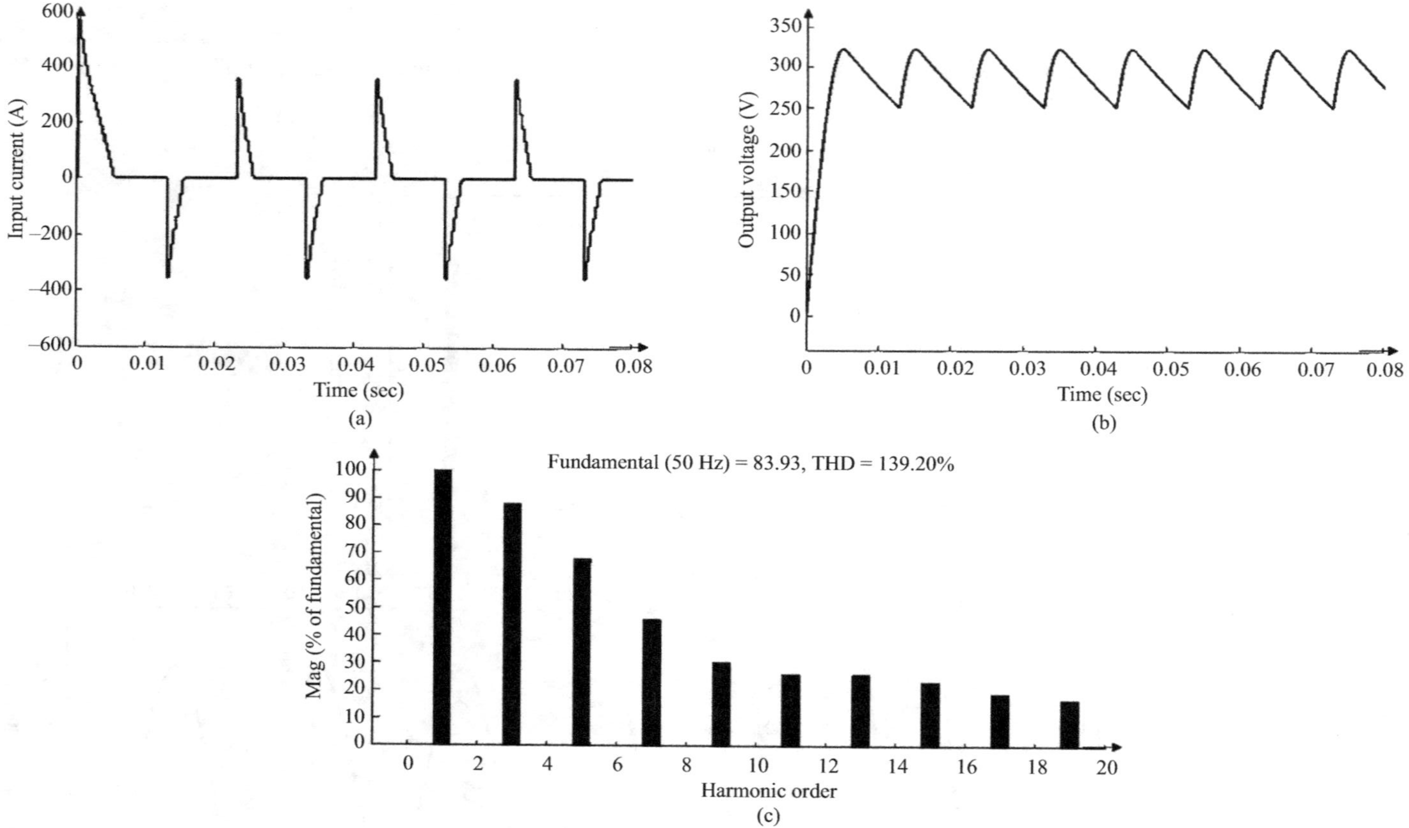

Figure 4.12 *Waveform of input current of diode rectifier with $R = 10\ \Omega$ and a filter capacitor of $3000\ \mu F$. (a) Input current, (b) output voltage and (c) harmonic spectrum of input current.*

4.4.3 Total Demand Distortion (TDD)

Total demand distortion [6,17] is the ratio of the rms value of harmonics component of current at the PCC to the fundamental component of maximum demand. It can be expressed as:

$$TDD_i = \frac{\sqrt{\sum_{h=2}^{n} I_{sh}}}{I_{MD}} \tag{4.29}$$

where I_{MD} is the maximum demand current at the PCC.

Multiplying with $\frac{I_{s1}}{I_{s1}}$

$$TDD_i = \frac{\sqrt{\sum_{h=2}^{n} I_{sh}}}{I_{MD}} \quad \frac{I_{s1}}{I_{s1}}$$

$$TDD = THD \times \frac{I_{s1}}{I_{MD}} \tag{4.30}$$

Consider a situation as shown in Figure 4.13, where linear loads (Load-2 and Load-3) are added with a non-linear load (Load-1) at PCC. The current source in this case is the addition of each load (Load-1 + Load-2 + Load-3). Though the distortion in Load-1 is high, addition of Load-2 and Load-3 results in less distortion in i_s and v_{PCC}.

The distortion in input current waveform depends on the ratio of linear load to non-linear load. As this ratio increases distortion in waveform increases or vice versa. To understand the effect of this ratio on THD, a linear load of $R = 1\ \Omega$ is added in parallel with the diode rectifier simulated in Figure 4.10. The ratio of linear load to non-linear load in this case is 3.87. The resultant input current waveform shown in Figure 4.14(a) with peak fundamental value 404.1 A and THD 28.1% as shown in Figure 4.14(b). Figure 4.14(c) indicates a sinusoidal waveform by increasing linear to non-linear current to 387.49. As the ratio increases the waveform becomes closer to sinusoidal waveform.

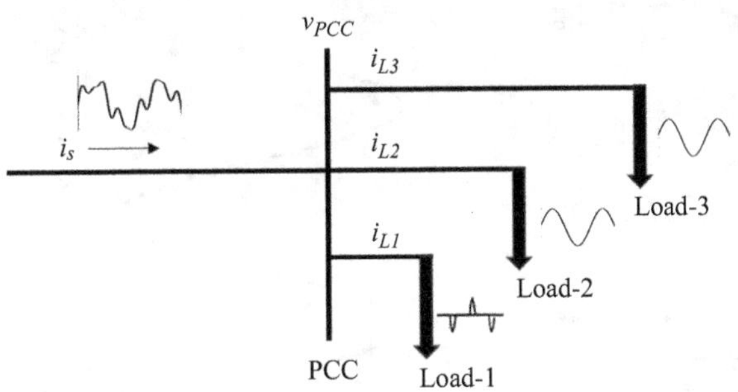

Figure 4.13 Transmission line connected to linear and non-linear load at PCC

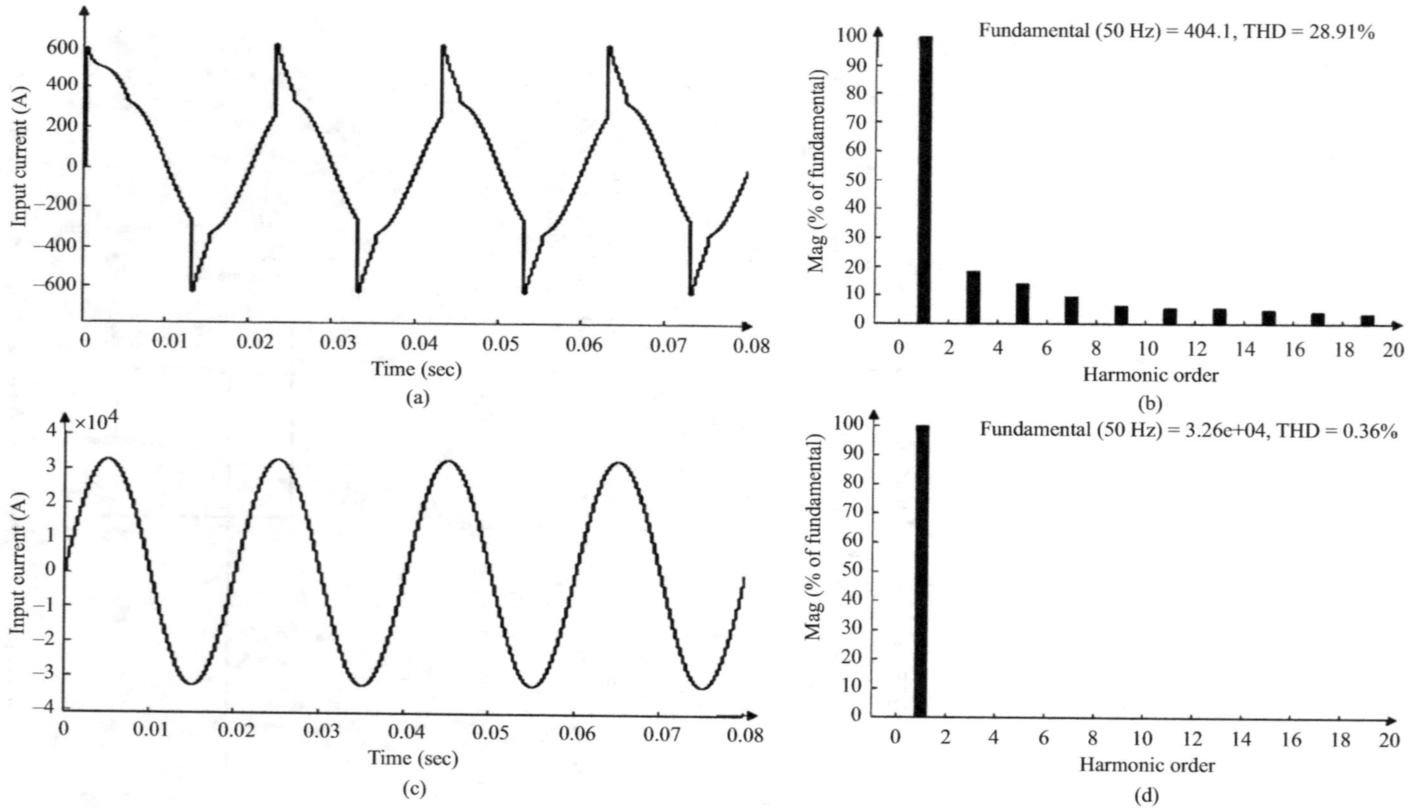

Figure 4.14 *Waveform of input current with linear load to non-linear load. (a) Input current, (b) harmonic spectrum with linear load to non-linear load ratio = 3.87, (c) input current and (d) harmonic spectrum with linear load to non-linear load ratio = 387.49.*

4.4.4 Telephonic Interference Factor (TIF)

Interference is a major concern of the harmonics produced by non-linear systems on adjacent signal lines. It is more pronounced on long telecommunication lines running close to the power transmission lines, especially where soil resistivity is relatively high. The magnetic fields set up by the harmonic currents flowing in power line induce voltages in nearby metallic conducting lines due to mutual coupling between them. Even a small harmonic level in power line can cause detrimental effects on the adjoining signal transmission systems viz. noise on data transmission, disturbance on telecommunication, maloperation protective relaying, measuring and communication equipment, service interruptions, etc. It may also affect the safety of operation and maintenance staff due to the induced voltage on the adjacent low-voltage signal line. Thus, considering the effect of harmonics on the telecom system, limit of harmonics is developed for trouble free operation the communication system. Other ways to reduce telephone interference are by electromagnetic shielding and capacitance balances among cables and the terminal equipment [18].

TIF is defined as the ratio of the square root of the sum of the squares of the weighted rms values of all the sinewave components to the root-mean-square value of the entire wave.

TIF in voltage (TIF_v) and current (TIF_i) are given by:

$$TIF_v = \frac{\sqrt{\sum_{h=1}^{\infty} (w_h V_h)^2}}{\sqrt{\sum_{h=1}^{\infty} V_h^2}} \tag{4.31}$$

$$TIF_i = \frac{\sqrt{\sum_{h=1}^{\infty} (w_h I)^2}}{\sqrt{\sum_{h=1}^{\infty} I_h^2}} \tag{4.32}$$

The weighted factor (w_h) is based on human ear response to noise at different frequencies

$$w_h = C_h 5 h f_1 \tag{4.33}$$

where, f_1, fundamental frequency; C_h, C-message weighting factor corresponding to order of harmonics developed by telephone industries in North American countries to reflect dependence of the ear on frequency and other parameters of the telephone instrument. Weighted factor (w_h) is based on C-message detailed in IEEE 519-2014.

Another similar term for harmonic interference is the telephone harmonic form factor (*THFF*). It is commonly used in European countries, that is defined in terms of weighting factor F_h, instead of *TIF* weighting coefficients (w_h). It is given by:

$$THFF = \frac{\sqrt{\sum_{h=1}^{\infty} (F_h V_h)^2}}{\sqrt{\sum_{h=1}^{\infty} V_h^2}} \tag{4.34}$$

$$F_h = \frac{P_h h f_1}{800} \tag{4.35}$$

where P_h is called the Psophometric weighting factor.

It may be noted both TIF and THEF are dimensionless quantities and are similar except that TIF and THFF differ by a constant of 4000 (i.e., $5\,h\,f_1$ and $h\,f_1$ / 800). However, both are single weighted frequency factors.

4.4.5 Telephone Interference (V–T and I–T indices)

Telephone Interference is the product of root-mean-square voltage magnitude (V), and its TIF. Similarly, I–T index is the product of the root-mean-square current magnitude (I), in amperes, and its TIF.

The relationship between V–T and I–T can be obtained as:

$$V - T = TIF_v \times V_{rms} \tag{4.36}$$

$$I - T = TIF_i \times I_{rms} \tag{4.37}$$

4.4.6 Distortion index (DIN)

THD symbolizes the magnitude of harmonics and is expressed as a percentage of the fundamental. If the signal is of the form $\sin(5\omega t) + \sin(7\omega t) + \sin(11\omega t) + \sin(13\omega t) + \cdots$, THD becomes infinite since no fundamental component exists in it. This situation occurs, for example, in PWM drives where the power frequency voltages and currents are modulated electronically to develop switching strategies. In such a case, Distortion Index (DIN) offers a viable solution to the above issue [19]. It is defined as the ratio of the square root of the sum of the squares of individual harmonics excluding the fundamental to square root of sum of square of individual harmonics including fundamental.

$$DIN = \frac{\sqrt{\sum_{i=2}^{\infty} I_i^2}}{\sqrt{\sum_{i=1}^{\infty} I_i^2}} \tag{4.38}$$

Thus, the amplitude of the fundamental of THD is replaced by the amplitude of the total signal in the denominator. Or the two indices are related by

$$DIN = \frac{THD}{\sqrt{1 + THD^2}} \tag{4.39}$$

$$THD = \frac{DIN}{\sqrt{1 - DIN^2}} \tag{4.40}$$

It may be noted from the above equations that DIN gives the same information as THD, and for low distortion cases, DIN and THD are about the same.

4.5 Computer-based systems to power supply disturbances

As discussed, most of the utilization equipment in the medical, fire protection, process control, communication and information technology sector are generally rated for nominal low voltage in the range from 120 to 240 V ±10% under steady state conditions, depending on the installation. Depending on its power requirement, it is connected either to single or three phase supply system, which thus subjected to its input power supply disturbance. At higher than nominal voltages, it may cause insulation failure or malfunction of the equipment. On the other hand, at lower voltage, the equipment may drop out of the supply, or performance may be affected. Thus, there is a range of voltage for its satisfactory operation. Over the past few decades, the use of computers in the critical area of automation and information technology has increased at a faster rate. Failure of computers, especially in real-time applications, causes interruption of the process or huge data loss. Thus, it may affect critical areas of our daily life and the economy.

4.5.1 CBEMA curve

Computer Business Equipment Manufacturer's Association USA (CBEMA), way back in 1977, developed a plot of voltage versus time to describe power supply tolerance for mainframe computers. It indicates the magnitude of voltage on the vertical (*Y*) axis and duration on the horizontal (*X*) axis, showing regions of supply voltage for the computer manufactured by its association members. After a revision, IEEE adapted it as a standard [20], and it is called the CBEMA curve (Figure 4.15). This plot has three regions formed by two curves, viz., higher and lower voltage from the rated voltage. One region, formed within both curves that extends the upper voltage range up to 106% and lower voltage range up to 87% on sustained basis. Voltage disturbance that lies within the area or zone is considered

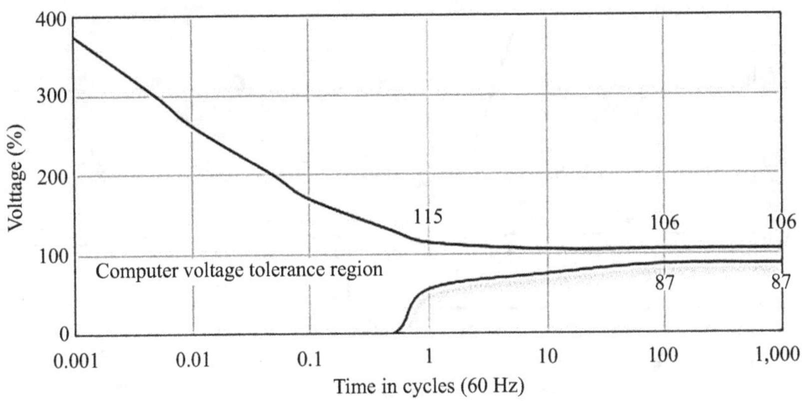

Figure 4.15 CBEMA curve

safe for the operation of the computer. Operation above the safe or voltage acceptability curve may adversely affect the electrical insulation of conducting parts or wires, junction voltage of semiconductor devices, capacitors, power supply units etc., by its failure. Operation below the safe zone, that is, under voltage, may affect the energy storage capacity of capacitors, the internal power supply unit diode that requires a certain minimum voltage for its operation, which may lead to a shutdown of the computer. Thus, operation of the computer above the overvoltage and below the undervoltage curves may be affected and hence not safe.

4.5.2 Information Technology Industry Council (ITIC) curve

Today, most of the modern automation processes or equipment in medical, industrial, business, transportation, office, or consumer electronic items, such as washing machine, microwave oven, etc., are based on Programmable Logic Controller (PLC) or microprocessor. Additionally, Information Technology Equipment (ITE), for example, laptops, servers, data centres, network or communication equipment, printers, copiers, etc., was nonexistent when the CBEMA curve for mainframe computers is formulated. As the above equipment is also highly susceptible to power supply disturbances, The Information Technology Industry Council (ITIC) adopted a modified version of CBEMA curve that can be tolerated by ITE and consumer electronics item in 1996. Figure 4.16 shows the curve and is called the New ITIC (CBEMA) curve [21].

Both the curves cover the transient and steady state conditions of AC input supply that can be tolerated by the equipment. Compared to the CBEMA curve shown in Figure 4.12, points in the New ITIC (CBEMA) curve are joined by lines

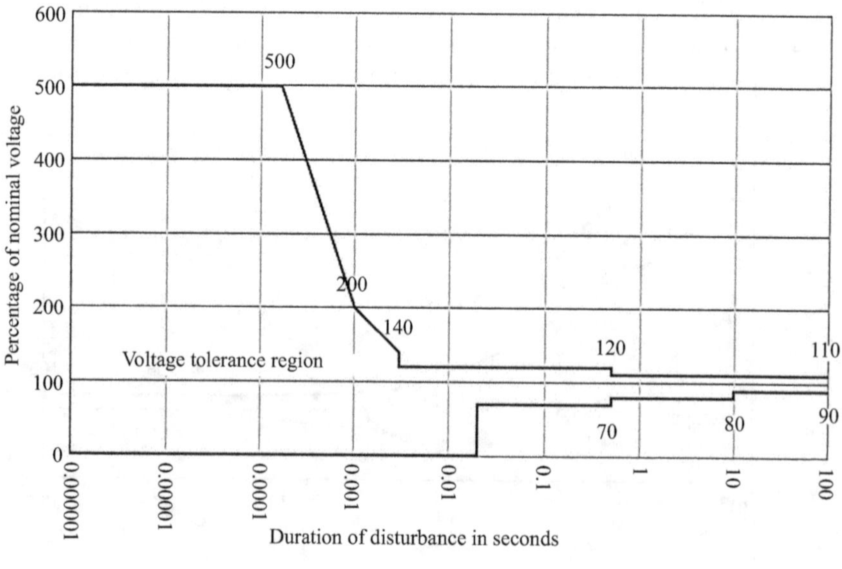

Figure 4.16 New ITIC (CBEMA) curve

rather than curves. Also, the safe zone extends the upper voltage range up to 110% and the lower voltage range up to 90% on a sustained basis. Since the points are well defined by line rather than dots formed by curves, equipment conformity can easily be established by testing.

4.6 Power supply interruptions and reliability measures

Electrical networks are subjected to various types of supply interruptions resulting loss of power to the consumer. While some of the interruptions may be due to faults in the networks (unplanned), others may be due to activities such as maintenance of the existing system equipment or upgradation to meet new requirements, such as increased fault level or expansion of the system to meet the increased customer demand (planned). Faults in the system are generally categorized as transient, semi-permanent or permanent and are isolated by the interrupting device (e.g., circuit breaker in high-voltage system or by fuse in low-voltage system). Based on the duration, the supply interruptions may be categorized as:

(a) Momentary: Faults generally cleared by automatic means within a brief period, usually less than 5 min.
(b) Sustained: Events with manual intervention. Duration of interruption is usually more than 5 min.

4.6.1 Reliability measures in power generation

Reliability is a measure of the availability of power to the consumer on a continuous basis. Measures taken to achieve continuity of electrical power in today's interconnected system involving multiple generating stations of different types are covered in Chapter 1. As detailed in the section on interconnected power systems, in case of nonavailability of any generation unit or transmission lines due to any technical issues or periodical maintenance, or even its failure, power flows to the consumer from the other connected stations.

4.6.2 Reliability measures in transmission system

Extra high-voltage overhead transmission power lines are subjected to more transient faults due to lightning. High or low voltage lines, on the other hand, are subjected to other temporary faults such as contact with live conductors with foreign objects such as birds, monkeys, trees, etc. Falling of a tree or its branch online may cause a semi-permanent fault or a permanent fault. Faults of this nature are more common during the rainy season and in forest areas. While the immediate tripping of the circuit breaker clears the transient fault, a delayed trip could burn away a semi-permanent fault. Auto-Reclose (AR) technique using AR relay in the protection scheme is generally used to re-energize the line after a fault trip of the Circuit Breaker (CB). In this scheme, the line is re-energized after a brief time interval to avoid re-striking the fault arc. It is observed that almost 80% of overhead line faults are transient, and the remaining are semi-permanent or permanent. Auto reclosure with single open-close or multiple open-close operation of CB with adjustable timing to restore the power supply after a transient or semi-permanent fault. The operation of a typical single AR scheme with time is shown in Figure 4.17. It may be noted that parameters such as opening time, arcing time, closing time, etc., depend on the type of protection provided on the line, type of arc quenching medium such as SF6 gas, oil or air in CB, load and system stability. The implementation of an AR scheme reduces the duration of supply interruption and improves the system stability. Interruptions arising out of auto reclosing of the CB are an example of momentary interruption of power supply.

In case of a permanent fault, after multiple (generally 3) unsuccessful AR operations, the associated CB would be locked out to open position. Faults of this nature are to be physically located and rectified before the power supply can be restored on the transmission line. Hence, in the case of radial lines, power supply interruption may be momentary or for a longer duration, resulting in sustained interruption. Power sharing measures followed in today's interconnected system during steady state, transient and contingency conditions such as loss of transmission lines are detailed in Chapter 1.

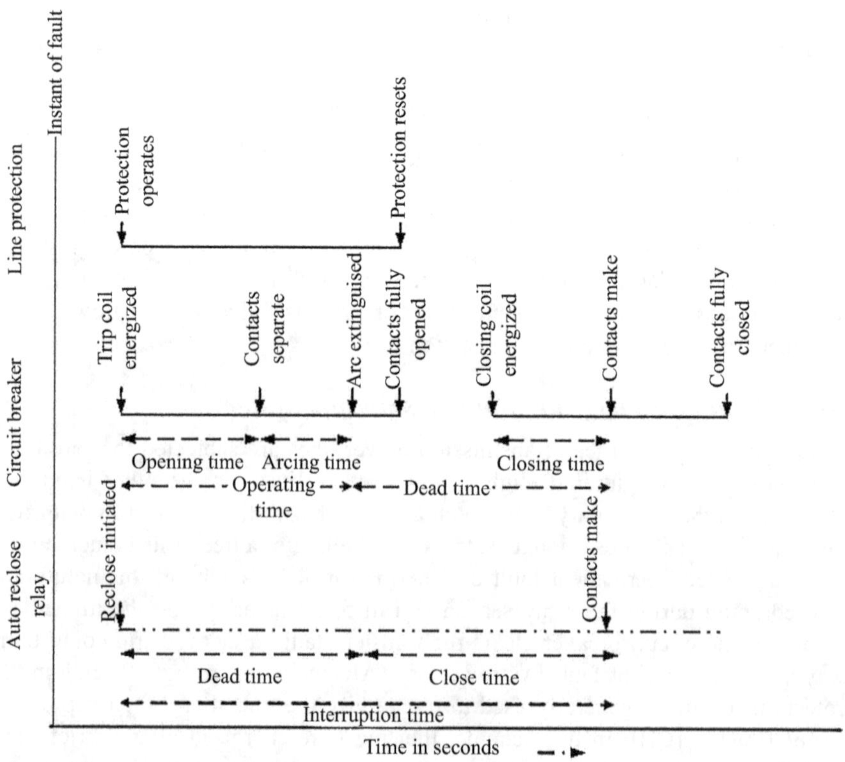

Figure 4.17 Operation sequence of auto reclose scheme on transmission line

4.6.3 Reliability measures in distribution system

On the consumer front, critical loads, such as railways, hospitals, airports, refineries, steel plants, defence installations, and mines, are usually connected to the grid with redundant transmission lines, possibly with different sources of power generation. Power supply interruptions are more common in distribution systems where radial lines are often used. This is especially true where the supply system to consumers is somewhat compromised, possibly due to various cost-cutting measures. However, in today's customer-centric era, several performance indices have been developed by industry and academia to measure the reliability from the end user point of view. The definitions of these indices and methods for calculating them can be found in several documents, most notably in the standard IEEE 1366 [22]. Globally, many countries have also implemented these indices as a monitoring tool for their distribution system and have shown encouraging results.

4.6.4 Distribution system reliability indices

In recent times, with the deregulation and cost optimization in electric power supply, more and more customers are concerned with the issues of power interruption. This has led the utilities and regulators to ensure that the power supply does not deteriorate beyond a certain acceptable value. Accordingly, the reliability indices are customer-based (number of customers) or load-based (connected load) and measured quantitatively. It tracks the performance of the distribution companies.

Supply reliability indices are calculated by frequency and duration of power interruptions. And on time scale for momentary or sustained interruptions based on the parameters indicated in Table 4.1.

For accuracy of computation, these parameters must be captured directly from the feeder devices, such as digital protection relays or smart meters, without any manual interventions. It is most often measured at the entire system, although it can also be measured for specific circuits. There are many terms and definitions used across the world. The most common and important terms for the distribution system are discussed in the following sections.

Table 4.1 Interruption parameters

i	Interruption event
ri	Restoration time for each interruption event
N_i	Number of interrupted customers for each sustained interruption
Ns	Total number of served customers
N_d	Number of distinct of interruption experiencing customers
L_i	Connected load interrupted
L_s	Total connected load served
M_i	Number of momentary interruptions
N_{mi}	Number of interrupted customers for each momentary interruption event

4.6.4.1 Customer-based sustained interruption indices

These indices give equal weight to all customers. For example, a residential customer with a few kW demand is considered at par with the steel industry with MW demand. Hence, for the computation of indices, only the number of consumers is required.

System Average Interruption Frequency Index (SAIFI)

One of the frequently used customers-based, sustained interruption reliability index is *SAIFI*. It indicates how many sustained interruptions an average customer experiences over a predefined period, typically a year. It is defined as,

$$\text{SAIFI} = \text{Total number of customers interrupted } (Ni)/\text{Total number of served customers } (Ns)$$

$$\text{i.e., SAIFI} = \frac{\sum N_i}{N_s} \tag{4.41}$$

System Average Interruption Duration Index (SAIDI)

It is a measure of the duration of sustained interruptions in hours or minutes, which an average customer experiences during a predefined period, typically a year. It is expressed as,

$$\text{SAIDI} = \text{Total durations of interruption } (r_i x N_i)/\text{Total number of customers served } (N_s)$$

where r_i is the duration of each interruption event.

$$\text{SAIDI} = \frac{\sum r_{i \times N_i}}{N_s} \tag{4.42}$$

Customer Average Interruption Frequency Index (CAIFI)

It is a measure of how many sustained interruptions an affected customer will experience over a predefined period. This index differs from SAIFI only by the denominator; instead of using the total number of customers served (N_s), the number of distinct customers experiencing interruption (N_d) is used. The customer is counted only once, regardless of the number of times the supply is interrupted. Mathematically,

$$\text{CAIFI} = \text{Total number of customer interruptions } (Ni)/\text{Total number of distinct customers interrupted } (N_d)$$

$$\text{CAIFI} = \frac{\sum N_i}{N_d} \tag{4.43}$$

Customer Average Interruption Duration Index (CAIDI)

It is a measure of how long an average interruption will last or the time required for restoration. This index differs from *SAIDI* only by the denominator; instead of using the total number of customers served; total number of customers interrupted is used. As such, *CAIDI* is a measure of restoration time. Mathematically, it is expressed as,

$$\text{CAIDI} = \text{Sum of all unplanned customer interruption durations}/\text{Total number of interrupted customers}$$

$$\text{CAIDI} = \frac{\sum r_{i \times N_i}}{\sum N_i} \tag{4.44}$$

$$\text{CAIDI} = \frac{\sum r_{i \times N_i}}{N_s} / \frac{\sum N_i}{N_s} = \frac{\text{SAIDI}}{\text{SAIFI}} \tag{4.45}$$

Thus, it is possible that measured CAIDI could deteriorate despite improved performances in both SAIDI and SAIFI.

4.6.4.2 Load-based sustained interruption indices

It may be seen from customer-based indices that they do not focus on the size of the interrupted load. Similarly, they don't differentiate between residential, commercial, or industrial customers. As the cost of supply interruption on each of them is different, there is a need to prioritize them. This is incorporated in load-based indices. These indices give weightage to the customer's connected load rather than considering all the customers on equal terms. Therefore, a more detailed knowledge of the consumers and their connected loads is required.

Average System Interruption Frequency Index (ASIFI)

ASIFI is the ratio of total load interrupted to the total load served. Compared to *SAIFI*, computation is based on load rather than the number of customers affected. In case all consumers are having equal load, *ASIFI* would be the same as *SAIFI*. Mathematically, it is expressed as,

$$\text{ASIFI} = \text{Total connected load interrupted } (Li)/\text{Total connected load served } (L_t)$$

$$\text{ASIFI} = \frac{\sum L_i}{L_t} \tag{4.46}$$

Average System Interruption Duration Index (ASIDI)

ASIDI is the ratio of total load interruption duration to the total load served. Compared to *SAIDI*, computation is based on load rather than the number of customers affected. Mathematically, it is expressed as,

$$\text{ASIDI} = \text{Total connected load duration of interruption}/\text{Total connected load served } (Lt)$$

$$\text{ASIDI} = \frac{\sum r_i L_i}{L_t} \tag{4.47}$$

4.6.4.3 Customer-based momentary interruption indices

With the growing automation and increased use of computers, even momentary power interruptions can lead to significant economic losses.

Momentary interruption indices are like sustained interruption. However, the duration of interruption in this case is momentary, unlike that of sustained. One of the frequently used customers-based, momentary interruption reliability index is explained in the next sub-section.

Momentary Average Interruption Frequency Index (MAIFI)

With the increasing use of digital technology and the use of computers, even momentary power interruptions can lead to significant loss of data or economic losses. *MAIFI* account above on performance of the distribution reliability term. *MAIFI* is like *SAIFI* with the duration of interruption limited to momentary (brief period, auto reclosure of circuit breakers) instead of sustained. It indicates how many momentary interruptions an average customer will experience over a pre-defined period, typically a year. It is defined as,

$$\text{MAIFI} = \text{Total number of customers interrupted momentarily}/\text{Total number of customers served } (N_s)$$

$$\text{MAIFI} = \frac{\sum Nm_i M_i}{N_s} \tag{4.48}$$

4.6.4.4 Factors affecting system reliability indices

The electric supply system interruption and hence reliability indices are volatile and depend on several factors, such as

(a) Environmental conditions: Predictable and unpredictable conditions such as temperature, fire, storm, lightning, floods, earthquake, etc.
(b) Type of customers: Agricultural, residential, commercial, educational and research, industrial, transportation sector such as rail and airports, emergency health care, communication, establishments of fire and defence, etc.
(c) Territories, region or terrain of service: Distribution system in urban area often use cable incurring heavy cost compared to rural overhead lines. Thus, supply interruption due to touching of trees, birds or foreign objects are avoided and hence reliability higher.
(d) Work practice of the distributor: Power supply system design, equipment used, installation, testing, commissioning, operation & maintenance, deployment of automation systems such as SCADA and smart meters, restoration procedure in case of interruption, conformity with the standards, etc.
(e) Condition of supply infrastructure: Adequacy of power, work practice and consequently on performance, competition among distributors and investment enhances reliability. But it is usual to balance the supply reliability with the cost to the customers.
(f) Social and economic policy: Industrial and economic development, regulatory requirements, monitoring system, compliance, etc.

Having realized the importance of a reliable electric power supply to the consumers, system reliability has been monitored in many countries. Countries in Europe, USA, Canada, Australia, New Zealand, etc., have established regulatory arrangements. Utilities in many of these countries are penalized or rewarded based on the measured reliability and established benchmark indices. In Ontario, Canada, distributors were given incentives based on *SAIDI*, *SAIFI* and *CAIDI* as early as 2000 [23]. In San Diego, an award or penalty of $ 2,50,000 were given for each variation of 0.01 below the target value of *SAIFI* 0.59 or above 0.63. It was based on 12,00,000 customers and the cost of outage $ 20.83 per customer[1] [24].

4.6.4.5 Reliability assessment benefits

(a) Assess the reliability of the distribution system in changing generation, transmission and load scenarios.
(b) Enable performance evaluation among distribution agencies and competition in the supply distribution system.
(c) Enable utilities to review their weakness and cause of interruptions, nature and type of existing generation, transmission and loads, viz. RE energy sources, power electronics in transmission and non-linear loads in utilization. Validate the electrical parameters with the installed system, forecast the trends, check adequacy, etc.
(d) Upgradation of existing distribution systems and procedures.
(e) Minimize power supply interruptions and hence improve power quality.
(f) Adopt new technology and innovation in distribution systems like SCADA, smart meters, DSP-based protection relays, underground cabling, condition monitoring.
(g) Enhance industrial growth and customer satisfaction.

In India, indices like SAIFI, SAIDI, CAIDI, and MAIFI are specified by the electricity regulatory authorities.

4.7 Problems and solution

Problem 4.1 A single-line diagram of a non-linear load is connected at the PCC as shown in Figure 4.18. The fundamental component of non-linear load current is 100 A (rms), and the percentage harmonic component present in terms of the fundamental component is given in Table 4.2. The maximum current demand at PCC is 1000 A. (a) Calculate the THD in the load current. (b) Plot the current waveform from the given information using any software, and comment on the neutral current in the system.

[1]System Reliability Regulation: A Jurisdictional Survey, May 2010, Pacific Economics Group Research, LLC, USA.

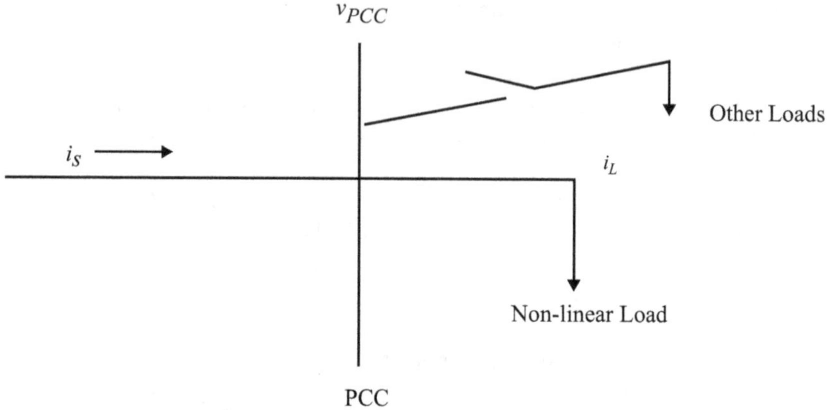

Figure 4.18 Problem 4.1 – PCC with non-linear and linear load

Table 4.2 Data for harmonic component of load – Problem 4.1

Harmonic order	Magnitude (% of fundamental)
1	100
5	18.36
7	11.33

Solution:

(a) Using (4.9) *THD* in load current is given by:

$$THD_i = \frac{\sqrt{\sum_{h=2}^{\infty} I_{sh}^2}}{I_{s1}} = \frac{\sqrt{18.36^2 + 11.33^2}}{100} = 0.2157$$

Therefore, *THD* in load current = 21.57%

(b) The waveform of the current drawn by the load is obtained on MATLAB platform as shown in Figure 4.19.

Neutral current, $I_n = 3I_{h3}$

Refer to Table 4.2, since triple order of harmonics is absent, neutral current will be zero. Hence, neutral conductors need to be sized in a normal way. That is need not be oversized.

Problem 4.2 In Problem 4.1, by adding 2nd order harmonics with their magnitude 50% of the fundamental component, how would the waveform appear? Justify it using the waveform of resultant current plotted using any software.

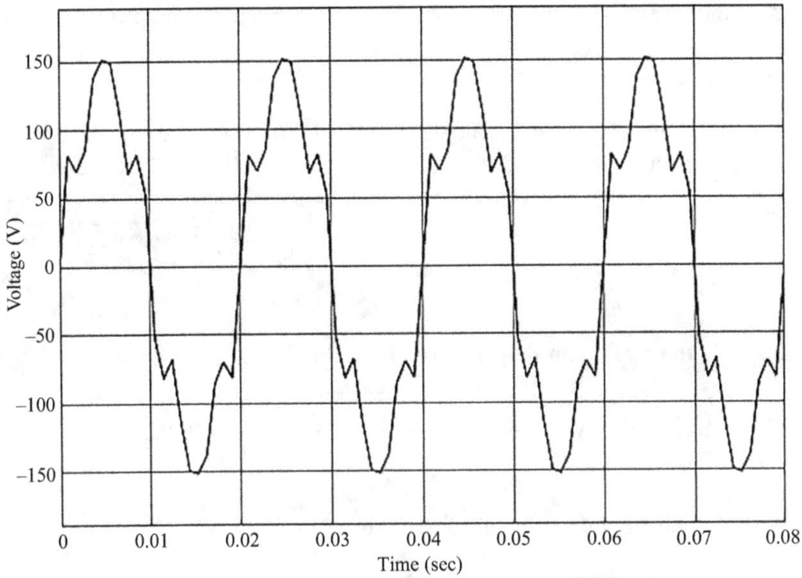

Figure 4.19 Problem 4.1 – waveform of the resultant current obtained using MATLAB platform

Solution:

The waveform is plotted using MATLAB software, and the resultant current waveform is shown in Figure 4.20.

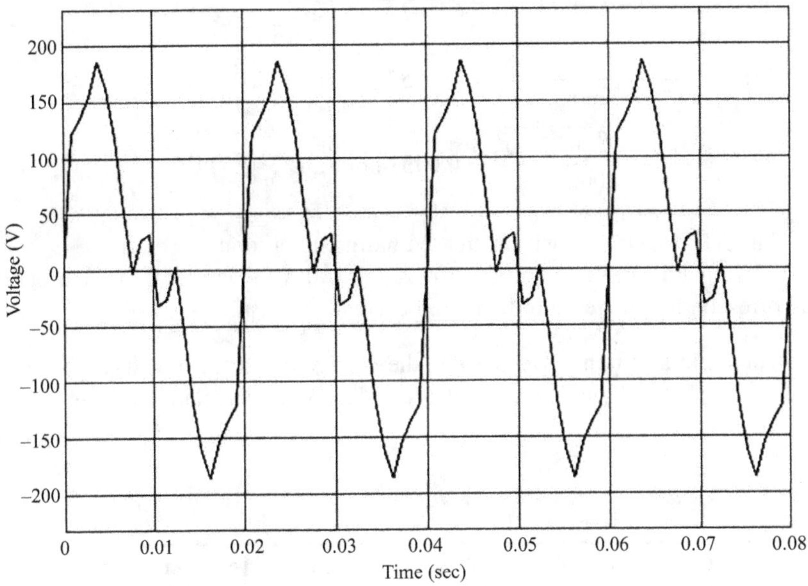

Figure 4.20 The resultant current waveform, for example, 4.2

(a) From Figure 4.20, it is clear that the waveform is unsymmetrical, having no DC offset. (b) In this case, neutral current $I_n = 3I_{h3}$ is zero as there are no triplen order harmonics present.

Problem 4.3 For Problem 4.1, calculate TDD. If the fundamental component of load current is increased to 200 A, calculate the new THD and TDD values.

Solution:

$$TDD_i = \frac{\sqrt{\sum_{h=2}^{n} I_{sh}}}{I_{MD}}$$

where I_{MD} is the maximum demand current at the PCC.

Therefore, TDD, with 1000 A $= \frac{\sqrt{18.36^2 + 11.33^2}}{1000} = 0.02157$ or $TDD = 2.157\%$.

With fundamental component of load current as 100 A,

$$THD_i = \frac{\sqrt{\sum_{h=2}^{\infty} I_{sh}^2}}{I_{s1}} = \frac{\sqrt{18.36^2 + 11.33^2}}{100} = 0.2157 \text{ or } THD \text{ in load}$$

current = 21.57%

When fundamental component of the load current is increased to 200 A,

$$I_1 = 200\,A, \quad I_5 = 36.72\,A, \quad I_7 = 22.66\,A.$$

Substituting above in $THD_i = \frac{\sqrt{\sum_{h=2}^{\infty} I_{sh}{}^2}}{I_{s1}}$,

$$THD = \frac{\sqrt{36.72^2 + 22.66^2}}{200} = 0.2157 \text{ or THD} = 21.57\%$$

$$TDD = \frac{\sqrt{36.72^2 + 22.66^2}}{1000} = 0.04314 \text{ or } TDD = 4.314\%.$$

Thus, TDD increased with increased harmonic current.

THD remains the same for both 100 A and 200 A fundamental load current due to a corresponding increase in harmonic current.

Problem 4.4 In a certain power system, the voltage measured in volts at the PCC is listed in Table 4.3. Calculate the voltage THD. Also, calculate the 5th-order harmonic distortion.

Table 4.3 Voltage at PCC – Problem 4.4

Fundamental component	**5th**	**7th**	**11th**	**13rd**	**17th**	**19th**	**23rd**	**25th**
440 V	88	62	40	25	20	10	2	5

Solution:
Voltage THD is given by:

$$Voltage\ THD = \frac{\sqrt{\sum_{h=2}^{25} V_h^2}}{V_1}$$

$$= \frac{\sqrt{88^2 + 62^2 + 40^2 + 25^2 + 20^2 + 10^2 + 2^2 + 5^2}}{440} = 0.2721$$

Therefore, $\%Voltage\ THD = 27.21\%$.

And % Fifth order harmonic distortion, $HD_5 = \frac{88}{440} = 20\%$.

It may be seen that the harmonics spectrum, apart from the fundamental, has frequency components at 5, 7, 11, 13, 17, 19, 23 and 25. It is following the pattern of $kq \pm 1,\ k = 1,\ 2,\ 3..$ and $q = 6$. Also, the harmonic component is reducing with increasing frequency. Thus, it is forming characteristic harmonics produced by a 3-ϕ, 6 pulse rectifier circuit.

Problem 4.5 Harmonic current in a typical 6-pulse converter is shown in Table 4.4. Assuming the fundamental current to be 500 A. Analyse the harmonic spectrum and suggest recommendations on possible interference on nearby communication lines.

Table 4.4 Harmonic current of typical 6-pulse converter – Problem 4.5

Harmonic (order)	1	5	7	9	11	13	15	17	19
Maximum current (I_n/I_1) %	100	20	14.28	11.11	1.05	0.75	0.57	0.44	0.35
Magnitude of current	500	100	71.4	55.55	5.25	3.75	2.85	2.2	1.75
w_i	0.5	225	650	–	2260	3360	–	5100	5630

Solution:

Using equation, $$TIF_i = \frac{\sqrt{\sum_{n=1}^{\infty} (w_n I_n)^2}}{\sqrt{\sum_{n=1}^{\infty} I_n^2}}$$

$$TIF_i = \frac{\sqrt{(0.5\times500)^2+(225\times100)^2+(650\times71.4)^2+(2260\times5.25)^2+(3360\times3.75)^2+(5100\times2.2)^2+(5630\times1.75)^2}}{\sqrt{500^2+100^2+71.4^2+55.55^2+5.25^2+3.75^2+2.85^2+2.2^2+1.75^2}}$$

$= 108.92$

Interpretations and observations:

$$I(\text{rms}) = \sqrt{500^2 + 100^2 + 71.4^2 + 55.55^2 + 5.25^2 + 3.75^2 + 2.85^2 + 2.2^2 + 1.75^2}$$

$$= 517.92\,\text{A}$$

IEEE Standard 519, the *I–T* product is $517.92 \times 108.92 = 56,415$ where I is the rms current in amperes, and T is TIF. It is greater than 25000, a level III category, and hence may cause interference on exposure to communication signal. Hence, precautions need to be taken.

Problem 4.6 It is observed that in a 3-ϕ, 60 kV, 60 Hz substation, the 3rd and 5th harmonic components of voltage have a magnitude of 1000 V and 500 V, respectively. The fundamental load power is 20 MVA at a unity power factor. Determine the TIF and THD of both the voltage and current. Also, determine *V–T* and *I–T* products Weighing factor is given in the Table 4.5.

Table 4.5 TIF weighting coefficients (wh) – Problem 4.6

Harmonic order	**1**	**3**	**5**
w_i	0.5	30	225

Given data:

$$V_1 = 60\text{ kV} \quad V_3 = 1000\text{ V} \quad V_5 = 500\text{ V}$$

Fundamental load power, $S_1 = 20$ MVA power factor $= 1$

$$P_1 = 20\text{ MW}$$

Solution:

Fundamental load power, $P_1 = \sqrt{3}V_1 I_1 \cos\phi_1 = 20$ MW

$$\cos\phi_1 = 1 \quad \therefore I_1 = \frac{P_1}{\sqrt{3} \times V_1 \times \cos\phi_1} = \frac{20 \times 10^6}{\sqrt{3} \times 60 \times 10^3 \times 1} = 192.45\text{ A}$$

$$\text{Resistance, } R = \frac{V_1/\sqrt{3}}{I_1} = \frac{60,000/\sqrt{3}}{192.45} = 180\ \Omega$$

$$I_3 = \frac{V_3}{R} = \frac{1000/\sqrt{3}}{180} = 3.21\text{ A}$$

$$I_5 = \frac{V_5}{R} = \frac{500/\sqrt{3}}{180} = 1.60 \text{ A}$$

Calculated values of harmonic component in voltage and current along with its THD are given in Table 4.6.

Table 4.6 Solution to Problem 4.6

V_1 (V)	V_3 (V)	V_5 (V)	I_1(A)	I_3(A)	I_5(A)	V_{THD}	I_{THD}
60,000	1000	500	192.45	3.21	1.60	1.86	1.86

$$TIF_i = \frac{\sqrt{\sum_{n=1}^{\infty} (w_n I_n)^2}}{\sqrt{\sum_{n=1}^{\infty} I_n^2}} = \frac{\sqrt{(0.5 \times 192.45)^2 + (30 \times 3.21)^2 + (225 \times 1.60)^2}}{\sqrt{192.45^2 + 3.21^2 + 1.60^2}}$$

$$= 1.99$$

$$TIF_v = \frac{\sqrt{\sum_{n=1}^{\infty} (w_n V_n)^2}}{\sqrt{\sum_{n=1}^{\infty} V_n^2}} = \frac{\sqrt{(0.5 \times 60{,}000)^2 + (30 \times 1000)^2 + (225 \times 500)^2}}{\sqrt{60{,}000^2 + 1000^2 + 500^2}}$$

$$= 2.003$$

$$I - T = \sqrt{\sum_{n=1}^{\infty} (w_n I_n)^2}$$

$$= \sqrt{\left((0.5 \times 192.45)^2 + (30 \times 3.21)^2 + (225 \times 1.60)^2\right)} = 384.88 \text{ A}$$

$$V - T = \sqrt{\sum_{n=1}^{\infty} (w_n V_n)^2}$$

$$= \sqrt{(0.5 \times 60{,}000)^2 + (30 \times 1000)^2 + (225 \times 500)^2} = 120{,}234 \text{ V}$$

Problem 4.7 Calculate DIN for Problem 4.1.

Solution:

Distortion index (DIN) is the ratio of the square root of the sum of square of individual harmonics excluding fundamental to square root of sum of square of individual harmonics including fundamental. Hence

$$DINi = \sqrt{\frac{\sum_{n=2}^{\infty} In^2}{\sum_{n=1}^{\infty} In^2}} = \sqrt{\frac{18.36^2 + 11.33^2}{100^2 + 18.36^2 + 11.33^2}} = 0.2108$$

THD is the ratio of the rms value of all harmonic content to the rms value of the fundamental quantity. THD in current (THD_i) is expressed as:

$$THD_i = \frac{\sqrt{\sum_{h=2}^{\infty} I_{sh}^{2}}}{I_{s1}}$$

$$THD_i = \frac{\sqrt{\sum_{h=2}^{\infty} I_{sh}^{2}}}{I_{s1}} = \frac{\sqrt{18.36^2 + 11.33^2}}{100} = 0.2157$$

Thus, THD and DIN differ marginally due to the change in the denominator among them. DIN gives a lower ratio compared to THD as DIN use rms value instead of fundamental in case of THD.

Problem 4.8 A transmission and distribution system with feeder details and the connected industrial units is shown in Figure 4.21. On 15.01.2023, a 3-ϕ fault on feeder 1, resulted in the opening of circuit breaker CB1 and interruption of industrial consumers on this feeder for 1 hour. Subsequently, a second fault on the distribution line appeared on 28.01.2023, resulting in the auto-reclosure operation of CB-TB twice. On failure of it, CB-TB locked out to the OPEN position, resulting in a loss of supply for 2 hours. A third fault on feeder 6 happened on 01.04.2023, resulting in an interruption of 30 minutes due to the opening of circuit breaker CB6. Calculate the indices SAIDI, SAIFI, CAIDI and *MAIFI* for the year 2023.

Solution:
Calculation for *SAIDI, SAIFI, CAIDI* and *MAIFI* for the year 2023 is done in tabular form, Tables 4.7–4.10, respectively.

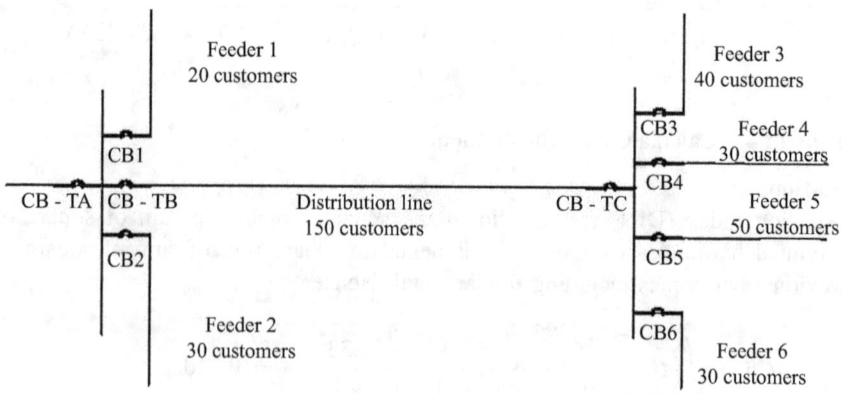

Figure 4.21 Single-line diagram of distribution system

Table 4.7 Solution to Problem 4.8 SAIDI

System Average Interruption Duration Index (SAIDI)

Total minutes every customer was without power due to sustained outage/Total number of customers

Sl. No.	Month	Number of consumers experienced a sustained interruption (Ni)	Restoration time for each interruption event (min) (r_i)	Total number of consumers served (Ns)	(r_i *Ni)	Minutes of sustained outages per customer per year. $SAIDI = \frac{\sum r_i \times N_i}{N_s}$
	(a)	(b)	(c)	(d)	(e)	(e)/(d)
a.	January CB1	20	60	200	1200	
b.	January CBTB	150	120	200	18,000	
c.	April CB6	30	30	200	900	
	Total (Year)	200		200	20,100	100.5

Table 4.8 Solution to Problem 4.8 SAIFI

System Average Interruption Frequency Index (SAIFI)Total number of customers experienced sustained interruption/Total number of customers

Sl. No.	Month	Number of consumers experienced a sustained interruption (Ni)	Total number of consumers served (Ns)	Number of sustained outages per customer per year. $SAIFI = \frac{\sum N_i}{N_s}$
1	(a)	(b)	(c)	(c)/(d)
a	January – CB1	20	200	0.1
b	January – CBTB	150	200	0.75
c	April – CB6	30	200	0.15
	Total (Year)	200	200	1

Table 4.9 Solution to Problem 4.8 CAIDI

Customer Average Interruption Duration Index (CAIDI)

Sl. No.	SAIDI	SAIFI	CAIDI = SAIDI / SAIFI Average time required to restore service to a customer
	(a)	(b)	(a)/(b)
1	100.5	1	100.5

Table 4.10 Solution to Problem 4.8 MAIFI

Momentary Average Interruption Frequency Index (MAIFI)Total number of customers experienced momentary outage/Total number of customers					
Sr. No.	**Month**	**Number of consumers experienced a momentary interruption (*Nmi*)**	**Number of momentary interruptions**	**(Nm_iM)**	**Number of momentary outages per customer per year.** $MAIFI = \frac{\sum Nm_iM_i}{N_s}$
1	**(d)**	**(e)**	**(f)**	**(g)**	**(c)/(d)**
(a)	January – CB1	0	0	0	0
(b)	January – CBTB	150	2	300	1.5
(c)	April – CB6	0	0	0	0
	Total (Year)	150	2	300	1.5

References

[1] Central Government notification New Delhi, 31st December 2020, Electricity (Rights of Consumers) Rules, 2020. Ministry of Power https://powermin.gov.in › sites › files › uploads.

[2] IEEE Std 1159-2019, IEEE Recommended Practice for Monitoring Electric Power Quality.

[3] LED Lighting Flicker and Potential Health Concerns: IEEE Standard PAR1789 Update Arnold Wilkins Dept. of Psychology University of Essex; UK Jennifer Veitch National Research Council Canada Ottawa; Canada Brad Lehman Dept. Elect. & Comp. Eng Northeastern University; USA, IEEE Energy Conversion Congress and Exposition (ECCE), 2010 IEEE. (https://www.researchgate.net/publication/273660125)

[4] IEC 61000-3-3, Electromagnetic Compatibility (EMC), Limitation of voltage, fluctuations and flicker in low-voltage Supply Systems for Equipment with Rated Current less than 16 A per phase.

[5] IEC Standard 61000-4-15, Flicker meter – Functional and Design Specifications.

[6] IEEE Std 519, IEEE Recommended Practice and Requirements for Harmonic Control in Electric Power Systems.

[7] Bo Zeng, and Zhaosheng Teng, "Parameter Estimation of Power System Signals Based on Cosine Self-Convolution Window with Desirable Side-Lobe Behaviors", *IEEE Transactions on Power Delivery*, Vol. 26, No. 1, 2011, pp. 250–257.

[8] Surya Santoso, Edward J. Powers, W. Mack Grady, and Peter Hofmann, "Power Quality Assessment via Wavelet Transform Analysis", *IEEE Transactions on Power Delivery*, Vol. 11, No. 2, 1996, pp. 924–930.

[9] Norman Mortensen, and Gary L Jognson, "A Power System Digital Harmonic Analyzer", *IEEE Transactions on Instrumentation and Measurement*, Vol. 37, No. 4, 1988, pp. 537–540.

[10] Jun-Zhe Yang, Member, IEEE, Chi-Shan Yu, and Chih-Wen Liu, Senior Member, IEEE, "A New Method for Power Signal Harmonic Analysis", *IEEE Transactions on Power Delivery*, Vol. 20, No. 2, 2005, pp. 1235–1239.

[11] Roger C. Dugan, Mark F. McGranaghan, and H. Wayne Beatyd,"Electric Power System and Quality". Third edition, 2012, McGraw-Hill.

[12] G. T. Heydt, "Electrical Power Quality". First edition, 1991, Stars in a Circle.

[13] Jos Arrillaga, Naville R. Watson, and S. Chen, "Power System Quality Assessment". 2000, John Wiley and Sons Ltd.

[14] W. Shepherd, and P. Zan, "Energy Flow and Power Factor in Non-Sinusoidal Circuits". 1979, Cambridge University Press.

[15] Bhim Singh, Ambrish Chandra, and Kamal Al-Haddad, "Power Quality: Problems and Mitigation Techniques". First edition, 2015, John Wiley and Sons Ltd.

[16] Sincy George, and Vivek Agarwal, "Optimum Control of Selective and Total Harmonic Distortion in Current and Voltage Under Nonsinusoidal Conditions", *IEEE Transactions on Power Delivery*, Vol. 23, No. 2, 2008, pp. 937–944.

[17] Ned Mohan, Tore M. Undeland, and William P. Robbins, "Power Electronics: Converters, Applications and Design", Second Edition, 1995, John Wiley & Sons (Asia) Pte Ltd.

[18] Indian standard IS 16071: 2013 / IEC/TR 62001: 2003 high-voltage direct current (HVDC) Systems—guidebook to the specification and design evaluation of AC Filters.

[19] G. T. Heydt, and W. T. Jewell "Pitfalls of Electric Power Quality Indices", *IEEE Transactions on Power Delivery*, Vol. 13, No. 2, 1998, pp. 570–576.

[20] IEEE Std 446, IEEE Recommended Practice for Emergency and stand by power system for Industrial and Commercial Applications, 1995.

[21] IEEE Std 1100, IEEE Recommended Practice for powering and grounding sensitive Electronic Equipment, 2005.

[22] IEEE Std 1366, IEEE Guide for Electric Power Distribution Reliability Indices, 2012.

[23] https://www.oeb.ca/oeb/_Documents/EB-2010-249/PEG_OEB_Service_Quality_Report.pdf

[24] *System Reliability Regulation: A Jurisdictional Survey*, 2010, Pacific Economics Group Research, LLC.

Chapter 5
Power factor under sinusoidal and non-sinusoidal conditions

Power factor is a key parameter for describing power quality and is the ratio of average power used in watts to the apparent power in volt-amperes delivered to the circuit.

Goals of the chapter

This chapter discusses another important aspect of power systems, the power factor. The term is not reflected in power quality standards, and they often believe power quality means power factor. It is, therefore, even more important to understand the technical definition and meaning of the term "power factor". This chapter aims to review the basic concepts of power factor under linear loads and its changing definition under the voltage and current distortion due to the growing presence of nonlinear electronic loads in recent times. It also briefly discusses the adverse effects of low power factor and ways to improve it under linear load conditions, as well as the impact of the conventional compensation technique under non-sinusoidal conditions. Finally, the need for new techniques for harmonics elimination, thereby improving the power factor, is emphasized. A few case studies using MATLAB®/Simulink® are also included to make the concept easy to understand.

5.1 Introduction

The growing demand for AC electrical power presents a unique challenge compared to DC systems, due to the appearance of the power factor. This is because most of the AC utilization equipment requires reactive power in addition to active power. This reactive or quadrature component is required mainly for the conversion of power through electric (capacitor) or magnetic (inductive) fields. The vast majority of the loads are inductive in nature. Induction motors, steel furnaces, transformers, etc., are a few among them. These loads require two kinds of power: i) active or working component of power P measured in watts (W) to perform the real work and ii) reactive volt-ampere Q measured in volt-ampere reactive (VAR) for transfer of power between electromagnetic fields. The vector sum of the active power and reactive volt-ampere makes up the apparent volt-ampere S, measured in volt-amperes (VA). The presence of harmonics due to nonlinear loads results in non-sinusoidal waveforms in the system, which further increases the apparent volt-ampere. This increase in S results in a reduction in power factor and system capacity. The basic power factor equations under sinusoidal and non-sinusoidal conditions are derived in the next sections.

5.1.1 Power factor under sinusoidal conditions

As we know, under sinusoidal conditions, only the fundamental frequency of current flows due to the fundamental frequency voltage present in the system. Under these conditions, let the instantaneous values of voltage, $v(t)$ and current $i(t)$ are given by:

$$v(t) = V_m \sin \omega t \tag{5.1}$$

$$i(t) = I_m \sin(\omega t \pm \phi) \tag{5.2}$$

where, V_m and I_m are the maximum values of the voltage and current, ω the angular frequency given by $\omega = 2\pi f$, f is the frequency of supply voltage and ϕ is the phase angle between voltage and current. For inductive load current lag($-\phi$) the voltage, and for capacitive load current leads ($+\phi$) the voltage. As the majority of loads are inductive, a lagging power factor is considered for analysis. However, similar analysis holds for leading current too. The average power (P), reactive volt-ampere (Q) and apparent volt-ampere (S) are given by:

$$P = V_{rms}\, I_{rms} \cos\phi \tag{5.3}$$

$$Q = V_{rms}\, I_{rms} \sin\phi \tag{5.4}$$

$$S = V_{rms}\, I_{rms} \tag{5.5}$$

Where V_{rms} and I_{rms} are the rms value of voltage and current, respectively, and are given by $V_{rms} = \dfrac{V_m}{\sqrt{2}}$ and current $I_{rms} = \dfrac{I_m}{\sqrt{2}}$

The power factor (PF) is given by [1,2]:

$$\text{Power factor, } PF = \frac{\text{Average power}}{\text{Apparent voltampere}} \tag{5.6}$$

Hence, using Eq. (5.3 and 5.5),

$$PF = \frac{P}{S} = \cos\phi \tag{5.7}$$

Thus, under sinusoidal supply and linear load conditions, PF defined by (5.7) is due to the displacement angle (ϕ) between voltage and current; hence, it is also called the displacement power factor.

From (5.7), it can also be noted that PF is a measure of the portion of the input apparent volt-ampere converted to real power. When both voltage and current are

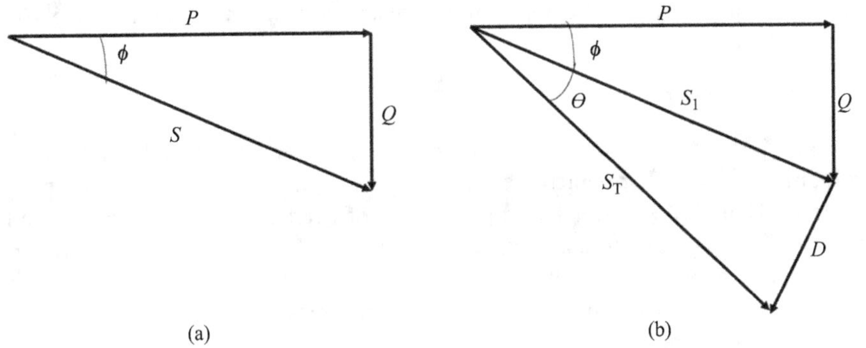

Figure 5.1 Power angle diagram. (a) Under sinusoidal condition and (b) under non-sinusoidal conditions.

sinusoidal in shape and in phase with each other, the PF is unity. Under this condition, the apparent volt-ampere drawn from the supply system is the same as the active power demanded by the load. This is true when the load is resistive. Ideally, PF should be unity for an efficient operation of the system. But for inductive-type loads, for their magnetic field, the current lags with respect to voltage. Similarly for capacitive types of loads, due to the associated electrostatic field, current leads the voltage. A phasor diagram with an inductive load under sinusoidal supply is given in Figure 5.1(a). The phasor diagram under non-sinusoidal conditions is shown in Figure 5.1(b) for comparison, and it is detailed in the next section.

5.1.2 Power factor under non-sinusoidal conditions

Basic concepts of power flow remain the same under both sinusoidal and non-sinusoidal conditions. However, under non-sinusoidal conditions, the power flow corresponding to the harmonic component present in voltage and current will also be added to the fundamental frequency components corresponding to the sinusoidal system. Similarly, the basic definition of power factor given in Eq. (5.6) is silent on the frequency components of supply voltage and/or load current waveforms. However, the amount of P, Q and S, and hence the PF of the system, is getting affected under non-sinusoidal conditions. The growing application of nonlinear devices in recent times, from generation to distribution systems, and in utilization from mobile phone adapters to laptops, electric drives to locomotives, etc., presents complexity in voltage and current waveforms. As discussed in Chapter 3, nonlinear equipment injects harmonics into the power system and alters the sinusoidal nature of voltage and current waveforms in addition to the phase shift ϕ between them. When voltage or current or both waveforms are distorted by the harmonics, the power factor deteriorates further [1,2]. The volt-ampere due to harmonics is called distortion volt-ampere (D), which is quadrature to apparent volt-ampere (S_1) due to the fundamental component of voltage and current, as shown in Figure 5.1(b). The resultant of P, Q and D indicates the total apparent volt-ampere S_T due to harmonics as well as the fundamental component of voltage and current. From Figure 5.1(b), S_T is written as:

$$S_T^2 = P^2 + Q^2 + D^2 \tag{5.8}$$

Thus, under non-sinusoidal conditions, power factor also depends on distortion volt-ampere due to harmonics present in the waveforms in addition to the phase angle ϕ. Though the power factor is the ratio of average power to apparent volt-ampere, it is not equal to $\cos\phi$ under non-sinusoidal conditions. In this case, the fundamental apparent volt-ampere S_1, which is equal to S in Figure 5.1(a) in the denominator, is replaced by the total volt-ampere, resulting in further reduction in power factor [3,4].

$$\text{PF} = \frac{P}{S_T} \neq \cos\phi \tag{5.9}$$

Issues with low PF and its need for improvement are briefly given in the following sections.

5.2 Need for improving power factor

We know that the current drawn from the supply depends on the type of load. For a resistive load, the power factor is close to unity, and hence the current drawn and apparent power are minimum. On the other hand, inductive and capacitive loads have a low power factor, higher load current and correspondingly higher apparent power. Thus, with a high power factor, there is a reduction in load current and apparent power for the same amount of active power. Thus, it reduces kVA loading on the system connected, and the system is provided with additional capacity otherwise. The generation units can use this saved capacity to serve an increased number of consumers.

Transmission and distribution systems transfer or transmit the electricity economically and reliably from generating stations to consumers, geographically located at faraway places. A large length of conductors of overhead and insulated cables carries the current. Transformers are used to change the voltage level from low to high (for reduction in current) at generating stations for transmission and from high to low voltage at distribution or utilization equipment ends, along with protective devices like circuit breakers and other switching devices which are also placed at strategic locations throughout the system. Thus, after generation, power is transmitted and distributed over a wide network involving several pieces of equipment. Reduction in current (I), reduces voltage drop ($I \times Z$) and energy loss ($I^2 \times R$) in all these transmission and distribution (T&D) components. Reduction in losses within equipment reduces its operating temperature and hence reduces the temperature-related ageing of dielectric material. Thus, reduction in equipment losses increases its useful life. By reduction in voltage drop, line voltage regulation improves and rated voltage at the terminal of the equipment is maintained. Alternatively, to maintain voltage at the equipment terminal within limits, we must use a higher conductor cross-section, that increases the material cost of the equipment and transmission system.

Many electricity distribution utilities' tariffs for electricity supply are based on a power factor. Some utilities charge penalties in case PF is lower than a stipulated value. While others give bonuses in their tariff in case of higher PF. Thus, utility charges for the active energy, maximum demand, and reactive volt-ampere. Hence, the electricity tariff is reflected on the power factor. Thus, there is a technical and economic benefit of maintaining PF close to unity. This is done by providing necessary compensating reactive volt-amperes locally so that current flowing in the upstream network is reduced. Fixed or switched shunt capacitors or reactors, overexcited synchronous machines, etc., are commonly used for providing reactive compensation. As capacitors are traditionally used for compensation of lagging reactive volt-amperes in systems everywhere, we shall discuss their use under various load-generation profiles. As power factor and THD are related to each other, THD is also analysed.

5.3 Power factor improvement using conventional compensation

Consider single-line diagram of a sinusoidal power supply system with linear and nonlinear loads connected through a transmission line as presented in Figure 5.2. When only linear loads (i_{L1}) are connected by closing circuit breaker (CB) S_1, transmission line voltage drop ($iX_{LINE} = i_{L1}X_{LINE}$) remains a sinusoidal and hence the voltage (v_r) at the receiving end bus or the point of common coupling (PCC) is

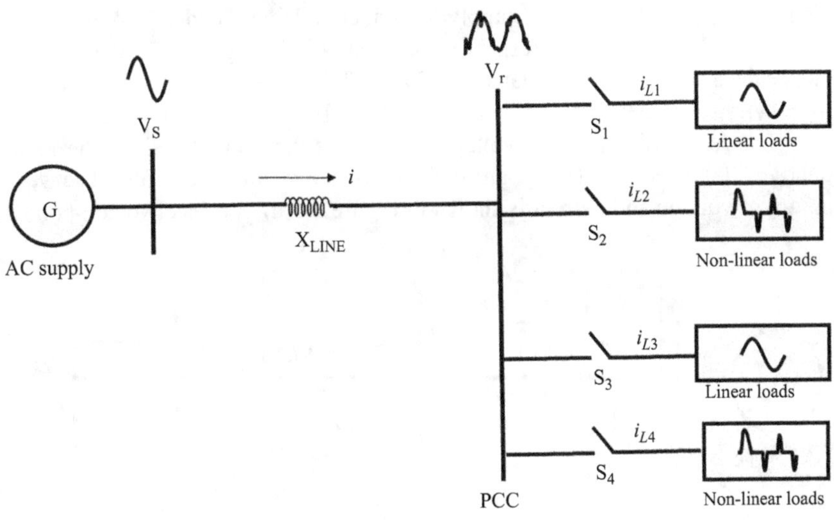

Figure 5.2 Single-line diagram of power supply system with linear and nonlinear loads

also sinusoidal, though there may be a dip in voltage depending on the load conditions. When nonlinear loads (i_{L2}) are added by closing CB S_2 to i_{L1}, v_r and i may get distorted depending upon the ratio of the amount of nonlinear load to linear load current $\left(\frac{i_{L2}}{i_{L1}}\right)$ through the transmission line. If the ratio is small, the transmission line current i and voltage at PCC may not be distorted considerably. However, if the ratio is high, voltage drops across the transmission line, iX_{LINE} and v_r at PCC may get distorted due to distortion in i. The distorted current and voltage increase THD in current and voltage and reduce the power factor of the system. To such a system when a linear, nonlinear or sensitive load is connected, it may affect the performance of the load. Hence, all such cases are considered and analysed as under.

Case 1: Both supply voltage and load current are sinusoidal
Case 2: Supply voltage is sinusoidal, and load current is non-sinusoidal
Case 3: Supply is non-sinusoidal, and load is linear
Case 4: Both supply and current are non-sinusoidal

5.3.1 *Case 1: PF and THD when supply voltage and load current are sinusoidal*

When only passive elements such as resistors, inductors and capacitors are connected in the system to the sinusoidal voltage supply, the current flowing is also sinusoidal. If the load is resistive, such as an incandescent lamp and heating loads, which is supplied with sinusoidal voltage, the load current is sinusoidal and in phase with voltage, and hence the power factor is unity. Whereas power factor is less than one in the case of capacitive or inductive loads. To analyse PF and THD in various conditions, let us consider a radial three-bus system with two feeders in which the 3-ϕ, 50 Hz source is supplying inductive loads (load 1 (i_{L2}) and 2 (i_{L3})) connected to buses 2 and 3, respectively, as shown in Figure 5.3. Various parameters such as P, Q, D and S related to PF and THD under various conditions of the power system are derived and analysed below. Let the instantaneous current in feeder 1 is $i(t) = i_1(t)$ and the voltage at bus 1 is $v(t) = v_1(t)$. Since the load is inductive, $i(t)$ lags the $v(t)$ by an angle ϕ as shown in Figure 5.4. The vector diagram of voltage and current is shown in Figure 5.5(a). I_{rms} has two components:

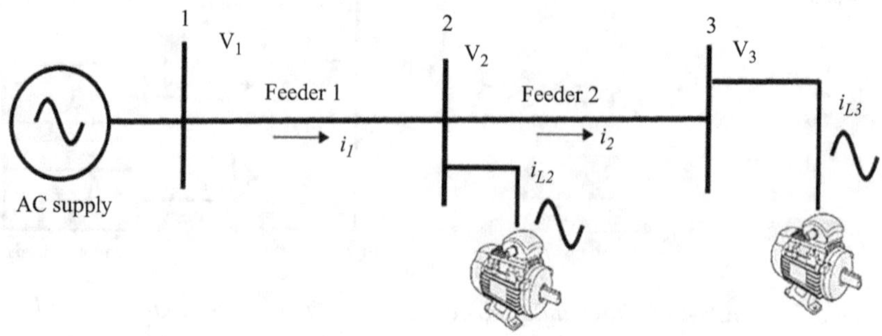

Figure 5.3 Single-line diagram of a radial three-bus system

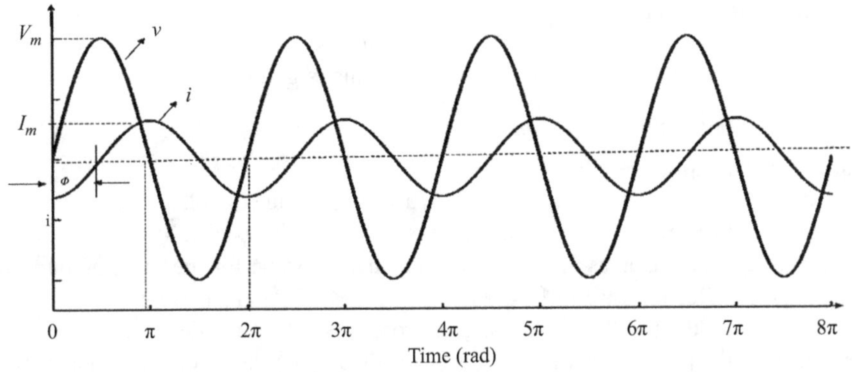

Figure 5.4 Waveforms of voltage and current with an RL load connected to the sinusoidal voltage source

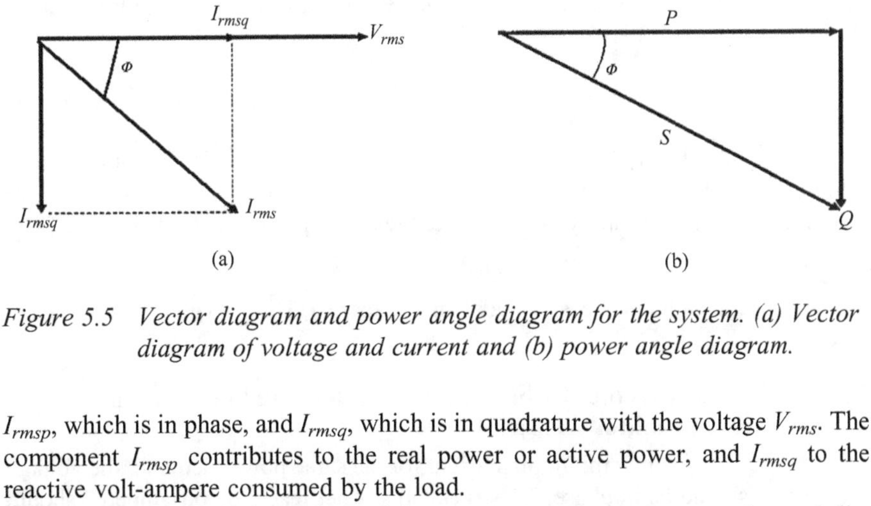

Figure 5.5 Vector diagram and power angle diagram for the system. (a) Vector diagram of voltage and current and (b) power angle diagram.

I_{rmsp}, which is in phase, and I_{rmsq}, which is in quadrature with the voltage V_{rms}. The component I_{rmsp} contributes to the real power or active power, and I_{rmsq} to the reactive volt-ampere consumed by the load.

To understand the behaviour of the system, average power, reactive and distortion volt-amperes, PF and THD are calculated as given below.

Average power *(P)*

Average power is defined as the integral of the product of instantaneous voltage and current over a period of one cycle. Thus, the active power over a time T (2π radians) can be written as:

$$P = \frac{1}{T}\int_0^T v(t) \times i(t) = \frac{1}{2\pi}\int_0^{2\pi} V_m \sin \omega t \times I_m \sin(\omega t - \phi) d\omega t \quad (5.10)$$

By solving the above, we get Eq. (5.3), that is, $P = V_{rms} \quad I_{rms}\cos\phi$. Thus, the average power, P is the product of rms value of voltage and the component of current in phase with the voltage.

Reactive volt-ampere *(Q)*

Similarly, reactive volt-ampere is the product of rms voltage and the component of current that is in quadrature with the voltage. This is given by:

$$Q = V_{rms} I_{rms} \sin\phi \; = V_{rms} I_{rmsq} \tag{5.11}$$

Apparent volt-ampere *(S)*

The apparent volt-ampere, which is the product of rms value of voltage and current, is given by Eq. (5.5), that is, $S = V_{rms} I_{rms}$. It is seen that as the value of ϕ increases, I_{rmsq} and hence I_{rms} also increases, while I_{rmsp} remains the same for the same amount of active power. This increase in feeder current causes a large increase in transmission loss (I^2R) in addition to the feeder voltage drop. The value of ϕ depends on the load connected to the power system, which is normally lagging due to the large number of inductive utilization equipment, and the power diagram of the system is shown in Figure 5.5(b). Using *P* and *S*, power factor of the system is calculated as follows.

Power factor (PF)

As power factor is defined as the ratio of average power *P* to apparent volt-ampere *S*, PF in Case 1 is given by:

$$\text{PF} = \frac{V_{rms} I_{rms} \cos\phi}{V_{rms} I_{rms}} = \cos\phi$$

Thus, under linear load conditions, the power factor deviates from its ideal value of unity, by $\cos\phi$ due to inductive or capacitive load, which draws reactive volt-amperes from the supply in addition to the active power demanded by the load.

THD

As current and voltage are sinusoidal in this case, THD in voltage and current is zero.

5.3.1.1 PF improvement when supply voltage and load current are sinusoidal – Case: 1

As stated before, to draw minimum current for the same power demand, the voltage and current should be in phase. That is ϕ and so "net reactive volt-ampere" should be zero. This can be achieved if the reactive volt-ampere required by the load is supplied by a local compensator circuit so that the generator needs to supply only the line reactive power, which depends on the reactance of the transmission line and the active power demanded by the load. Such an arrangement using a capacitor bank connected in parallel with the load as a compensator is shown in Figure 5.6.

Assume that the capacitor draws leading current ($I_c = I_{c2} + I_{c3}$) from the system so that the reactive component of current from the feeder is reduced to $I_{rmsq}' = I_{rmsq} - I_c$, which reduces the total current from I_{rms} *to* I_{rms}' as shown in Figure 5.7(a). From the figure, the minimum current drawn from the system is $I_{rms\text{-}upf}$ when the power factor angle ϕ is zero. This is obtained when the reactive power demanded by the load is fully supplied by the capacitor bank. Under this condition $I_c = I_{rmsq}$, and unity power factor (UPF) is achieved. The power triangle diagram given in Figure 5.7(b) also indicates that with capacitor bank

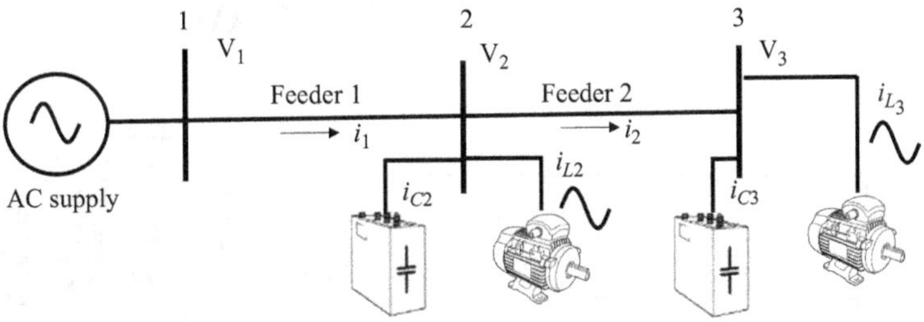

Figure 5.6 Single-line diagram of a system with a capacitor bank as a compensator

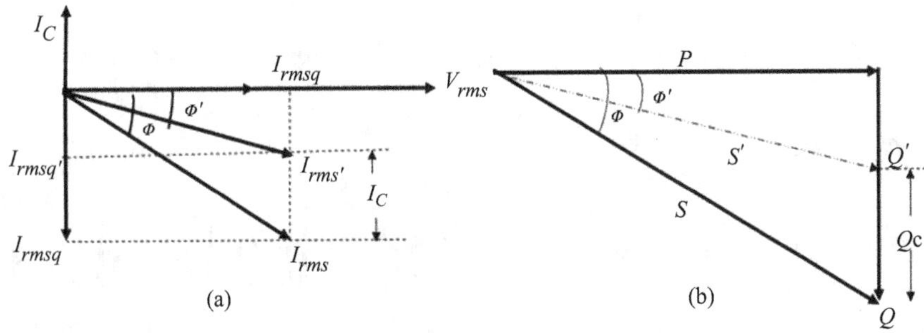

Figure 5.7 Vector diagram and power triangle diagram of Figure 5.6. (a) Vector diagram of voltage and current and (b) power angle diagram.

compensation (Q_C), power factor is improved [3]. Q, S and ϕ before compensation are changed to Q', S' and ϕ'' respectively, while P remains the same before and after compensation. If 100% compensation is done, $P = S$, and hence unity power factor (UPF) is achieved.

Induction motors are among the largest inductive loads in the power system that draw lagging current with a typical power factor ranging from 0.8 to 0.9. To improve power factor at load, a shunt capacitor bank of adequate capacity is installed close to the equipment so that the capacitor bank acts as a compensating source for reactive volt-ampere.

5.3.2 Case 2: PF and THD when supply voltage is sinusoidal and current is non-sinusoidal

As discussed in previous sections, nonlinear loads draw non-sinusoidal current from the power supply. Consider a system as in Figure 5.3 where voltage at PCC is stiff and linear loads (loads 1 and 2) are replaced with nonlinear loads by introducing motor drives for the induction motors as shown in Figure 5.8.

Assume voltage at all the buses is sinusoidal ($v = v_1$). To such a system, when drives are connected, a non-sinusoidal current, $i = i_1 = i_{L2} + i_{L3}$ flows as shown in Figure 5.9.

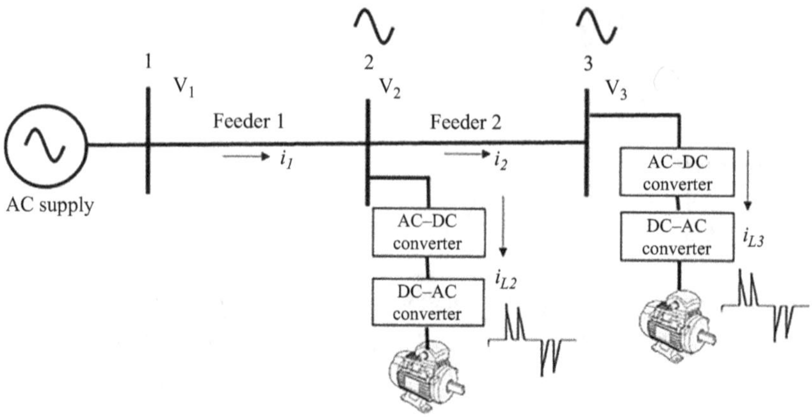

Figure 5.8 Sinusoidal power supply with nonlinear load – Case 2

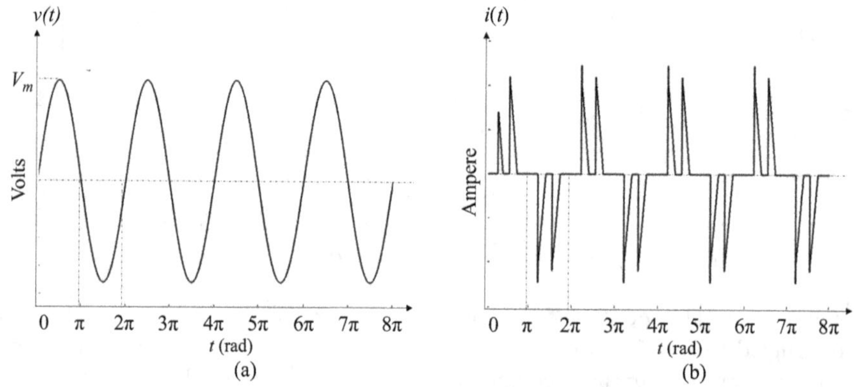

Figure 5.9 Sample waveform of (a) sinusoidal v(t) and (b) non-sinusoidal current i(t) – Case 2

The resulting non-sinusoidal load current *i(t)* can be written as:

$$i(t) = \sum_{h=1}^{n} I_{mh} \sin(h\omega t - \phi_h) \tag{5.12}$$

where ω is the supply voltage of frequency, I_{mh} is the maximum value of hth-order harmonic component of load current and ϕ_h is its phase angle, and "n" is the order of the highest harmonic component present in the load current. Note that the fundamental is the only harmonic component present in both voltage and current, whereas higher-order harmonic components are present in current in Case 2.

Voltage and current THD

As voltage is sinusoidal in Case 2, voltage THD is zero, while current THD depends on the magnitude of harmonics present in it. As the current contains harmonics, the equation for $i(t)$ can be divided into two parts to represent the fundamental component and harmonic component of the current as follows:

$$i(t) = i_1(t) + i_h(t) = I_{m1}\sin(\omega t - \phi_1) + \sum_{h=2}^{n} I_{mh}\sin(h\omega t - \phi_h) \tag{5.13}$$

The rms value of current (I_{rms}) can be determined by integrating the current Eq. (5.13) for a complete cycle and is given by:

$$I_{rms} = \sqrt{\frac{1}{T}\int_0^T i^2(t)dt} \tag{5.14}$$

Thus, the rms value of the load current can be written as:

$$I_{rms} = \sqrt{I_{rms1}^2 + \sum_{h=2}^{n} I_{rmsh}^2} \tag{5.15}$$

where, I_{rmsh} represents the rms value of the "h" order harmonic component of current and is given by $I_{rmsh} = \frac{I_{mh}}{\sqrt{2}}$. The second term under the square root gives distortion component ($I_{rms-dis}$) of current, which measures the THD of the waveform as below:

$$I_{rms-dis} = \sqrt{\sum_{h=2}^{n} I_{rmsh}^2} \tag{5.16}$$

THD in current is the ratio of rms value of the distortion current ($I_{rms-dis}$) to the fundamental component of the current(I_{rms1}). So, the percentage of THD in current (I_{THD}) is given by:

$$I_{THD} = \frac{I_{rms-dis}}{I_{rms1}} \times 100 \tag{5.17}$$

$$I_{THD} = \frac{\sqrt{I_{rms}^{\ 2} - I_{rms1}^{\ 2}}}{I_{rms1}} \times 100 \tag{5.18}$$

$$I_{THD} = \sqrt{\sum_{h=2}^{n} \frac{I_{rmsh}^{\ 2}}{I_{rms1}^{\ 2}}} \times 100 \tag{5.19}$$

It is clear from (5.19) that the THD depends on the order of harmonics and their magnitudes with respect to the fundamental component of current. As distortion in the waveform increases, THD also increases.

Average power

Using (5.1) and (5.12), average power in case 2 is given by:

$$P = \frac{1}{2\pi}\int_0^{2\pi} V_m \sin\omega t \times \sum_{h=1}^{n} I_{mh}\sin(h\omega t - \phi_h) \tag{5.20}$$

$$P = \frac{1}{2\pi}\int_0^{2\pi} V_m I_{m1} \sin \omega t \times \sin(\omega t - \phi_1) + V_m I_{m2} \sin \omega t \times \sin(2\omega t - \phi_2) + \cdots + V_m I_{mh} \sin \omega t \times \sin(h\omega t - \phi_h) \tag{5.21}$$

Since voltage $v(t)$ is with only the fundamental component and current $i(t)$ has higher-order harmonic components along with the fundamental component, integration of all the terms except the first term is zero. This means that when voltage is sinusoidal and current is non-sinusoidal, the average power or power delivered to the load by the source is only due to fundamental component of voltage and current. Thus, the power will be transmitted only due to the similar order frequency components present in voltage and current. Thus, active power in this condition is given by:

$$P = V_{rms} \times I_{rms1} \times \cos \phi_1 \tag{5.22}$$

where, V_{rms} is the rms value of the sinusoidal supply voltage which contains only the fundamental component, $I_{\text{rms}1}$ is the rms value of the fundamental component of load current and $\phi_1$is the phase angle between the supply voltage and the fundamental component of current. Other higher-order harmonic components of current will not contribute to real power transfer.

Apparent volt-ampere

As we know, the apparent power in any circuit is the product of rms value of voltage and current as given by:

$$S = V_{rms} \times I_{rms} = \frac{V_m}{\sqrt{2}} \times \sqrt{\sum_{h=1}^{n} I_{rmsh}^2} \tag{5.23}$$

$$\therefore S^2 = V_{rms}{}^2 \times I_{rmsh}^2 = V_{rms}{}^2 \sum_1^n I_{rmsh}^2 \tag{5.24}$$

$$S^2 = \frac{V_m^2}{2}\left(I_{rms1}^2 + I_{rms2}^2 + \cdots + I_{rmsn}^2\right) \tag{5.25}$$

Thus, it is seen that the apparent power increases with distortion in the current waveform.

Power factor

Recalling the definition of power factor, we can write:

$$\text{PF} = \frac{P}{S} = \frac{V_{rms} \times I_{rms1} \times \cos \phi_1}{V_{rms} I_{rms}} = \frac{I_{rms1} \times \cos \phi_1}{\sum_{h=1}^{n} I_{rmsh}^2} \tag{5.26}$$

$$\text{PF} = \frac{I_{rms1}}{I_{rms}} \times \cos \phi_1 \tag{5.27}$$

The term $\frac{I_{rms1}}{I_{rms}}$ is the ratio of rms value of the fundamental component of current to the rms value of the resultant current and is termed the distortion factor. Hence, *PF* in case 2 is written as:

$$\text{PF} = \text{distortion factor} \times \text{displacement factor} \tag{5.28}$$

Note that if there is no distortion in the waveforms of voltage and current, the distortion factor turns out to be unity. Displacement factor is due to the displacement between voltage and current. It is the cosine of the angle between voltage and the fundamental component of current, i.e., $\cos(\phi_1)$. When current contains harmonic components, *PF* becomes less than unity, and it reduces further with more harmonic content [1]. This means nonlinear load further reduces the *PF* than linear load on the system.

Reactive volt-ampere

The reactive volt-ampere (Q) arises due to the fundamental component of current which is 90° out of phase with the voltage. Thus:

$$Q = V_{rms\times} I_{rms1} \times \sin\phi_1 \tag{5.29}$$

$$\therefore Q^2 = {V_{rms}}^2 \times {I_{rms1}}^2 \times \sin^2(\phi_1) \tag{5.30}$$

$$Q^2 = {V_{rms}}^2 \times {I_{rms1}}^2 \times \left(1 - \cos^2(\phi_1)\right) \tag{5.31}$$

$$Q^2 = {V_{rms}}^2 \times {I_{rms1}}^2 \times \left(1 - \text{Displacement factor}^2\right) \tag{5.32}$$

Thus, the reactive volt-ampere can be represented in terms of displacement factor. When both voltage and current are sinusoidal and in phase with each other, the displacement factor is unity, and the reactive volt-ampere is zero. This is the case with the resistive load. However, when a nonlinear load like a rectifier with a DC filter capacitor is connected to a sinusoidal supply, the input current becomes non-sinusoidal. The reactive volt-ampere is zero, as the fundamental component of voltage and the fundamental component of current are in phase. Though $Q = 0$, PF is less than unity due to harmonics present in current or due to distortion of volt-amperes as explained below.

Referring to Figure 5.1(b), the apparent volt-ampere (S_T) in terms of active, reactive and distortion volt-amperes can be written as:

$${S_T}^2 = P^2 + Q^2 + D^2 = {V_{rms}}^2 \times {I_{rms}}^2 \tag{5.33}$$

From Eq. (5.33), distortion power is written as:

$$D^2 = {V_{rms}}^2 \times {I_{rms}}^2 - \left(P^2 + Q^2\right) = {V_{rms}}^2\left({I_{rms}}^2 - {I_{rms1}}^2\right) \tag{5.34}$$

$$D^2 = {V_{rms}}^2\left(\sum_{h=1}^{n} {I_{rmsh}}^2 - {I_{rms1}}^2\right) \tag{5.35}$$

$$D^2 = V_{rms}^{\ 2}\left(\sum_{h=2}^{n} I_{rmsh}^{\ 2}\right) \tag{5.36}$$

From the above expression of *D*, distortion volt-ampere, voltage contains only the fundamental component of voltage, and the summation series gives the rms value of the higher-order harmonic component of current. This means that *D* is the product of voltage and current of dissimilar harmonic components. The average value of distortion volt-ampere will be zero, as the integration of dissimilar orders of harmonics over one complete cycle is zero. Hence no power associated with it. However, it contributes to the apparent volt-ampere to the system.

The distortion volt-ampere can be expressed in terms of the distortion factor as below:

$$D^2 = V_{rms}^{\ 2}\left(I_{rms}^{\ 2} - I_{rms1}^{\ 2}\right) = V_{rms}^{\ 2} I_{rms}^{\ 2}\left(1 - \frac{I_{rms1}^{\ 2}}{I_{rms}^{\ 2}}\right)$$

$$D^2 = S^2\left(1 - \text{Distortion factor}^2\right) \tag{5.37}$$

When both voltage and current are sinusoidal, the distortion factor will be unity and distortion volt-ampere will be zero. However, as the nonlinearity of the current increases, the distortion factor decreases and the distortion volt-ampere increases. Thus, in the case of diode rectifier circuit, though the reactive volt-ampere *Q* is absent, the distortion volt-ampere *D* is present, which will contribute to an increase in the volt-ampere of the system. This in turn reduces the power factor of the system.

5.3.2.1 PF improvement when supply is sinusoidal and current is non-sinusoidal – Case 2

Consider capacitor *C* is connected in parallel to the load as shown in Figure 5.10 for compensating volt-ampere demanded by the load as in case of 1. Let *v(t)* be the voltage and $i(t) = i_1 = i_{L2} + i_{L3} + i_{C2} + i_{C3}$ be the current at bus 1. Figure 5.11 shows its equivalent circuit.

Total active power *P* supplied from the source is given by:

$$P = V_{rms} \times I_{R-1} \tag{5.38}$$

where, I_{R-1} is the resistive component of the non-linear load current.

Similarly, reactive volt-ampere (Q_C), by the capacitor that draws leading current is given by:

$$Q_C = V \times I_C \tag{5.39}$$

where rms value of capacitive current I_C, $I_C = V\omega C$.

Capacitor is a linear element that draws current I_C, in proportion to the magnitude and frequency of each component, including fundamental, harmonic and Interharmonic, present in the supply voltage. Also, each order of harmonic component of capacitive current leads the respective harmonic voltage by 90°. However, in case 2, as voltage is sinusoidal, capacitor current also is sinusoidal

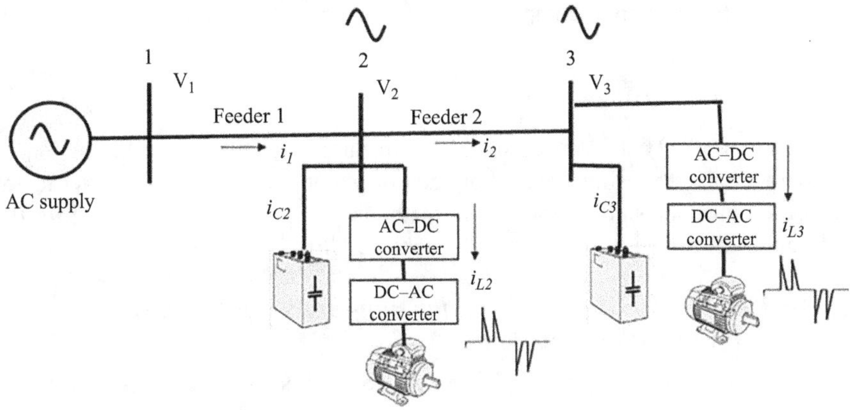

Figure 5.10 Power factor improvement using a capacitor for the system – Case 2

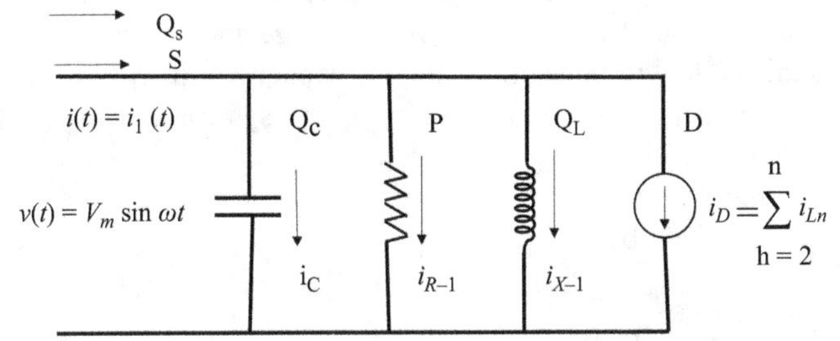

Figure 5.11 Equivalent circuit – sinusoidal supply, non-sinusoidal load, capacitor compensation

which leads the voltage by 90°. Thus, Q_C can compensate reactive volt-ampere due to Q_L in the load current i_L. Thus I_C can be made equal to $i_{C2} + i_{C3}$ by suitably selecting the capacitor or $Q_C = Q_L$ and thus $Q_s = 0$. Thus, complete compensation for load reactive volt-ampere is possible, as in Case 1. However, the capacitor contribution to distortion volt-ampere is zero. Thus, distortion volt-ampere (D), which is due to unlike frequency components in voltage and load current, will be present in the system:

$$D = \mathrm{V}_{rms} \times \sqrt{\sum_{h=2}^{n} I_{rmsh}{}^2} \tag{5.40}$$

Thus, with capacitor compensation, the total apparent power (S) is written as:

$$S = V_{rms} \times I_{rms} = \mathrm{V}_{rms}\sqrt{\sum_{h=1}^{n} I_{rmsh}{}^2} = \sqrt{P^2 + D^2} \tag{5.41}$$

Thus, after capacitive compensation of reactive volt-ampere,

$$PF = \frac{P}{\sqrt{P^2 + D^2}} \tag{5.42}$$

Equation (5.42) shows that under nonlinear load conditions, elimination of reactive volt-ampere using a capacitor bank does not result in a unity power factor because of harmonics. Therefore, other methods [5–8] need to be explored for the elimination of distortion volt-ampere to achieve UPF.

5.3.3 Case 3: PF and THD under non-sinusoidal supply voltage with linear load

Consider the same 3-ϕ system (Figure 5.3) considered in case 1. Assume that a nonlinear load connected at bus 2 results in a non-sinusoidal voltage at bus 2 and bus 3 as shown in Figure 5.12(a). When a linear load is connected to a non-sinusoidal voltage, the current drawn by it will have harmonics corresponding to supply voltage harmonics [9] with the phase angle depending on the type of load. If the load is resistive, the phase angle between voltage and current will be zero. If it is inductive, the angle will be lagging, and for capacitive, it will be leading. Let us assume that an inductive load is connected to the system as shown in the block diagram in Figure 5.12(b).

Let the instantaneous non-sinusoidal voltage, $v(t)$ and corresponding load current be $i(t)$ are given by:

$$v(t) = \sum_{h=1}^{n} V_{mh} \sin(h\omega t + \alpha_h) \tag{5.43}$$

$$i(t) = \sum_{h=1}^{n} I_{mh} \sin(h\omega t + \alpha_h - \phi_h) \tag{5.44}$$

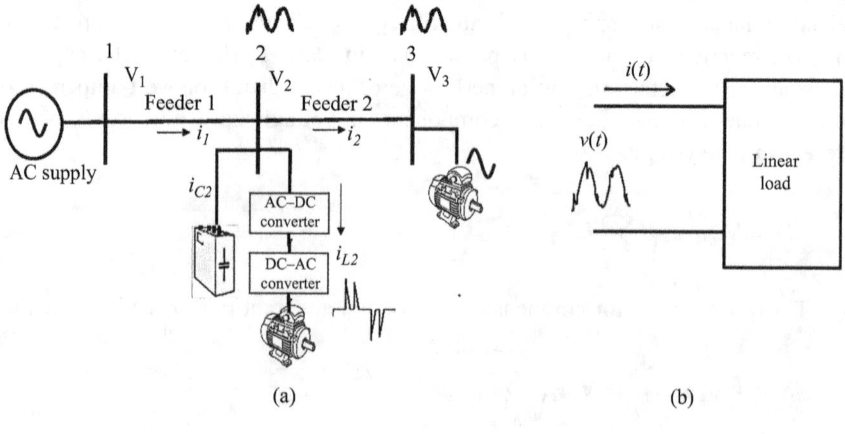

Figure 5.12 Single-line diagram of a linear load connected to a non-sinusoidal supply: Case 3. (a) Non-sinusoidal supply with linear load and (b) block diagram.

where,

α_h is the phase angle of "h" order harmonic component of voltage with a reference frame.
ϕ_h is the phase angle between "h" order harmonic component of the voltage and current.

Total harmonic distortion (THD)
Though a linear load is connected to the system, the current is distorted due to the distortion in voltage. Thus, harmonics exist in the system, and hence voltage (V_{THD}) and current THD (I_{THD}) can be expressed as:

$$V_{THD} = \sqrt{\sum_{h=2}^{n} \frac{{V_{rmsh}}^2}{{V_{rms1}}^2}} \tag{5.45}$$

$$I_{THD} = \sqrt{\sum_{h=2}^{n} \frac{{I_{rmsh}}^2}{{I_{rms1}}^2}} \tag{5.46}$$

Note that the distortion in current increases with distortion in voltage.

Active power
Average power can be calculated as

$$P = \frac{1}{2\pi}\int_0^{2\pi} \left\{ \sum_{h=1}^{n} V_{mh}\sin(h\omega t + \alpha_h) \times \sum_{h=1}^{n} I_{mh}\sin(h\omega t + \alpha_h - \phi_h) \right\} d\omega t \tag{5.47}$$

It is seen that integration of products of non-similar order of harmonic components of voltage and current over one complete cycle will be zero. In other words, integration of the product of similar orders of harmonic components of voltage and current will give the non-zero average power of the system. Thus, the power will be transmitted to the load only due to the similar order of harmonic components of voltage and current. The active power is given by:

$$P = V_{rms1}I_{rms1}\cos(\alpha_1 - \phi_1) + V_{rms2}I_{rms2}\cos(\alpha_2 - \phi_2) + \cdots + V_{rmsn}I_{rmsn}\cos(\alpha_n - \phi_n) \tag{5.48}$$

$$P = \sum_{h=1}^{n} V_{rmsh}I_{rmsh}\cos(\alpha_h - \phi_h) \tag{5.49}$$

It may be noted that in case 3, all harmonic components of voltage contribute to power flow, as current contains similar-order harmonic components due to its linear nature.

Apparent volt-ampere
Since voltage is non-sinusoidal, its rms value is given by

$$V_{rms} = \sqrt{V_{rms1}^2 + V_{rms2}^2 + \cdots + V_{rmsn}^2} = \sqrt{\sum_{h=1}^{n} V_{rmsh}^2} \tag{5.50}$$

As the load is linear, current also contains similar order of harmonic components. Thus, rms value of the current is written as:

$$I_{rms} = \sqrt{I_{rms1}^2 + I_{rms2}^2 + \cdots + I_{rmsn}^2} = \sqrt{\sum_{h=1}^{n} I_{rmsh}^2} \tag{5.51}$$

Apparent volt-ampere is given by:

$$S = V_{rms} I_{rms} = \sqrt{\sum_{h=1}^{n} V_{rmsh}^2}\sqrt{\sum_{h=1}^{n} I_{rmsh}^2} \tag{5.52}$$

$$S^2 = V_{rms}{}^2 I_{rms}{}^2 = \sum_{h=1}^{n} V_{rmsh}^2 \sum_{h=1}^{n} I_{rmsh}^2 \tag{5.53}$$

As the harmonic component in voltage increases, the apparent power also increases, which reduces the power factor of the system for a given active power transmission.

Power factor

Using Eqs. (5.6), (5.49) and (5.53), the power factor is given by:

$$PF = V_{rms1} I_{rms1} \cos(\alpha_1 - \phi_1) + \cdots V_{rmsn} I_{rmsn} \cos\frac{(\alpha_n - \phi_n)}{V_{rms} I_{rms}} \tag{5.54}$$

$$= \frac{1}{V_{rms} I_{rms}} \sum_{h=1}^{n} V_{rmsh}\, I_{rmsh} \cos(\alpha_h - \phi_h) \tag{5.55}$$

Thus, the power factor when both voltage and current contain similar-order harmonic components depends on individual voltage and current harmonics.

Reactive volt-ampere

The reactive volt-ampere in case 3 is given by:

$$Q = V_{rms1} I_{rms1} \sin(\alpha_1 - \phi_1) + V_{rms2} I_{rms2} \sin(\alpha_2 - \phi_2) + \cdots + V_{rmsn} I_{rmsn} \sin(\alpha_n - \phi_n) \tag{5.56}$$

$$Q = \sum_{h=1}^{n} V_{rmsh} I_{rmsh} \sin(\alpha_h - \phi_h) \tag{5.57}$$

The distortion volt-ampere D can be determined using (5.34), (5.49), (5.53) and (5.57) and is given by:

$$D^2 = V_{rms}{}^2 I_{rms}{}^2 - \sum_{h=1}^{n} V_{rmsh}^2 I_{rmsh}^2 \tag{5.58}$$

5.3.3.1 PF improvement under non-sinusoidal supply voltage with linear load – Case 3

Under non-sinusoidal supply voltage conditions, if the load is resistive, the power factor will be unity, though there is distortion in both voltage and current.

When inductive load is connected to a non-sinusoidal voltage, a low power factor due to reactive volt-ampere can be eliminated using capacitor; however, it cannot be made unity. This is due to a distortion that cannot be compensated for using capacitor banks.

5.3.4 *Case 4: PF and THD when supply voltage and current are non-sinusoidal*

Consider a scheme shown in Figure 5.13(a) where nonlinear loads on bus 2 result in non-sinusoidal voltage at buses 2 and 3. Under such a situation, a nonlinear load is connected at bus 3. The block diagram of the system is shown in Figure 5.13(b). This situation is quite often present in recent power distribution system.

When a nonlinear load is connected to a non-sinusoidal supply, the current will be distorted due to the distortion in voltage as well as due to the nonlinear characteristic of the load. Thus, there will be similar orders of harmonics and dissimilar orders of harmonics in the load current. Let the supply voltage contain a set of harmonic components k to n_1 that produces similar order harmonics in the load current and l to n_2 that do not result in corresponding components in the load current. Then, the instantaneous voltage, $v(t)$ and its V_{rms} can be written as:

$$v(t) = \sum_{h=k}^{n_1} V_{mh}\sin(h\omega t + \alpha_h) + \sum_{h=l}^{n_2} V_{mh}\sin(h\omega t + \alpha_h) \tag{5.59}$$

$$V_{rms}^2 = \sum_{h=k}^{n_1} V_{rmsh}^2 + \sum_{h=l}^{n_2} V_{rmsh}^2 \tag{5.60}$$

Let the load current, i_L contains a group of harmonic components k to n_1, which are of similar order as the voltage, and m to n_3, which are of dissimilar order

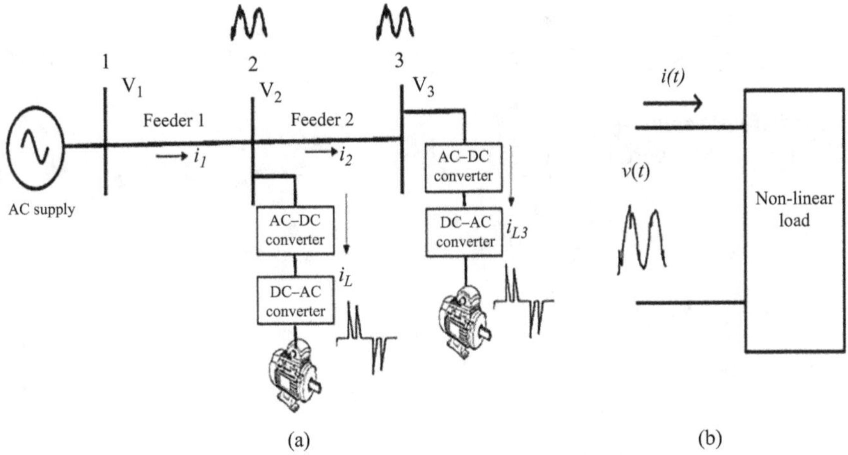

Figure 5.13 Single-line diagram of nonlinear load connected to non-sinusoidal supply: Case 4. (a) Non-sinusoidal supply with nonlinear load and (b) block diagram.

of harmonics in voltage and current. The instantaneous current, $i_L(t)$ and rms value of current, I_{rms} are given by:

$$i_L(t) = \sum_{h=k}^{n_1} I_{mh}\sin(h\omega t + \alpha_h - \phi_{Lh}) + \sum_{h=m}^{n_3} I_{mh}\sin(h\omega t + \alpha_h - \phi_{Lh}) \quad (5.61)$$

$$I_{rms}^2 = \sum_{h=k}^{n_1} I_{rmsh}^2 + \sum_{h=m}^{n_3} I_{rmsh}^2 \quad (5.62)$$

Total harmonic distortion (THD)

The voltage and current THD can be expressed as:

$$V_{THD} = \sqrt{\sum_{h=k\neq 1}^{n_1} \left[\frac{V_{rmsh}^2}{V_{rms1}}\right]^2 + \sum_{h=l\neq 1}^{n_2} \left[\frac{V_{rmsh}^2}{V_{rms1}}\right]^2} \quad (5.63)$$

$$I_{THD} = \sqrt{\sum_{h=k\neq 1}^{n_1} \left[\frac{I_{rmsh}^2}{I_{rms1}}\right]^2 + \sum_{h=m\neq 1}^{n_3} \left[\frac{I_{rmsh}^2}{I_{rms1}}\right]^2} \quad (5.64)$$

Average power

As indicated earlier, active power flow is due to a similar order of harmonics present in voltage and current and is given by:

$$P = \sum_{h=k}^{n_1} V_{rmsh} I_{rmsh} \cos(\alpha_h - \phi_h) \quad (5.65)$$

Reactive volt-ampere

The reactive volt-ampere Q is also due to similar-order harmonic components of voltage and current given by:

$$Q = \sum_{h=k}^{n_1} V_{rmsh} I_{rmsh} \sin(\alpha_h - \phi_h) \quad (5.66)$$

Distortion volt-ampere

Distortion volt-ampere using (5.34) as given by:

$$D^2 = V_{rms}{}^2 I_{rms}{}^2 - \sum_{h=k}^{n_1} V_{rmsh}^2 I_{rmsh}^2 \quad (5.67)$$

Apparent volt-ampere

Since voltage and current are non-sinusoidal in case of 4, rms value of voltage and current can be written as:

$$V_{rms} = \sqrt{\sum_{h=k}^{n_1} V_{rmsh}^2 + \sum_{h=l}^{n_2} V_{rmsh}^2} \quad (5.68)$$

$$I_{rms} = \sqrt{\sum_{h=k}^{n_1} I_{rmsh}^2 + \sum_{h=m}^{n_3} I_{rmsh}^2} \tag{5.69}$$

Thus, reactive volt-ampere is given by:

$$S^2 = V_{rms}{}^2 I_{rms}{}^2 = \left[\sum_{h=k}^{n_1} V_{rmsh}^2 + \sum_{h=m}^{n_2} V_{rmsh}^2\right]^2 \left[\sum_{h=1}^{n_1} I_{rmsh}^2 + \sum_{h=1}^{n_3} I_{rmsh}^2\right]^2 \tag{5.70}$$

Power Factor

Power factor in the case of non-sinusoidal supply voltage and load current, as in case 4, can be determined using Eq. (5.6) as below:

$$PF = \frac{P}{S} = \frac{\sum_{h=k}^{n_1} V_{rmsh} I_{rmsh} \cos(\alpha_h - \phi_h)}{\sqrt{\left[\sum_{h=k}^{n_1} V_{rmsh}^2 + \sum_{h=l}^{n_2} V_{rmsh}^2\right]^2 \left[\sum_{h=1}^{n_1} I_{rmsh}^2 + \sum_{h=1}^{n_3} I_{rmsh}^2\right]^2}} \tag{5.71}$$

5.3.4.1 PF improvement when both voltage and current are non-sinusoidal – Case 4

As in cases 2 and 3, when a capacitor bank is used for power factor improvement, it can eliminate only reactive power due to a similar order of harmonics in voltage and current. However, volt-ampere due to dissimilar harmonics, which are classified as distortion volt-ampere, remain in the system. Thus, in case 4 too, improvement in power factor using a capacitor bank will not result in UPF like cases 2 and 3.

5.4 Conclusion

As highlighted, traditionally power factor refers to $\cos\phi$ or displacement factor. The reactive volt-ampere (VAR) produced by a load is compensated by supplying an equal and opposite reactive volt-ampere using a shunt capacitor $Q_C = V^2(\omega C)$ for a lagging (inductive) load, or a shunt reactor $Q_L = V^2/(\omega L)$ for a leading (capacitive) load. However, in today's power generation and load profile, the voltage and current waveforms have departed from their ideal sine wave of one frequency to waveforms composed of multiple frequencies. Consequently, the PF is also changed to PF = distortion factor × displacement factor. We also have seen that under non-sinusoidal supply conditions, distortion volt-amperes are produced by the unlike frequency components in voltage and current. Since a capacitor is a linear and passive device, it cannot neutralize distortion volt-ampere, D. Also, under harmonic conditions, the capacitor may create series and parallel resonances, which are discussed in previous chapters. Thus, other techniques for compensating harmonics are required in cases 2, 3 and 4. These are discussed in the next chapters.

5.5 Problems and solutions

Problem 5.1 A 1000 W resistive load is connected to a 1-ϕ, 50 Hz, 230 V system.

(i) Calculate the rms value of current drawn from the source and the power factor.

(ii) In another load condition, it is found that the system is drawing a current of 6 A, though the active power is still 1000 W. Calculate the power factor.

(iii) What could be the reason for the low power factor in case (ii) though the active power is the same in both cases? Draw the current waveforms for each case and justify

Solution:

Given parameters:

$P = 1000$ W
$V = 230$ V, 50 Hz

(i) Calculate rms value of the current drawn and the power factor at the source (Figure 5.14).

Since the load is resistive,

$$\text{rms value of load current} = \frac{P}{V} = 1000/230 = 4.35 \text{ A}$$

Apparent volt-ampere $S = V \times I = 230 \times 4.35 = 1000$ VA

As $P = S$, the power factor is unity. Otherwise also we know that for a resistive load connected to a sinusoidal supply, the current will be sinusoidal and in phase with the voltage, and hence PF = 1.

(ii) In another load condition, it is found that the system is drawing 6 A, though the active power drawn is still 1000 W. Calculate the power factor.

Since the active power demanded by the load in part (ii) is the same, the active component of current remains 4.35 A only. Since the system is drawing 6 A current,

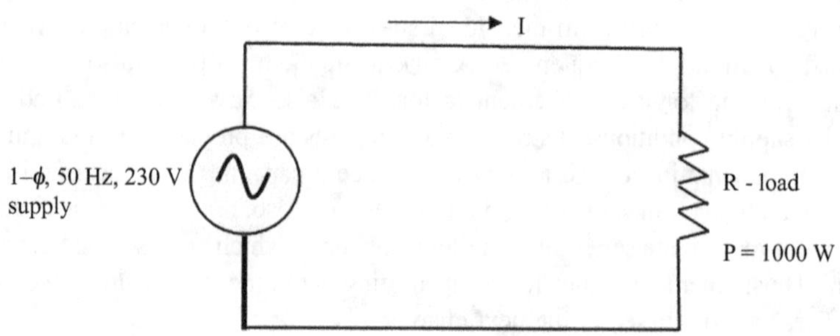

Figure 5.14 Circuit diagram for problem 5.1 case (i)

it means reactive power is also being drawn by the load. However, it is given that the load is resistive. This indicates that it is drawing distortion power. That may be due to the introduction of power electronic-based equipment with resistive load, as shown in Figure 5.15.

Apparent power (S) is given by:

$S = VI = 230 \times 6 = 1380$ VA

With the rectifier, PF is given by: $PF = \dfrac{P}{S} = \dfrac{1000}{1380} = 0.72$

(iii) What could be the reason for the low power factor in case (ii) though the active power is the same in both cases? Draw the current waveform for each case and justify

In case (i) a linear load is connected to supply, which consumes 1000 W power whereas in case (ii) a rectifier with a resistor that forms, a nonlinear load is connected, which consumes the same power as in case (i); however, it consumes distortion power, which results in a low power factor. A simulation is carried out using MATLAB. The waveform of supply voltage and input current for the same output power of 1000 W in case (i) and case (ii) is shown in Figure 5.16. A diode rectifier

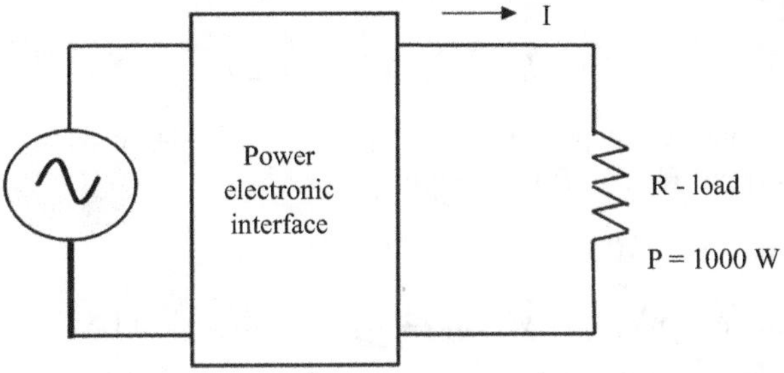

Figure 5.15 Circuit diagram for problem 5.1 case (ii)

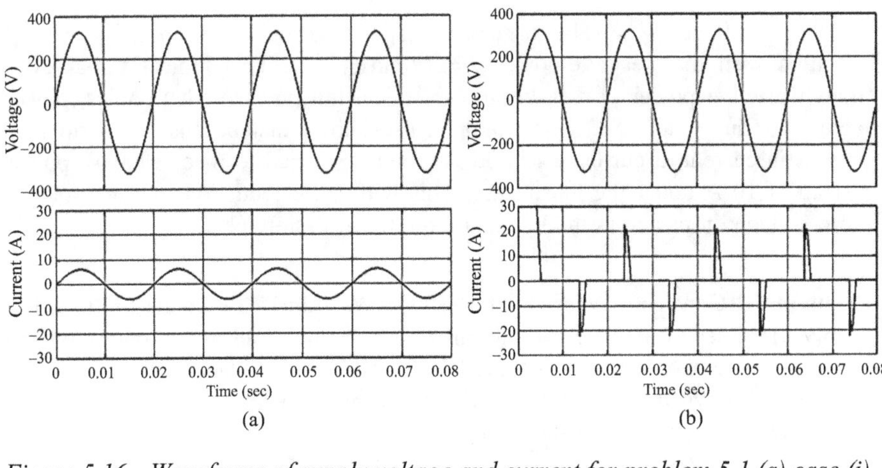

Figure 5.16 Waveforms of supply voltage and current for problem 5.1 (a) case (i). (b) case (ii).

is considered as a power electronic interface in case (ii). It is clear that the peak value of current in case (i) is 6 A and sinusoidal, and in case (ii) it is 20 A, discontinuous and peaky. However, the fundamental component of current in both cases is the same, as both loads consume the same active power of 1000 W.

A larger deviation of power factor from unity to 0.72 is due to the peak nature of the load current or distortion of power consumption by the load. In other words, the load draws more current than required to meet the same active power demand in case (ii).

Problem 5.2 Consider a system with sinusoidal voltage of 100 V feeding a non-linear load which results in a non-sinusoidal current with a fundamental component of 10 A, a 5th harmonic component 5 A and a 7th harmonic component of 2 A. The phase angle between the fundamental component of voltage and current is zero. Show that $\frac{P}{S}$ and $\cos\phi_1$ are not equal. Justify the answer. Neglect harmonic orders greater than 7.

Solution:
Given values:

$$V = 100\ \text{V}\ I_{rms1} = 10\ \text{A}\ I_{rms5} = 5\ \text{A}\ I_{rms7} = 2\ \text{A}$$

We know that active power flow is due to a similar order of harmonics in voltage and current. Hence, P is given by:

$$P = VI_1 \cos\phi_1 = 100 \times 10 \times 1 = 1000\ \text{W}$$

$$\text{Resultant current}, I_{rms} = \sqrt{I_{rms1}{}^2 + I_{rms5}{}^2 + I_{rms7}{}^2} = 11.35\ \text{A}$$

$$\text{Apparent power},\ S = VI_{rms} = 100 \times 11.35 = 1135.00\ \text{VA}$$

$$PF = \frac{\text{Average power}}{\text{Apparent voltampere}} = \frac{P}{S} = \frac{1000}{1135.00} = 0.88$$

As $\phi_1 = 0$, $\cos\phi_1 = 1$. But the calculated power factor is 0.88. Thus, $\frac{P}{S} \neq \cos\phi_1$.

It is seen that reactive power $Q = VI_1 \sin\phi_1 = 0$. This indicates that even though reactive power $Q = 0$, because of distortion power or harmonics, power factor is reduced to 0.88. This example also shows that because of harmonics present in current, a current of 11.35 A (rms) is needed to meet 1000 W power instead of 10 A under the linear case. This will cause reduction in power factor, more voltage drops across the line and more transmission loss.

Problem 5.3 Consider a linear load of 1000 kW with 0.7 PF connected to a 1-ϕ supply of 11 kV. Calculate the current drawn by the load and supplied by the source. If the power factor is improved to 0.9 by using a capacitor bank, calculate the current drawn from the supply. Verify the results using simulation software. Also draw a line diagram of the arrangement.

Solution:

Given data

Power consumed by the load, $P = 1000$ kW

Power factor of the load = 0.7, supply voltage = 11 kV, assuming ϕ_1 as angle between supply voltage and current without capacitor compensation,

$$\text{Current drawn by the load, } I_L = \frac{P}{V\cos\phi_1} = \frac{1000 \times 1000}{11 \times 1000 \times 0.7} = 129.87 \text{ A}$$

The current supplied by the source also will be 129.87 A, as there is no device connected in between the supply and load.

A MATLAB simulation is done to verify the results. The waveforms for supply voltage and load current are shown in Figure 5.17. The power factor observed in simulation is 0.699 ($\approx$0.7), and the supply current is 129.9 A (rms). Figure 5.18 shows the single-line diagram and phasor diagram.

When the power factor is improved to 0.9, current drawn from the source (I_s), by assuming ϕ_2 angle between supply voltage and current with capacitor compensation,

$$I_s = \frac{P}{V\cos\phi_2} = \frac{1000 \times 10^3}{11 \times 10^3 \times 0.9} = 101.01 \text{ A}$$

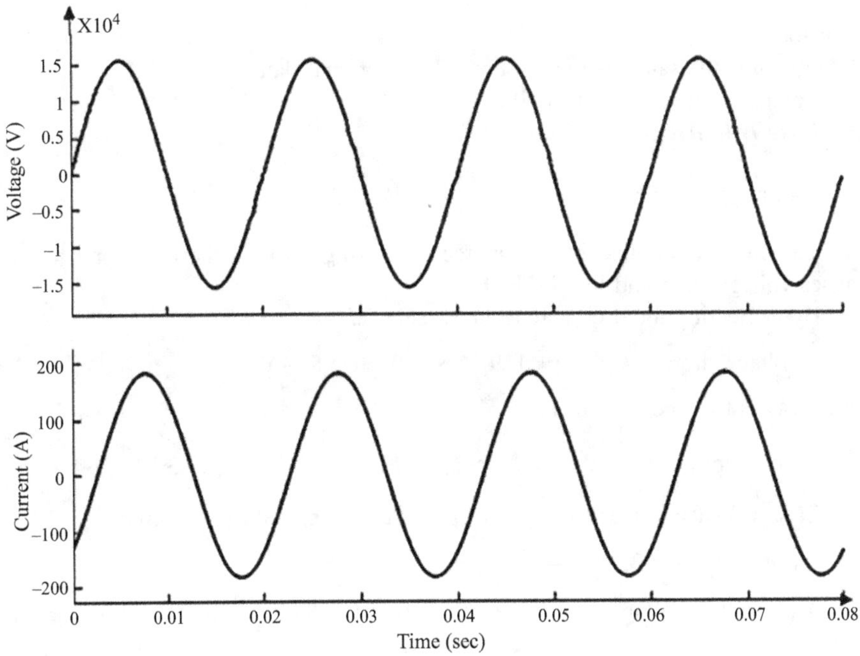

Figure 5.17 Waveforms of supply voltage and current of 100 kW, 0.7 pf, 11 kV system – Problem 5.3

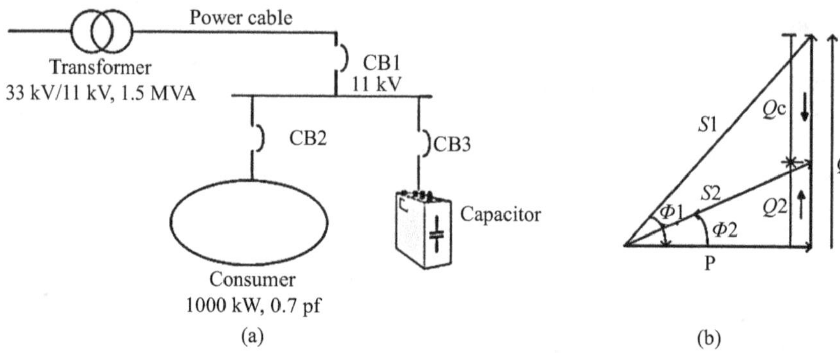

Figure 5.18 Single-line diagram of 11 kV system and phasor diagram – Problem 5.3. Power (a) supply arrangement and (b) phasor diagram of P, Q and S.

Thus, the supply current is reduced to 101.01 A from 127.87 A by providing a leading compensating device capacitor connected across the load.

Problem 5.4 Consider a 1-ϕ, 11 kV, 1000 kW load connected to a power supply using a power cable of impedance $0.02 + j0.02\ \Omega$. Calculate the power loss and voltage drop in the cable when the power factor of the load is case (i) 0.7 and case (ii) 0.9.

Solution:

Data given is the same as in problem 5.3. Additional data

$$\text{Cableimpedance} = 0.02 + j0.02$$

Case 1. When power factor is 0.7,

$$\text{As in problem 5.3,}\quad I = \frac{1000}{11 \times 0.7} = 129.87\ \text{A}$$

The current supplied by the source is flowing through the cable and hence, causes voltage drop and loss in the line.

By neglecting the drop across the resistance,

$$\text{Voltage drop} = I \times X_L = 129.87 \times 0.02 = 2.597\ \text{V}$$

where, X_L is the line reactance.

$$\text{Power loss in the cable} = I^2 \times R = 129.87^2 \times 0.02 = 333.3\ \text{W}$$

Case 2. When power factor is at 0.9, current supplied by the source,

$$\text{As in problem 5.3, } I = \frac{1000}{11 \times 0.9} = 101.01\ \text{A}$$

And hence, $\text{Voltagedrop} = I \times X_L = 101.01 \times 0.02 = 2.02\ \text{V}$ and loss in the cable is

$$I^2 \times R = 101.01^2 \times 0.02 = 204.06\ \text{W}$$

We have seen from the earlier examples that as the power factor improves, the current requirement for a given active power demand reduces. In this problem, it is proved that improvement in power factor helps in reducing losses and voltage drop across the transmission line.

Problem 5.5 A resistive load of 10 Ω is connected to a 1-ϕ, 230 V, 50 Hz supply. Calculate the real and reactive power supplied by the source and the power factor of the system. Comment on the THD of voltage and current.

Solution:
Given data:

$$V_{rms} = 230V$$

$$I_{rms} = \frac{V_{rms}}{R} = \frac{230}{10} = 23 \text{ A}$$

Real power, $P = \frac{V^2}{R} = I^2 \times R = V_{rms} I_{rms} \cos\phi = 230 \times 23 \times \cos 0 = 5290$ W

Reactive power
Since the connected load is resistive, the reactive power consumed by the load is zero, and hence only active power will be supplied by the source.

Power factor
Power factor of the circuit will be unity, as voltage and current are in phase.

THD
Since the resistive load is connected to a sinusoidal supply, both voltage and current are sinusoidal and in phase. Thus, THD in both voltage and current will be zero.

Problem 5.6 An R-L load of 10 + j10 Ω is connected to a 1-ϕ, 230 V, 50 Hz supply. Calculate the real power, reactive power, and power factor of the system. Comment on the THD of voltage and current.

Solution:
Given data:

$$V_{rms} = 230 \text{ V} \quad Z = (10 - j10)\Omega$$

$$I_{rms} = \frac{V_{rms}}{Z} = \frac{230}{10 + j10} = (11.5 - j11.5)A = 16.26\angle - 45^\circ$$

The current equation shows that current lags the voltage by an angle 45°. Hence, the power factor angle ϕ is 45°.

Active power

$$P_L = V_{rms} I_{rms} \cos\phi = 230 \times 16.26 \times \cos 45 = 2644.44 \text{ W}$$

Reactive power

$$Q_L = V_{rms}I_{rms}\sin\phi = 230 \times 16.26 \times \sin 45 = 2644.44 \text{ VAr}$$

Apparent power

$$S = V_{rms}I_{rms} = 230 \times 16.26 = 3739.8 \text{ VA}$$

Power factor

$$PF = \frac{P}{S} = \cos 45 = 0.707$$

THD

Since the load connected to the system is linear, the current waveform is sinusoidal, and hence, THD is zero.

Problem 5.7 In Problem 5.6, calculate the value of the capacitor required to improve the power factor to unity.

Reactive power to be supplied by the capacitive compensator (Q_c) to improve power factor to unity is:

$$Q_c = Q_L = 2644.44, \text{ to make PF unity}$$

Capacitor current required for full compensation, $I_c = \frac{Q_c}{V_{rms}} = \frac{2644.44}{230} =$ 11.4976 A

$$X_c = \frac{V_{rms}}{I_c} = \frac{230}{11.4976} = 20.004\ \Omega$$

$$C = \frac{1}{2 \times \pi \times f \times X_c} = \frac{1}{2 \times \pi \times 50 \times 20.004} = 159.12\ \mu\text{F}$$

159.12 μF capacitor to be connected in parallel with the load to make power factor unity.

Problem 5.8 Assume a supply voltage of $v = \sqrt{2} \times 230\sin\omega t$ supplying a nonlinear load, which results in a current of $i = \sqrt{2} \times 5\sin\omega t + \sqrt{2} \times 2\sin 5\omega t + \sqrt{2} \times 3\sin 7\omega t$.

Calculate active power, reactive power, distortion power, power factor and THD.

Solution:

Given data:

$$V_{rms} = 230 \text{ V}$$

$$I_{rms1} = 5 \text{ A} \quad I_{rms5} = 2 \text{ A} \quad I_{rms7} = 3 \text{ A}$$

$$\phi_1 = \phi_5 = \phi_7 = 0$$

$$P = VI_{rms1}\cos\phi_1 = 230 \times 5 \times \cos 0 = 1150 \text{ W}$$

$$Q = VI_{rms1}\sin\phi_1 = 230 \times 5 \times \sin 0 = 0$$

Distortion power

$$I_{rms} = \sqrt{I_{rms1}{}^2 + I_{rms5}{}^2 + I_{rms7}{}^2} = \sqrt{5^2 + 2^2 + 3^2} = 6.164 \text{ A}$$

$$S = V \times I_{rms} = 230 \times 6.614 = 1417.72 \text{ VA}$$

$$D = \sqrt{S^2 - (P^2 + Q^2)} = 829.11 \text{ VA}$$

Power factor

$$PF = \frac{P}{S} = \frac{1150}{1417.72} = 0.8112$$

Current THD

$$I_{THD} = \frac{\sqrt{\sum_{h=2}^{n} I_{rmsh}}}{I_{rms}} = 0.5849$$

Current THD is 58.49% and voltage THD is zero, as the voltage is purely sinusoidal.

Problem 5.9 A 1-ϕ, 50 Hz, sinusoidal supply of rms value 230 V, is supplying a nonlinear load. The current drawn from the source contains a fundamental component, a second-order and a third-order harmonic component with rms value of 20 A, 15 A and 10 A, respectively, and phase differences of −30°, 60° and 30°, respectively. Obtain the instantaneous values of voltage and current. Calculate active power, reactive power, distortion power, displacement factor, distortion factor and power factor. Assume that the capacitor compensation is used to improve the power factor; calculate the value of the capacitor of fundamental frequency to improve power factor. Comment on the results.

Solution:
Given data:

$$V_{rms} = 230 \text{ V}$$

$$I_{rms1} = 20 \text{ A} \quad I_{rms2} = 15 \text{ A} \quad I_{rms3} = 10 \text{ A}$$

$$\phi_1 = -30^\circ \quad \phi_2 = 60^\circ \quad \phi_3 = 30^\circ$$

Voltage and current equations are written as:

$$v = \sqrt{2} \times 230 \sin \omega t$$

$$i = \sqrt{2} \times 20 \sin(\omega t - 30) + \sqrt{2} \times 15 \sin(2\omega t + 60) + \sqrt{2} \times 10 \sin(3\omega t + 30)$$

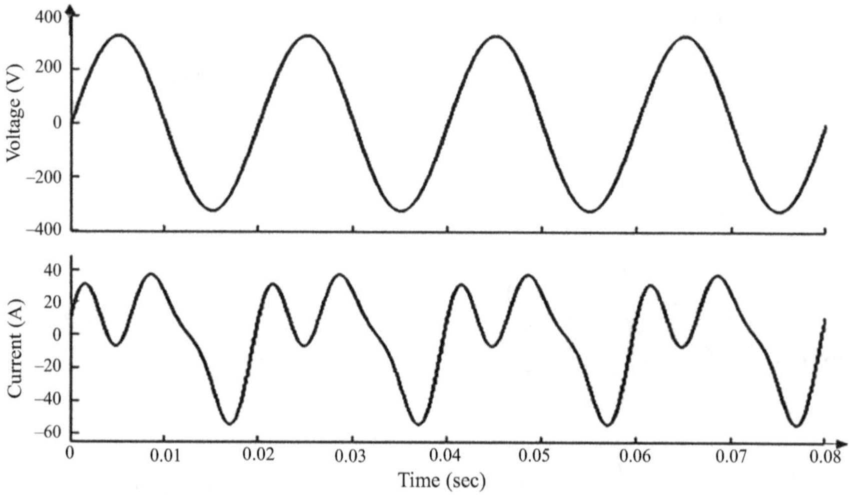

Figure 5.19 Waveform of voltage and current for Problem 5.9

The waveform of current is plotted using Simulink® to understand the distortion of current with harmonics. As it contains even-order harmonics, the waveform will not be symmetrical, as may be noted from the current waveform shown in Figure 5.19.

From the equation of voltage and current, the voltage contains only the fundamental component, while the current contains the 2nd and 3rd orders of harmonic components. Various other parameters are calculated as follows.

Active power

$$P = VI_{rms1}\cos\phi_1 = 230 \times 20 \times \cos 30 = 3983.717\ \text{W}$$

Reactive power

Similarly reactive power is given by:

$$Q = VI_{rms1}\sin\phi_1 = 230 \times 20 \times \sin 30 = 2300\ \text{VAr}$$

Power factor

The rms value of current is given by:

$$I_{rms} = \sqrt{{I_{rms1}}^2 + {I_{rms2}}^2 + {I_{rms3}}^2} = \sqrt{20^2 + 15^2 + 10^2} = 26.926\ \text{A}$$

Apparent volt-ampere is given by:

$$S = V \times I_{rms} = 230 \times 26.926 = 6192.98\ \text{VA}$$

$$PF = \frac{P}{S} = \frac{3983.717}{6192.98} = 0.643$$

THD

THD in voltage is zero, as voltage is sinusoidal. Since the current contains harmonics, current THD is given by:

$$\%I_{THD} = THD_i = \frac{\sqrt{\sum_{h=2}^{\infty} I_{sh}{}^2}}{I_{s1}} \times 100 = \frac{\sqrt{15^2 + 10^2}}{20} = 90.13\%$$

Problem 5.10 A six-pulse rectifier with a resistive load is connected to a 3-ϕ, 415 V, 50 Hz sinusoidal supply. The fundamental component of the resultant current is 1252 A (peak). Write down the equations of instantaneous voltage and current. The phase angles of the 5th and 7th harmonic components are 160° and 140°, respectively. Calculate P, Q, D, PF and THD in the circuit. Verify using simulation.

Solution:

The phase voltage equations are:

$$v_a(t) = \sqrt{2} \times \frac{415}{\sqrt{3}} \sin \omega t$$

$$v_b(t) = \sqrt{2} \times \frac{415}{\sqrt{3}} \sin (\omega t - 120°)$$

$$v_c(t) = \sqrt{2} \times \frac{415}{\sqrt{3}} \sin (\omega t - 240°)$$

For a six-pulse converter, current harmonics present will be 5, 7, 9, 11, ... with magnitudes of 1/5th, 1/7th, 1/9th, 1/11th ..., respectively.

$$i_a(t) = 1252\sin (\omega t) + \frac{1252}{5} \sin (5\omega t - 160°) + \frac{1252}{7} \sin (7\omega t - 140°)$$

rms values of phase quantities

$$V_a = \frac{415}{\sqrt{3}} = 239.6 \text{ V}$$

$$I_1 = \frac{1252}{\sqrt{2}} = 885.29 \text{ A}, \quad I_5 = \frac{1252}{5 \times \sqrt{2}} = 177.06 \text{ A}, \quad I_7 = \frac{1252}{7 \times \sqrt{2}}$$

$$= 126.471 \text{ A}$$

$$\text{Power per phase} = 239.6 \times 885.29 = 212115.48 \text{ W}$$

$$I_{rms} = \sqrt{I_1^2 + I_5^2 + I_7^2} = \sqrt{885.29^2 + 177.06^2 + 126.471^2} = 911.441 \text{ A}$$

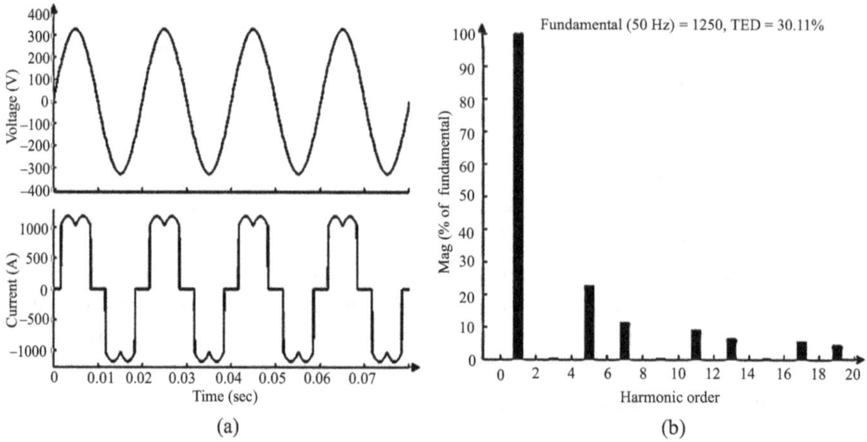

Figure 5.20 Waveforms of supply voltage, current and current harmonic spectrum. (a) Voltage and current waveforms and (b) harmonic spectrum of current.

Reactive power per phase, $S = VI = 239.6 \times 911.441 = 218,382.016$ VA

$$\text{Power factor} = \frac{P}{S} = \frac{212,115.48}{218,382.016} = 0.9713$$

$$THD_i = \frac{\sqrt{\sum_{h=2}^{\infty} {I_{sh}}^2}}{I_{s1}} = \frac{\sqrt{I_5^2 + I_7^2}}{I_1} = \frac{217.59}{885.29} = 0.2457$$

Current THD is 24.57%.

A simulation is done using MATLAB using the blocks and a load of $R = 0.55\ \Omega$. The harmonic spectrum of the input current along with the current waveform is shown in Figure 5.20. The maximum value of the fundamental component of current is observed to be 1252 A with a THD of 30.11. The calculated value of THD was 24.77. The difference is because in simulation higher-order harmonics up to 1000 Hz are considered, whereas in calculation only the 5th and 7th are considered.

Problem 5.11 A 1-ϕ, 230 V, 50 Hz sinusoidal supply is feeding a bridge diode rectifier circuit with $R = 100\ \Omega$ and $C = 6000\ \mu$F. Simulate the circuit and Compute the rms value of the source current, active power consumed by the rectifier, reactive power, distortion power and power factor of the system, and comment on the results.

Solution:

Figure 5.21 shows the supply voltage and current waveforms plotted using MATLAB simulation. The input current is peaky, as a high-value capacitor is connected for filtering.

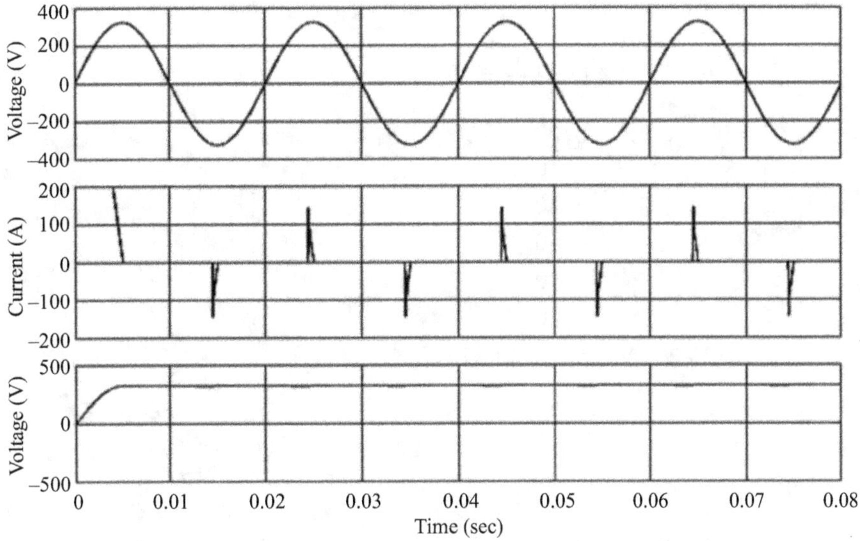

Figure 5.21 Waveform of supply voltage, current and, DC output voltage of single-phase diode rectifier

Simulation results:

$$V_{rms} = 230\text{ V} \quad I_{rms} = 13.91\text{ A} \quad P = 1049\text{ W}$$

$$Q = -58\text{ Var} \quad D = 3022\text{ VA} \quad THD = 2.878 \quad PF = 0.3277$$

Comment on the results:
From the waveforms it is clear that the fundamental component of the current is almost in phase with the supply voltage, which means the reactive power consumed by the load and hence the reactive power supplied by the source is almost zero. But it is observed that the power factor is very low, i.e., 0.3277 which indicates the presence of distortion power due to harmonics present in the current.

Problem 5.12 Consider a sinusoidal voltage

$$v = \sqrt{2} \times (100\sin(\omega t) + 50\sin(3\omega t - 20) + 30\sin(5\omega t - 10) + 70\sin(7\omega t - 20))$$

supplying a resistive load of 10 Ω. Plot the waveforms of voltage and current using any software. Determine the ratio of the individual harmonic component of the voltage to the current. Calculate the power factor and THD.

Solution:

$$V_{rms1} = 100\text{ V} \quad I_{rms1} = \frac{100}{10} = 10\text{ A}$$

$$V_{rms3} = 50\text{ V} \quad I_{rms3} = \frac{50}{10} = 5\text{ A}$$

$$V_{rms5} = 30\text{ V} \quad I_{rms5} = \frac{30}{10} = 3\text{ A}$$

$$V_{rms7} = 70\text{ V} \quad I_{rms7} = \frac{70}{10} = 7\text{ A}$$

$$V_{rms} = 135.277\text{ V} \quad I_{rms} = 13.52\text{ A}$$

Ratio of voltage to current

$$\frac{V_1}{I_1} = \frac{V_3}{I_3} = \frac{V_7}{I_7} = 10$$

Phase angle between voltage and current of the hth harmonic, $\alpha_h - \phi_h = 0$. The current equation is written as:

$$i = \sqrt{2} \times (10\sin(\omega t) + 5\sin(3\omega t - 20) + 3\sin(5\omega t - 10) + 7\sin(7\omega t - 20))$$

$$P_1 = 100 \times 10 = 1000\text{ W}$$

$$P_3 = 50 \times 5 = 250\text{ W}$$

$$P_5 = 30 \times 3 = 90\text{ W}$$

$$P_7 = 70 \times 7 = 490\text{ W}$$

$$P = 1000 + 250 + 90 + 490 = 1830\text{ W}$$

$$S = V_{rms} \times I_{rms} = 135.277 \times 13.52 = 1829.99\text{ VA}$$

Power factor, $PF = \frac{P}{S} = 1.$

It is seen that $P = S$, so, the power factor will be unity. However, harmonics present in the supply voltage may create problems for the load as well as other equipment connected to it. So, in addition to power factor, harmonics also need to be controlled.

The simulation results show active power and reactive volt-ampere to be nearly the same, and hence the power factor is unity. THD in voltage and current is 91.02%, which indicates that the harmonic order and the ratio of voltage to current in each order of harmonics are the same. The waveforms of voltage and current are shown in Figure 5.22.

It can be seen that to make power factor unity under any conditions, whether sinusoidal or non-sinusoidal, (i) a similar order of harmonic components needs to be present in both voltage and current; (ii) the phase angle between voltage and

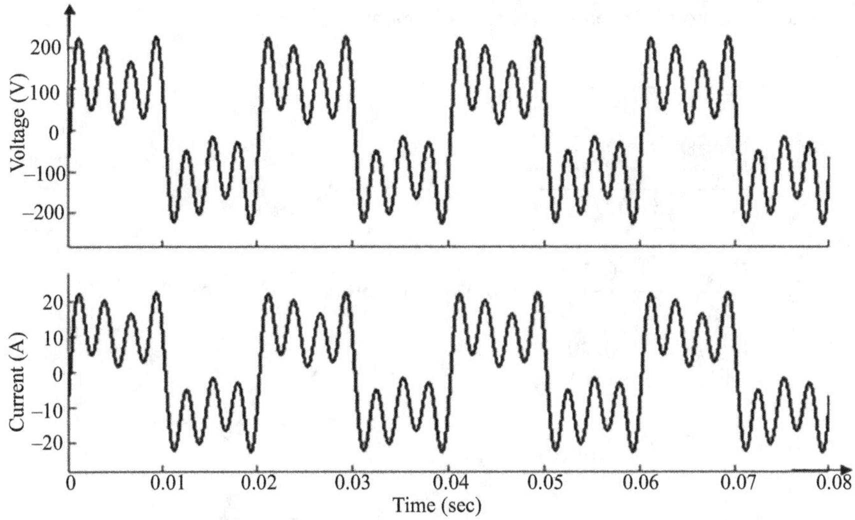

Figure 5.22 The waveforms of voltage and current in Problem 5.12

current at each order of harmonics needs to be same and (iii) the ratio of harmonic voltage component to similar order harmonic current component needs to be the same.

Problem 5.13 Consider a sinusoidal voltage,

$v(t) = 100 \times \sqrt{2}\sin(\omega t) + 50 \times \sqrt{2}\sin(3\omega t + 20) + 30 \times \sqrt{2}\sin(5\omega t + 10) + 70 \times \sqrt{2}\sin(7\omega t + 20)$ applied to a load of 10+j10 Ω at 50 Hz. Plot the waveforms of voltage and current using any software. Calculate (i) power factor, (ii) THD of voltage and current, and (iii) distortion power. Assume a 50 Hz supply.

Solution:
Given data:

$$V_{rms1} = 100 \text{ V} \quad V_{rms3} = 50 \text{ V} \quad V_{rms5} = 30 \text{ V} \quad V_{rms7} = 70 \text{ V}$$

$$Z_1 = 10 + J10 = 14.14\angle + 45\ \Omega$$

The impedance of the load will change depending upon the harmonic component present in the voltage; however, the resistance of the load will remain the same as it is independent of the frequency. The impedance of the load corresponding to the order of the harmonic component present in the voltage is given by:

$$Z_3 = 10 + J30 = 31.62\angle 71.56\ \Omega.$$

$$Z_5 = 10 + J50 = 50.99\angle 78.69\ \Omega.$$

$$Z_7 = 10 + J70 = 70.71\angle 81.86\ \Omega$$

The corresponding load current component is given by:

$$I_1 = \frac{V_{rms1}}{Z_1} = \frac{100\angle 0}{14.14\angle 45} = 7.07\angle -45^\circ \text{A}$$

$$I_3 = \frac{V_{rms3}}{Z_3} = \frac{50\angle 20}{31.62\angle 71.56} = 1.58\angle -51.56 \text{ A}$$

$$I_5 = \frac{V_{rms5}}{Z_5} = \frac{30\angle 10}{50.99\angle 78.69} = 0.58\angle -68.69 \text{ A}$$

$$I_7 = \frac{V_{rms7}}{Z_7} = \frac{70\angle 20}{70.71\angle 81.86} = 0.989\angle -61.86 \text{ A}$$

$$P = \sum_{h=1}^{n_1} V_{rmsh} I_{rmsh} \cos(\alpha_h - \phi_h)$$

$$P = 100 \times 7.07 \times \cos 45 + 50 \times 1.58 \times \cos 51.56 + 30 \times 0.58 \times \cos 68.69$$
$$+ 70 \times 0.989 \times \cos 61.86$$
$$= 588.01 \text{ W}$$

$$Q = 100 \times 7.07 \times \sin 45 + 50 \times 1.58 \times \sin 51.56 + 30 \times 0.58 \times \sin 68.69$$
$$+ 70 \times 0.989 \times \sin 61.86$$
$$= 639.06 \text{ VAr}$$

$$V_{rms} = \sqrt{{V_{rms1}}^2 + {V_{rms3}}^2 + {V_{rms5}}^2 + {V_{rms7}}^2} = \sqrt{100^2 + 50^2 + 30^2 + 70^2}$$
$$= 135.277 \text{ V}$$

$$I_{rms} = \sqrt{{I_{rms1}}^2 + {I_{rms3}}^2 + {I_{rms5}}^2 + {I_{rms7}}^2}$$
$$= \sqrt{7.07^2 + 1.58^2 + 0.58^2 + 0.989^2} = 7.335 \text{ A}$$

$$S = V_{rms} \times I_{rms} = 992.257 \text{ Var}$$

$$D = \sqrt{S^2 - (P^2 + Q^2)} = 480.02 \text{ VA}$$

$$PF = \frac{P}{S} = \frac{588.01}{992.257} = 0.593$$

$$V_{THD} = \sqrt{\sum_{h=3,5,,7} \frac{V^2}{{V_1}^2}} = \sqrt{\frac{V_3^2 + V_5^2 + V_7^2}{V_1^2}} = \sqrt{\frac{50^2 + 30^2 + 70^2}{100^2}} = 0.911$$

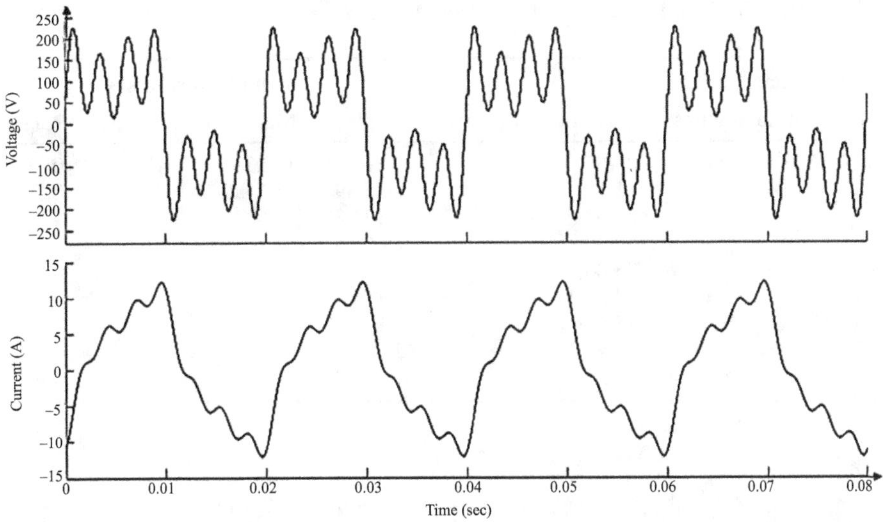

Figure 5.23 The waveforms of voltage and current in Problem 5.13

$$I_{THD}=\sqrt{\sum_{h=3,5,7}\frac{I_h{}^2}{I_1{}^2}}=\sqrt{\frac{I_3^2+I_5^2+I_7^2}{I_1^2}}=\sqrt{\frac{1.58^2+0.58^2+0.989^2}{7.07^2}}=0.2761$$

The waveforms of voltage and current drawn using MATLAB are shown in Figure 5.23.

It is seen that $\frac{V_1}{I_1}\neq\frac{V_3}{I_3}\neq\frac{V_7}{I_7}$ and phase angle between the voltage and current of the hth harmonic, $\alpha_h-\phi_h\neq 0$. Therefore, (i) the waveshapes of the voltage and current are different, (ii) the power factor is not unity and (iii) THD of voltage and current are different.

Problem 5.14 A 3-ϕ diode rectifier with a large value of capacitor filter is connected to a 3-ϕ supply, and the harmonic components measured in voltage and current are given in Table 5.1. Calculate per phase values of P, Q, S, PF and THD. Justify the results.

Solution:

$$V_{rms}=\sqrt{V_1^2+V_3^2}=\sqrt{162.2^2+35.4^2}=166.02\text{ V}$$

$$I_{rms}=\sqrt{I_1^2+I_3^2+I_5^2+I_7^2+I_9^2+I_{11}^2+I_{13}^2+I_{15}^2+I_{17}^2}$$

$$=\sqrt{23.69^2+12.38^2+28.02^2+5.63^2+25.06^2+17.57^2+15.43^2+24.4^2+3.63^2}$$

$$=57.57\text{A}$$

Table 5.1 Harmonic components – Problem 5.14

Current harmonics			Voltage harmonics	
Order of harmonics	**Magnitude (V rms)**	**Angle**	**Magnitude (V rms)**	**Angle**
1	23.69	−11.08	162.2	0
3	12.38	−25.39	35.4	0
5	28.02	−171.3	–	–
7	5.63	48.57	–	–
9	25.06	9.538	–	–
11	17.57	−156	–	–
13	15.43	−172	–	–
15	24.4	23.5	–	–
17	3.63	−35.18	–	–

$$P = V_1 \times I_1 \times \cos\phi_1 + V_3 \times I_3 \times \cos\phi_3$$

$$= 162.2 \times 23.69 \times \cos(11.08) + 35.4 \times 12.38 \times \cos(25.39) = 4166.814 \text{ W}$$

$$Q = V_1 \times I_1 \times \sin\phi_1 + V_3 \times I_3 \times \sin\phi_3$$

$$= 162.2 \times 23.69 \times \sin(11.08) + 35.4 \times 12.38 \times \sin(25.39) = 926.36 \text{ Var}$$

$$S = V_{rms} \times I_{rms} = 166.42 \times 57.57 = 9580.799 \text{ VA}$$

$$PF = \frac{P}{S} = \frac{4166.814}{9580.799} = 0.435$$

$$\%THD_v = \frac{V_3}{V_{rms}} \times 100 = \frac{35.4}{166.42} \times 100 = 21.27\%$$

$$\%THD_i = \frac{\sqrt{12.38^2 + 28.02^2 + 5.63^2 + 25.06^2 + 17.57^2 + 15.43^2 + 24.4^2 + 3.63^2}}{166.42} \times 100$$

$$= 343.2\%$$

Thus, if both voltage and current are distorted, the power factor will be very low, while the THD in both voltage and current will be very high.

Problem 5.15 The distorted single-phase power supply voltage and current are given by the equation below:

$$v(t) = 230\sqrt{2}\sin(\omega t) + 25\sqrt{2}\sin(3\omega t - 30°)\text{V}$$

$$i(t) = 2 + 15\sqrt{2}\sin(\omega t - 30°) + 10\sqrt{2}\sin(3\omega t - 60°)\text{A}$$

Now, based upon these instantaneous voltage and current equations, determine the active power, reactive power, apparent power and the power factor of the distorted power supply.

Solution:
The given supply voltage has a third harmonic component along with the fundamental, whereas the current has a DC component along with a third harmonic and fundamental component.

(a) Calculating active power:

$$P = V_{dc}I_{dc} + \sum_{n=1}^{\infty} V_n I_n \cos\phi_n$$

$$P = V_{dc}I_{dc} + V_1 I_1 \cos\phi_1 + V_3 I_3 \cos\phi_3$$

$$\phi_1 = \phi_3 = 30^\circ$$

Hence,

$$P = 0 \times 2 + 230 \times 15 \times \cos 30^\circ + 25 \times 10 \cos 30^\circ$$

$$P = 0 + 2987.79 + 216.51$$

$$P = 3204.3 \text{ Watts}$$

(b) Calculating reactive power:

$$Q = \sum_{n=1}^{\infty} V_n I_n \sin\phi_n$$

$$Q = V_1 I_1 \sin\phi_1 + V_3 I_3 \sin\phi_3$$

Thus,

$$Q = 230 \times 15 \times \sin 30^\circ + 25 \times 10 \times \sin 30^\circ$$

$$Q = 1725 + 125$$

$$Q = 1850 \text{ VAR}$$

(c) Calculating apparent power

$$S = V \times I$$

$$S = \left(\sqrt{V_{dc}^2 + V_1^2 + V_3^2}\right) \times \left(\sqrt{I_{dc}^2 + I_1^2 + I_3^2}\right)$$

$$S = \left(\sqrt{0^2 + 230^2 + 25^2}\right) \times \left(\sqrt{2^2 + 15^2 + 10^2}\right)$$

$$S = 231.35 \times 18.14$$

$$S = 4196.69 \text{ VA}$$

(d) Calculating power factor

The power factor now can be calculated by simply taking the ratio of active power (P) and apparent power (S), as apparent power now includes reactive as well as distortion/harmonic power. Therefore, the power factor (pf) is given by:

$$pf = \frac{P}{S} = \frac{3204.3}{4196.69} = 0.7635 \text{ (lagging)}$$

References

[1] W. Shepherd, and P. Zand, *Energy Flow and Power Factor in Non-Sinusoidal Circuits*, Cambridge University Press, 1979.

[2] N. Mohan, T. M. Undeland, and W. P. Robbins (eds.), *Power Electronics: Converters, Applications and Design*, Second Edition, 1995, John Wiley & Sons (Asia) Pte Ltd.

[3] L. X. Zhou, Z. D. Yin, X. N. Xiao, Z. Q. Wang, and L. Zheng, "Study on power factor through the similarity of waveform", *2008 3rd IEEE Conference on Industrial Electronics and Applications*. Singapore, 2008, pp. 1534–1537.

[4] Report on Power Quality of Electricity Supply to Consumers, August 2018, Forum of Regulators, Central Electricity Regulatory Commission, India. https://forumofregulators.gov.in/Data/Reports/Power07.pdf.

[5] IS 16071: 2013 IEC/TR 62001: 2003, Indian Standard high-voltage direct current (HVDC) systems — guidebook to the specification and design evaluation of AC Filters. Bureau of Indian Standards, 2013.

[6] A. I. Maswood, and F. Liu, "A Unity Power Factor Front-End Rectifier with hysteresis Current Control", *IEEE Transactions on Energy Conversion*, Vol. 21, No. 1, 2006, pp. 69–76.

[7] D. Stepins, "An Improved Control Technique of Switching-Frequency-Modulated Power Factor Correctors for Low THD and High Power Factor", *IEEE Transactions on Power Electronics*, Vol. 31, No. 7, 2016, pp. 5201–5214.

[8] Q. Wu, L. Wu, J. Zhou, Y. Li, Z. Di, and Z. Zhu, "A Novel Multiplier Circuit for PFC Controllers With Low Total Harmonic Distortion and High Power Factor", *IEEE Transactions on Power Electronics*, Vol. 40, No. 11, 2025, pp. 16105–16110.

[9] S. George, and V. Agarwal, "A Novel, DSP Based Algorithm for Optimizing the Harmonics and Reactive Power Under Non-Sinusoidal Supply Voltage Conditions", *IEEE Transactions on Power Delivery*, Vol. 20, No. 4, 2005, pp. 2526–2534.

Chapter 6

Passive techniques for control of voltage variation, reactive power and compensation for harmonics

Goals of the chapter

Reliable operation of the electrical system necessitates control of voltage and mitigation of reactive power and harmonics, apart from managing real power for doing the work at any instant.

This chapter discusses another important necessity of today's power system: control of voltage variation, reactive power and mitigation of harmonics. After going through the chapter, the reader shall get a broad idea about the contribution of the transmission system to reactive power and its effects on voltage variation, another vital issue of power quality. It also details various static compensation techniques for reactive power and a conventional method for the elimination of harmonics from the system using passive filters. At the end, the need for new techniques for harmonics elimination for today's power generation-utilization scenario is emphasized. A few problems and solutions are also included to make the concept easy to understand.

6.1 Introduction

An electrical power system instantaneously manages real power for doing the useful work and reactive power for facilitating it. As discussed in previous chapters, today, electrical energy is generated from various sources, from conventional to renewable, and transmitted over long distances on both overhead and underground cables of HVDC and HVAC to the various utilization equipment of different natures spread over the region. In this process, various types of electrical equipment, from conventional machines like generators, motors, transformers, overhead transmission lines, cables, heating and lighting equipment to converter-based power supplies, UPS and sensitive electronic equipment for telemetry, PLC, communication, information technology, SCADA, robotics, aerospace, industrial automation, healthcare, etc., are used. In this energy transformation process, reactive and distortion volt-amperes are produced by the fundamental and its multi-frequency components. These adversely affect the system and may cause disruption of services in a widespread area, resulting in substantial economic losses.

In the transmission system, reactive power directly affects the power system and voltage of the network; if not managed within limits, it may lead to large disturbances or even the collapse of the grid. The European Network of Transmission System Operators for Electricity, in its report titled "Technical background and recommendations for defense plans in the continental Europe synchronous area", indicates inadequate reactive power compensation as one of the prime reasons for large disturbances across the world [1].

The Indian grid also experienced major disturbances on July 30, 2012, and July 31, 2012. The report of the enquiry committee on this incident concluded "Inadequate dynamic reactive reserves" as one of the reasons for this large disturbance in the power grid. It is also recommended to provide dynamic reactive reserves at strategic locations in the power grid to take care of contingencies like high line loading, large load throw-offs, or disconnection and phase-earth or phase-phase faults [2].

Also, during the past few decades, voltage level and length of transmission lines have gone as high as 1200 kV AC. This has led to higher grid voltage during low line loading periods due to transmission line capacitance, necessitating isolation of lines to avoid overfluxing of the transformer and dielectric failure of equipment connected to the grid. It also led to lower voltages during peak inductive line loading. Voltage variation beyond the limit causes isolation of the affected portion of the system by the electrical protective scheme for its protection and isolation to avoid possible large-area disruption.

Traditionally, reactive power has been controlled by various methods such as excitation control of synchronous machines, changing transformer voltage ratio by online or offline tap changers, installation of reactors (inductance) during the period of low transmission line loading, and placement of shunt capacitors near to inductive loads or at distribution level to cater to the needs of groups of inductive loads. However, the increasing nonlinear impedances in recent times have changed the scenario of the system with the presence of multiple frequencies. We have seen that under such conditions, capacitors cannot neutralize distortion volt-ampere (D) produced by the unlike frequency components. Also, harmonics and inter-harmonics are harmful to the system and its equipment. Hence, we have to adhere to the limits of voltage and current harmonics imposed on them from time to time by various international standards and regulations. A few of the standards are detailed in chapter 3, Table 3.1. Hence, it is mandatory to control reactive power and voltage variation and limit harmonics in the system, as it affects both the system and its equipment.

We shall discuss solutions to both the issues of (i) reactive power and voltage variation, (ii) mitigation of harmonics using traditional techniques in this chapter. As reactive power is a major concern for variation of voltage in many countries from its specified value, the mechanism of its generation is briefly presented in the following section.

6.2 Inductive and capacitive reactive power

As we know, when current I_L flows through any conductor, a magnetic field or flux, $\Phi = LI$, is set up depending on its inductance L and I_L. This inductive current lags the supply voltage by a phase angle depending upon the reactance X_L given by ωL of the circuit, as shown in Figure 6.1.

Electrical equipment like induction motors, induction generators, transformers, etc., draws lagging current and thereby produces inductive reactive power (Q). Similarly, an energized conductor with a dielectric material between the conductors and/or ground produces an electric charge, $q = CV$, where C is the capacitance and V is potential difference across the dielectric material. This charge q result in an electric field in the dielectric medium. When the voltage is variable over a cycle, the stored charge and current $I_c = C\frac{dv}{dt}$ return over the next half cycle to the source. This capacitive current i_c lead the voltage by a phase angle, depending upon the reactance of X_c given by the $\frac{1}{\omega C}$ circuit. Figure 6.2 shows the representation of a capacitor and phasor diagram of its current and voltage. Capacitors, cables, overhead transmission lines, etc., are of this type.

Thus, the system components or its equipment produce inductive or capacitive reactive power (Q) that is measured in volt-ampere reactive (VAR).

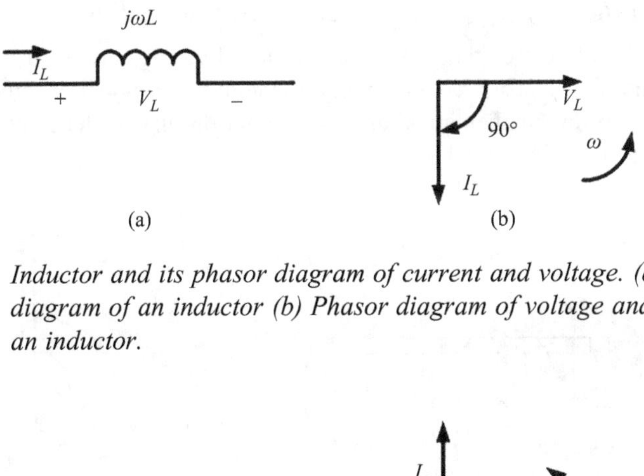

Figure 6.1 Inductor and its phasor diagram of current and voltage. (a) Line diagram of an inductor (b) Phasor diagram of voltage and current in an inductor.

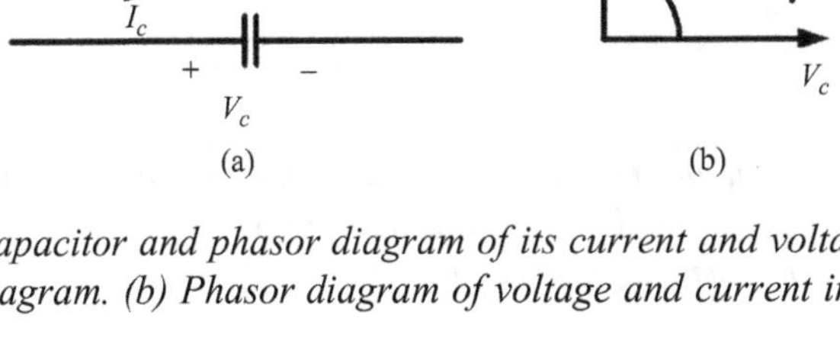

Figure 6.2 Capacitor and phasor diagram of its current and voltage. (a) Line diagram. (b) Phasor diagram of voltage and current in a capacitor.

6.2.1 Reactive power generation on the power transmission line

As detailed in the previous section, when a 3-ϕ overhead transmission line is connected to the supply source, the potential difference between the R-Y-B lines and/or ground creates a capacitance with the dielectric medium. Also, when the load is connected at the receiving end of the transmission line, current flows through the line, generating a magnetic field around it. Figure 6.3(a) and (b) show the electric field and magnetic field produced by a transmission line carrying a current. A similar condition exists in the case of an insulated cable.

Alternating voltage or current in lines creates an alternating electric field and magnetic field throughout the length of the line. A variable electric field results in charging current or leakage current between lines and/or ground, thus generating reactive power on the line. Similarly, a variable magnetic field around the conductor results in an induced voltage on the line or absorbs reactive power from the line. This is distributed inductance and shunt capacitance are represented as a lumped series inductor (L) and shunt capacitor (C) while modelling the line. These result in series impedance (Z) measured in ohms (Ω), represented by reactance (x) and resistance (r) along with shunt admittance (Y) measured in mho (℧), represented by conductance (g) and susceptance (b), respectively. Both line capacitance and series inductance result in variation of line voltage and thus impair power quality.

6.2.2 Voltage variation on power transmission line

To analyse the effects of these parameters on voltage variations, a model of the transmission line is considered. Generally, the nominal π model or T model is used for transmission line mathematical analysis. In our discussion, let us consider the π model as shown in Figure 6.4.

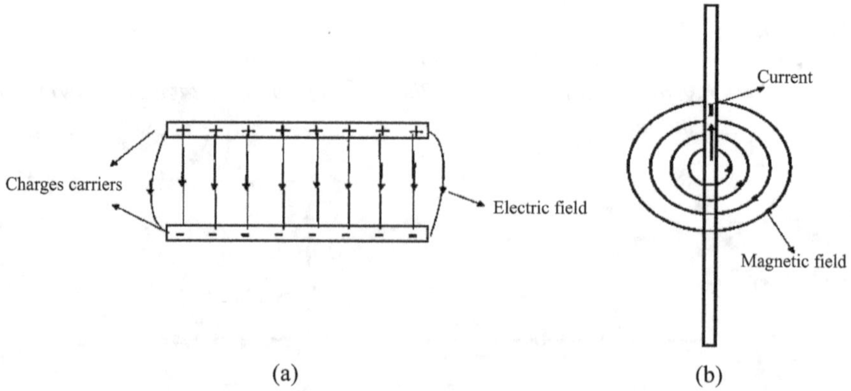

Figure 6.3 Electric field and magnetic field around a transmission line. (a) Electric field produced a line. (b) Magnetic field around the current-carrying conductor.

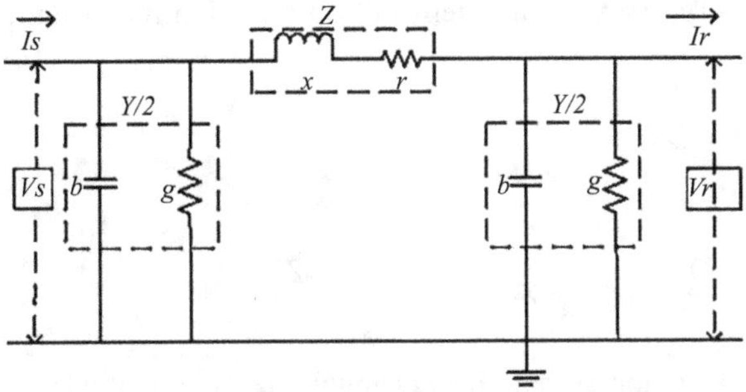

Figure 6.4 Nominal π model of a transmission line

In the π model of the line [3], shunt admittance Y is divided into two halves, with one half ($Y/2$) at the sending end and the other half at the receiving end. V_s and V_r are the sending-end and at the receiving-end voltages. I_s and I_r are corresponding currents.

The expression for the sending end voltage and current in terms of the receiving end voltages, current, and line parameters can be derived as detailed below.

At the receiving end:

$$\text{Current in shunt capacitance} = \frac{V_r Y}{2} \tag{6.1}$$

$$\text{and current in } Z = I_r + \frac{V_r Y}{2} \tag{6.2}$$

Thus, at the sending end, voltage is:

$$V_s = \left(\frac{V_r \cdot Y}{2} + Ir\right) Z + Vr \tag{6.3}$$

or

$$V_s = \left(\frac{Z \cdot Y}{2} + 1\right) Vr + Z \cdot Ir \tag{6.4}$$

And sending-end current is given by:

$$I_s = \frac{V_s \cdot Y}{2} + \frac{V_r \cdot Y}{2} + Ir \tag{6.5}$$

$$I_s = \left(\frac{Z \cdot Y}{4} + 1\right) Y \cdot Vr + \left(\frac{Z \cdot Y}{2} + 1\right) Ir \tag{6.6}$$

Thus, V_s and I_s are written in terms of transmission line parameters as:

$$V_s = AVr + BIr \tag{6.7}$$

$$I_s = CVr + DIr \tag{6.8}$$

where,

$$A = D = \frac{ZY}{2} + 1; B = Z \text{ and } C = \left(1 + \frac{ZY}{4}\right)Y$$

Power transmission system, being capital intensive, is implemented considering the anticipated power demand over a long period. With the line current and series reactance voltage drops, and with shunt capacitance, voltage rises along the length of the line. That is, line inductance consumes reactive power in proportion to the square of the load current ($I^2\omega L$) and capacitance generates or produces reactive power in proportion to the square of the line voltage ($V^2\omega C$). Thus, during off-peak demand, $V^2\omega C$ may exceed $I^2\omega L$, leading to a rise in voltage. That is to say, under a leading power factor, the receiving end voltage (V_r) is more than the sending end voltage (V_s). This phenomenon is called the *Ferranti effect* [3]. This is especially the case on long, lightly loaded extra-high-voltage lines. This may often necessitate off-loading of the line from the grid to increase the loading on other lines and to control the overvoltage within the rated voltage. Another issue associated with off-peak demand is if a circuit breaker (CB) opens at a far end of an energized line, CB end of the line may experience overvoltage, and it will be difficult to synchronize back the line or add a new generator to the system.

Similarly, while the line is loaded heavily (generally loads are inductive) due to seasonal or intraday variation of load, reactive power absorbed by the line inductance may exceed the generation by the line capacitance. Hence, the voltage reduces due to line inductance from the sending end along the line. This is especially the case where inductive reactance (X_l) dominates over capacitive reactance (X_c). Thus, with the varying power flow, high- or extra-high-voltage lines are prone to voltage instability. Figure 6.5(a) shows a single-line diagram of a radial line neglecting its resistance. Figure 6.5(b) and (c) shows the vector diagrams of the line voltages under lagging and leading conditions of load current. As discussed, under lagging conditions of current (i), the receiving end voltage (V_r) is less than the sending end voltage (V_s) by the voltage drop due to the line inductance (X_l). It is further reduced from the generator voltage (V_g) by the generator inductance (X_g) voltage drop. Similarly, under the leading condition, the resulting receiving-end voltage is more than the sending-end voltage.

The issues arising out of the transmission line distributed inductance and capacitance, along with load-side reactive power, can be summarized as follows:

1. Reactive power absorbed or generated by the line inductance or capacitance under varying load conditions affects the voltage profile – dip or rise along the length of the line.

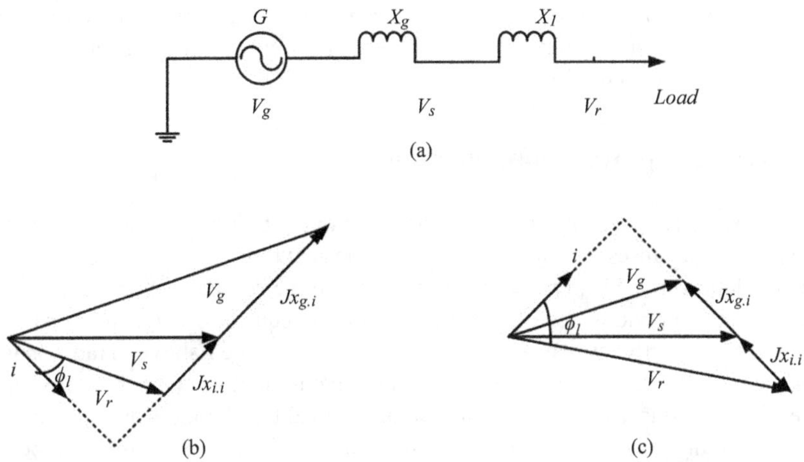

Figure 6.5 Radial transmission line. (a) Single-line diagram, (b) Vector diagram under pf lag, (c) Vector diagram under pf lead.

2. Variation of voltage beyond the permissible synchronizing limit prevents closing of the circuit breaker of a new line or adding a generating unit to the system.
3. A rise in voltage affects the performance of the equipment connected and increases reactive power generation of the line. It adversely affects the dielectric strength of insulation material, overfluxing of transformers and equipment breakdown.
4. Change of operating region of synchronous generators from lagging zone to capacitive-leading zone results in a reduction in generator excitation that may cause generator instability.
5. Electrical protective relays for undervoltage and overvoltage are designed to respond to voltage variations. Operation of these relays may cause isolation of lines, load shedding or large-scale system disturbances.
6. It affects the power factor of the system. It may go down to lagging during heavy load and to leading during light load conditions.
7. Power handling capability of the line may be compromised due to the additional line reactive power, in addition to the active and reactive power requirement of utilization equipment.
8. The variation of transmission line voltage is transferred by the voltage ratio of the transformers to low-voltage utilization equipment. This variation in voltage adversely affects equipment voltage withstand capability and enhances its ageing or requires frequent maintenance. Thus, a major challenge is the power quality issue.

Thus, due to the distributed capacitance and inductance of the transmission line, with the varying load current, the system voltage varies over a wide range and impairs power quality. Thus, we need to compensate for the reactive volt-ampere produced by the various electrical equipment, as well as the transmission line. These are briefly covered in the following sections.

6.3 Reactive power compensation

As the flow of reactive current in addition to the load current, apart from power quality issues, it causes additional power (I^2R) loss in transmission and distribution systems. Hence, it needs to be compensated near to the source of its generation. As detailed in the previous section, compensation is required for (i) inductive and (ii) capacitive reactive power to control various issues cited above. Traditionally, synchronous generators, over-excited synchronous motors (synchronous condensers), power transformers with on-load or off-load tap changers on transmission level, and shunt capacitors mainly on power distribution level for reactive power management or voltage control. Additionally, harmonic filters are also closely linked to the reactive power management. Based on system characteristics, the type of compensator for a particular location can be selected. However, with the development of electronics, notably high-power switching devices and controllers, other static and dynamic compensation techniques have been developed [3–5]. The following section details commonly used static var (SVC) compensation techniques. As the power transmission system itself contributes to reactive power and voltage variation, we shall start with ways of compensation in the transmission line.

6.3.1 Shunt compensation on the transmission system

The function of the shunt compensator is to provide the desired amount of reactive power absorbed or generated by the system components. It is represented as a current source with stored energy in an electric field (capacitors) or magnetic field (inductors) connected between the line and ground potential. Shunt capacitors are used for compensating undervoltage and inductors/reactors for compensating overvoltage conditions. Figure 6.6 depicts a single-line diagram of a transmission line showing distributed line inductance (d_{xl}), shunt line capacitor (d_{cl}), shunt compensator supplying reactive volt-ampere (VAR), and receiving end voltages (V_r).

In the case of a large interconnected multi-machine system, where sending and receiving end voltages are approximately fixed because of its stiffness, voltage

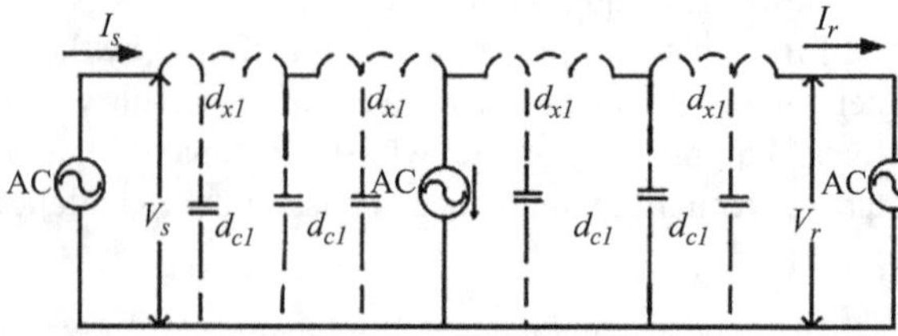

Figure 6.6 Single-line diagram of a transmission line with shunt compensation

variation (sag or swell) occurs along the line, peaking towards the middle [4]. Hence the ideal location of the compensator would be at the midpoint of the line where $V_s = V_m = V_r$. If the length of the line is longer, multiple compensators are required along the line at intermediate places so that an ideal profile can be achieved with maximum possible transfer of power. So, depending on power flow and length of the transmission line, the capacity of shunt reactors or shunt capacitors is decided. We shall consider a stiff system with midpoint compensation as shown in Figure 6.7.

Let the reactance of the line be X_l and the angle between the sending-end voltage V_s and receiving-end voltage V_r be δ. For analysis, let us consider a few points on the line, say sending end s, point at $\frac{X_l}{4}$, midpoint m at $\frac{X_l}{2}$, point at $\frac{3X_l}{4}$ and at the receiving end r at X_l. As the line is compensated fully, the magnitude of voltage and current at intermediate points (say, the midpoint) on the line is equal to the sending end, and receiving end.

$$|V_s| = |V_r| = |V_m| = V$$

$$I_s = I_r = I_m = I$$

Also, let the line be lossless. Thus, power and receiving-end power are equal. From the vector diagram shown in Figure 6.8,

$$V_{sm} = V\cos\frac{\delta}{4} \tag{6.9}$$

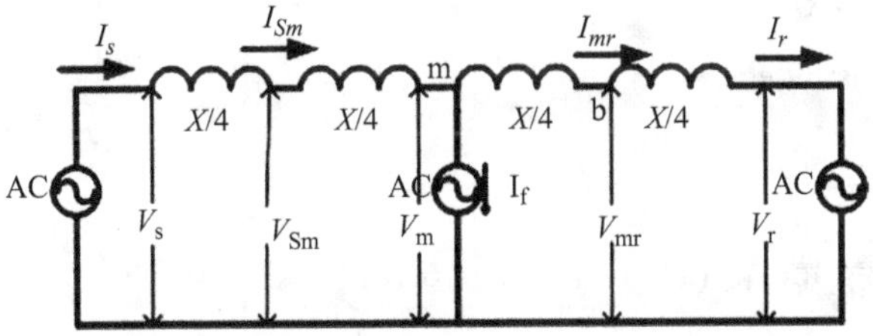

Figure 6.7 Single-line diagram of a transmission line with midpoint compensation

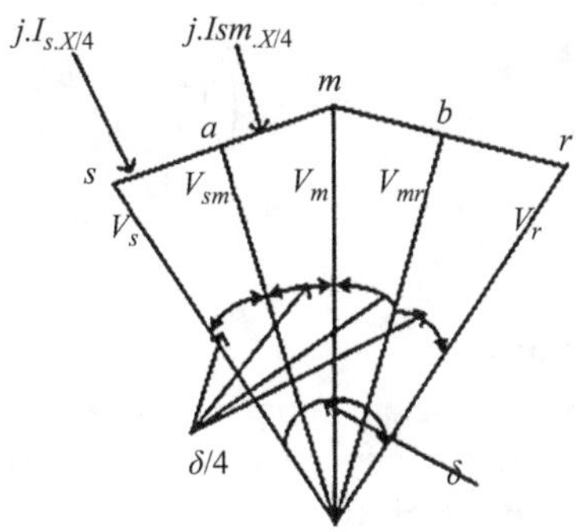

Figure 6.8 Phasor diagram of a transmission line with shunt compensation

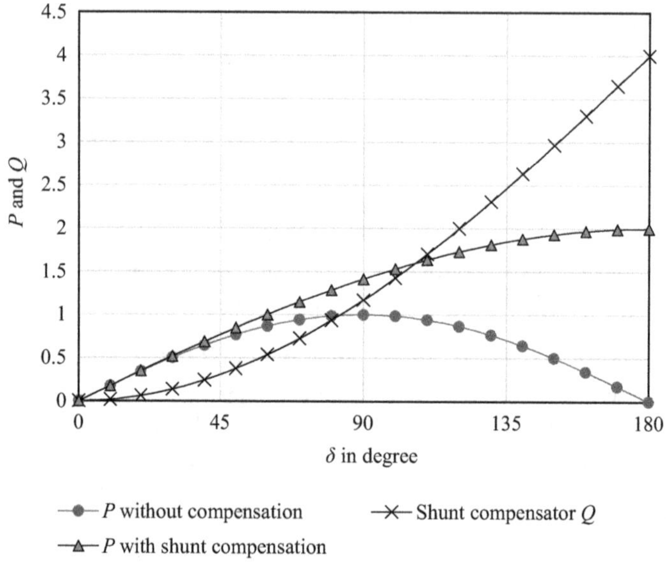

Figure 6.9 Waveforms of active and reactive power with mid-point compensation

and I_{sm}

$$I_{sm} = I = \frac{4V}{X_l}\sin\frac{\delta}{4} \tag{6.10}$$

Power transmitted,

$$P = V_{sm}I_{sm} = V_{mr}I_{mr} = VI\cos\frac{\delta}{4} \tag{6.11}$$

Substituting for I_{sm} (as given in eqn 6.10)

$$P = V\frac{4V}{X_l}\sin\frac{\delta}{4}\cos\frac{\delta}{4} = \frac{2V^2}{X_l}\sin\frac{\delta}{2} \tag{6.12}$$

Similarly,

$$Q = VI\sin\frac{\delta}{4} = \frac{4V^2}{X_l}\sin\frac{\delta}{4}\sin\frac{\delta}{4} = \frac{2V^2}{X_l}\left(1 - \cos\frac{\delta}{2}\right) \tag{6.13}$$

Using (6.11)–(6.13), waveforms of active and reactive power with mid-point compensation are drawn as Figure 6.9.

6.3.1.1 Shunt capacitors

Capacitor is one of the most efficient components used in a variety of configurations, along with other electrical components such as resistance and/or inductor, to solve power quality issues [5,6]. It is widely used in both 1-ϕ and 3-ϕ

configurations and from low-voltage utilization systems to high-voltage transmission systems. It consists of two electrodes separated by dielectric material (paper, plastic film, etc.) impregnated with oil or as a dry type (gas-filled, resin-filled, etc.). Traditionally, shunt capacitors are installed for various applications such as 1-ϕ induction motor start and run, reactive power compensation, power factor correction, harmonic mitigation, surge or transient suppressors, etc. For major loads, capacitors are installed near the loads, such as induction motors for pumps, blowers, etc. For larger loads such as industries, arc and induction furnaces, pump houses, etc., capacitors are placed at the distribution substations, catering to a group of loads. Railway traction loads are different from normal industrial ones. They are 1-ϕ loads and vary rapidly from no load to full load as the locomotive approaches the section of the substation that feeds the traction load. Locomotives fitted with controlled rectifiers with smoothing reactors for feeding the DC traction motors, apart from being unbalanced, introduce harmonics of 3rd, 5th, 7th, 9th, and 11th order of varying magnitude depending on the switching system. As capacitors are sensitive to frequency $\left(X_c = \dfrac{1}{2\pi fc}\right)$, due to harmonics, capacitive reactance reduces with an increase in f and, hence a large current flows through them. Also, as discussed in chapter 5, capacitors may form a resonating circuit with other frequency components in the power system. Hence, its effectiveness needs to be checked with the anticipated harmonic spectrum.

The following are the advantages of using shunt capacitors for reactive power compensation:

1. Provide compensation to the reactive power consumed by the line and inductive load, and thus improve the power factor of the system.
2. Reduce voltage drops across the line and thereby improve the voltage profile.
3. Reduce transmission and distribution losses and hence the temperature rise of equipment and components, also with a reduction in unit energy cost.
4. Enhance transmission line and transformer capacity, and hence the system can serve more active loads with the same reactive power.

When the system voltage is low, the shunt capacitor produces less reactive power, and when the voltage is high, it produces more reactive power as it generates reactive power in proportion to the square of the voltage ($V^2\omega C$). Also, overcompensation may lead to a rise in terminal voltage and a reduction in excitation power requirement at the generator side, leading to system instability. Thus, only adequate capacity for compensation needs to be provided. Accordingly, fixed- and switched-type capacitor banks are to be provided. Fixed capacitors (FC) are simple and low cost and are widely used for base load or steady load applications in low- and medium-voltage applications. However, it may lead to overcompensation and possibly operate at a leading power factor, resulting in overvoltage conditions. Manual switching in of the capacitor bank will result in transients due to capacitive inrush current. Hence, either a circuit breaker capable of handling inrush current or a thyristor as a switch is used. In thyristor switched capacitors (*TSC*), thyristors are used to switch ON and OFF the capacitor as required by the system, but not on a continuous basis. Hence, it supplies either zero or maximum, and an inject cannot

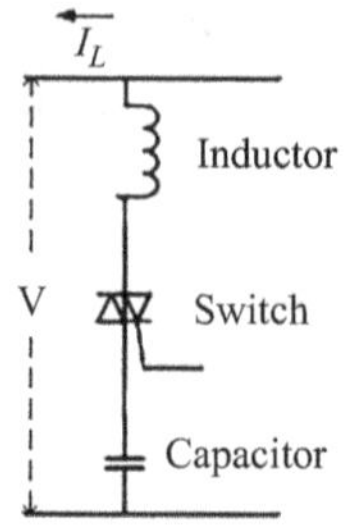

Figure 6.10 Block diagram of thyristor-switched capacitors

have a variable magnitude of reactive power. Figure 6.10 shows the block diagram of a *TSC* scheme. The circuit consists of a capacitor, a bidirectional thyristor, and an inductor. The inductor is used to limit capacitive surge current during turn-on and also to avoid resonance with system impedance.

6.3.1.2 Shunt reactor

As discussed in the preceding section, long extra high voltage (EHV- 66 kV and above) transmission lines often create overvoltage during light load or low power flow conditions due to higher capacitive generation than the inductive consumption of the line [5]. The severity of the problem increases with system voltage and length of the line. Hence, considering voltage stability, line loading needs to be kept below the natural or surge impedance loading of the line. This puts a restriction on the loading of the line. The installation of inductive equipment that stores energy in the form of a magnetic field (shunt reactor) for absorbing reactive power offers a satisfactory solution to these problems. Shunt reactors are like transformers used in transmission lines. However, their function is to absorb the reactive power generated by the line capacitance under no-load conditions, and hence they are different from transformers in design and construction. Its reactance must match the capacitance to avoid voltage fluctuations. To achieve higher impedance and a linear magnetization (*B-H*) curve to limit excessive harmonic current generation, the magnetic core is made with a gapped core (air gap with cold-rolled-gain-oriented or non-oriented laminations) or without any magnetic material in the core. It may be single phase or three-phase, oil-filled, and appear like a power transformer. It is normally in service or switched on/off type. Figure 6.11 shows a line diagram of commonly used reactor topologies. For transmission lines, reactors with an additional reactor called a neutral grounding reactor (NGR) are also connected between neutral and ground to increase the zero-sequence impedance. This facilitates a reduction in single-phase ground fault current and faster clearance of the fault by the interrupting circuit breaker. Thus, the arc at the fault location is extinguished for auto reclosing of the circuit breaker.

Keeping in mind the importance of quality and reliability, India has standardized the technical parameters and rating of shunt reactors. Standardized ratings of reactors above 245 kV class [7] are tabulated in Table 6.1.

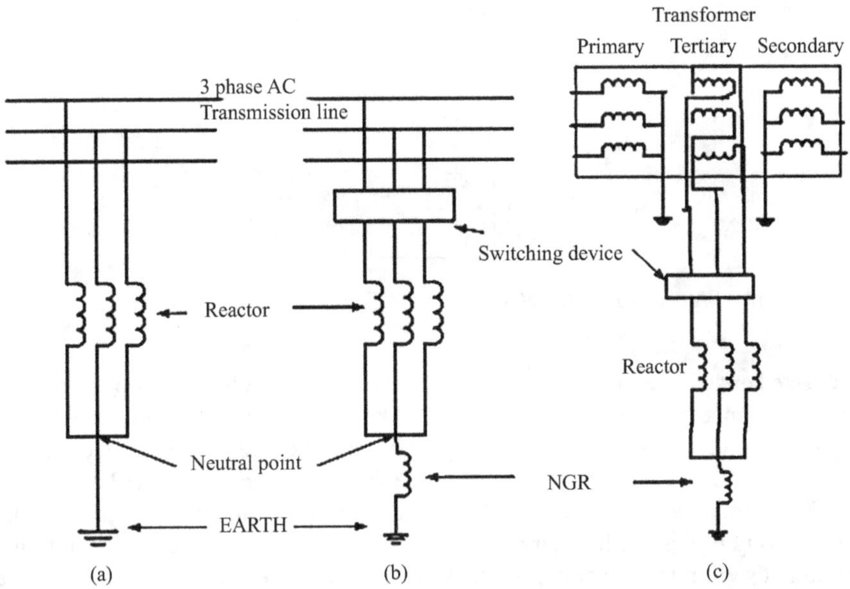

Figure 6.11 Line diagram of reactors. (a) Bus reactor, (b) line reactor, (c) Transformer – tertiary winding reactor.

Table 6.1 Standardized ratings for shunt reactors above 245 kV

Rated voltage (kV)	Capacity (MVA)	Type
$765/\sqrt{3}$	110/80	Single phase
420	125/80/63/50	Three phase
245	50/25	Three phase

The following is achieved using a shunt reactor:

1. Capacitive generation is balanced with an inductive power consumption of the line and shunt reactor.
2. It controls the voltage profile of the line under all operating conditions, such as normal and abnormal service conditions, viz. light load or no-load condition. Thus, it offers a solution to the receiving-end problem of no-load overvoltage, the Ferranti Effect.
3. Reactors close to generating stations enable increased generator excitation, thereby enhancing the stability of the system.
4. They enhanced the power transfer capacity of the line with respect to its thermal limit, leading to better utilization of the line.

Use of shunt reactors may result in reduced voltage under heavy load conditions and increased system losses. Thus, it necessitates a choice between a permanently

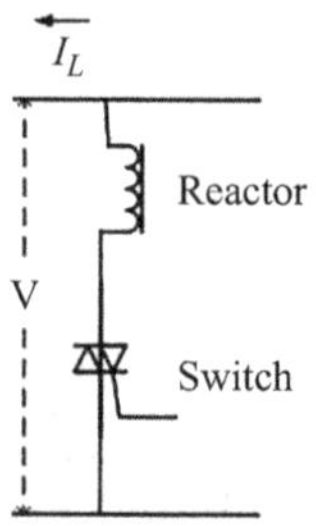

Figure 6.12 Block diagram of TCR with shunt reactor

connected or switched-in or switched-off type reactor for voltage control. In case of fixed reactance requirement, thyristor-switched reactor (*TSR*) with fixed firing angle, and for variable reactance, thyristor-controlled reactor (*TCR*) offers an efficient solution. A block diagram of the *TCR* is shown in Figure 6.12.

Thus, shunt compensation is effective in controlling the voltage profile on the line, making it close to line surge impedance loading limits, which helps in transmitting power up to its thermal limit. However, it is not effective in enhancing the power transfer limit set by the line voltage and angle (δ) between the sending end (V_s) and receiving (V_r) end. Series compensation described in the following section briefly explains ways to enhance the power transfer limit.

6.3.2 Series compensation on the transmission system

As detailed earlier, real power (P) is transferred from the sending end to the receiving end of a transmission line with line reactance (X_l) is given by:

$$P = \frac{V_s V_r}{X_l} \sin \delta \tag{6.14}$$

and the reactive power at the receiving end Q_r and sending end Q_s are given by:

$$Q_r = V_r \frac{(V_r - V_s \cos \delta)}{X_l} \tag{6.15}$$

$$Q_s = V_s \frac{(V_s - V_r \cos \delta)}{X_l} \tag{6.16}$$

Thus, by modification of the line reactance X_l for a given magnitude of voltages at the ends of the transmission line, transmittable real and reactive power can be altered. This modification can be done by introducing a series of passive elements of inductive ($\Delta X > 0$) or capacitive ($\Delta X < 0$) on the transmission line. Changing the value by ΔX, the effective impedance of the line is altered. Thus, a series capacitor reduces the impedance ($X_l - X_c$) and the reactor increases the impedance ($X_l + X_r$) of the line. Representation of a series compensation is shown in Figure 6.13. As the compensation unit is in series with the transmission line, current through it will be the same as load current, and the voltage drop across it

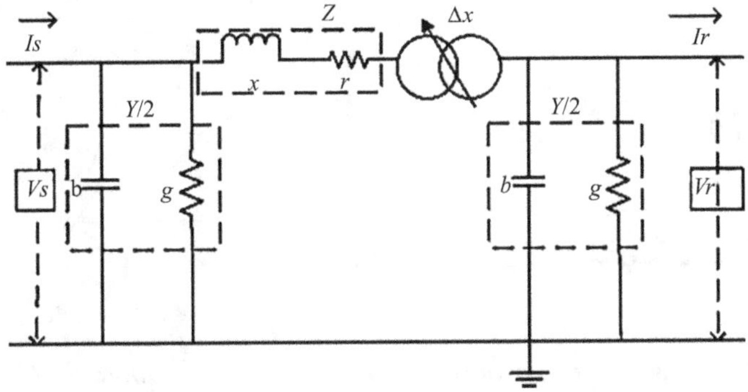

Figure 6.13 Equivalent circuit of a transmission line with series compensation

depends on the value of the elements connected as a compensator. Based on the voltage drop across the compensator, the net voltage drop across the transmission line is controlled, and hence voltage at the receiving end of the line. Thus, it can also be represented as a variable voltage source that varies with load current.

In an interconnected system, there may be different transmission routes or power transmission lines connecting the generation bus to the load bus. When there is a change in demand at the load bus, the current through the line depends on the voltage at the two ends and the reactance between the buses. If these points are connected by double circuit lines of the same length, power will be shared equally. If lines are of different lengths, the shorter line will carry more power compared to the longer line due to the impedance difference. This may overload the short line and underload the longer line. Thus, a series inductor is used to alter the power flow through an interconnected parallel line. To control power flow, the inductor is connected in series so that the net reactance of the line increases, and hence the power flow is altered.

6.3.2.1 Series capacitor compensation

As discussed in the previous section, by installing a series capacitor on the line, the reactance of the line X_l gets reduced by an amount equal to the series capacitive reactance (X_c) value. In other words, the line gets electrically shorter. Thus, the transmittable power gets modified to,

$$P = \frac{V_s V_r}{X_l - X_c} \sin\delta \tag{6.17}$$

where V_s is the voltage at the sending end, V_r at the receiving end.

Consider the model of transmission system as shown in Figure 6.14(a) with X_l and X_c split into two identical sections. Also, let the magnitude of $V_s = V = V_r$ and the degree of compensation be $k = \frac{X_c}{X_l}$. Thus, the reactance of the line after series capacitive reactive compensation,

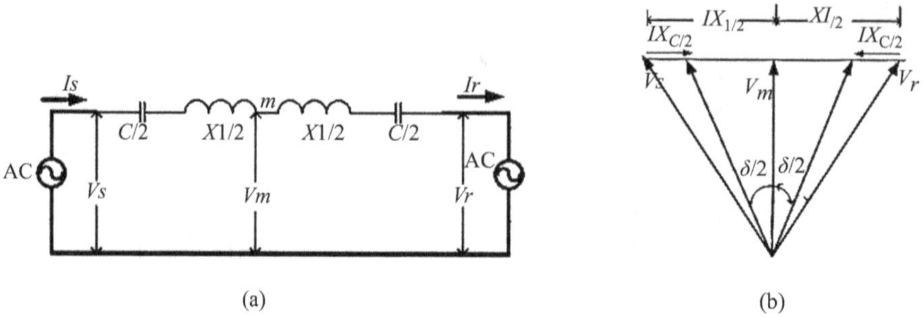

Figure 6.14 Power system with series capacitive compensation. (a) Model of the system with series capacitive compensation and (b) vector diagram.

$$X_l - X_c = X_l - kX_l = X_l(1-k) \tag{6.18}$$

From the vector diagram shown in Figure 6.14(b),

$$I = \frac{2V}{(X_l - X_c)} \sin\frac{\delta}{2} \tag{6.19}$$

since $V_m = V\cos\frac{\delta}{2}$; real power transferred,

$$P = V_m I = \frac{2V^2}{(1-k)X_l}\sin\frac{\delta}{2}\cos\frac{\delta}{2} = \frac{2V^2}{2(1-k)X_l}\sin\delta \tag{6.20}$$

$$P = \frac{V^2}{(1-k)X_l}\sin\delta \tag{6.21}$$

and capacitive reactive power,

$$Q_c = I^2 X_c = \left(\frac{2V}{(X_l - X_c)\sin\frac{\delta}{2}}\right)^2 X_c \tag{6.22}$$

$$Q_c = \frac{2V^2 k}{X_l(1-k)^2}(1-\cos\delta) \tag{6.23}$$

Thus, with a series capacitor, the transmittable power over the existing line increases. This is an added advantage of transmission of power in urban areas where space for the line corridor is limited. Figure 6.15 shows the elementary scheme of a thyristor-switched series capacitor (*TSSC*). In the actual scheme, depending on the compensation, a number of capacitor units are connected in series. In case of frequently varying compensation, a thyristor-controlled series capacitor (*TCSC*) is often employed.

Figure 6.16 shows the plot of active power P and reactive power Q transfer for a compensated and uncompensated line. It can be noted from the waveforms that with compensation, both active power and reactive power transfer increase. As the

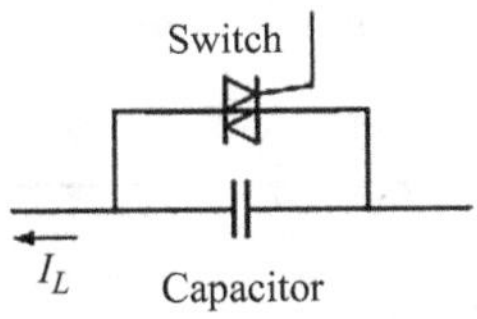

Figure 6.15 Basic arrangement of a thyristor-switched series capacitor

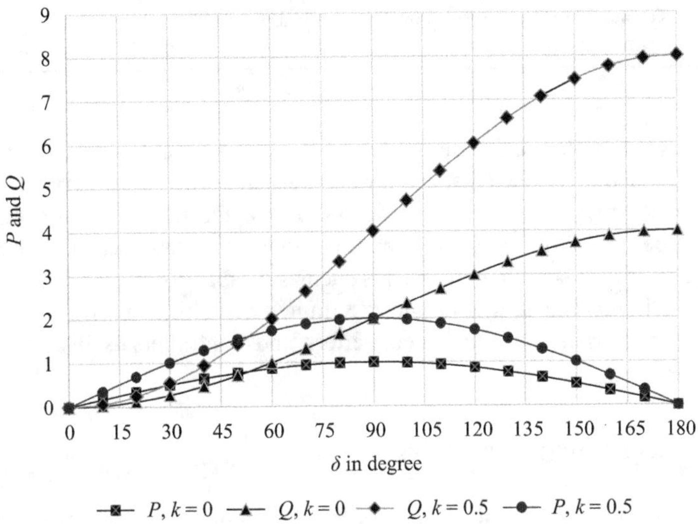

Figure 6.16 Waveform of active and reactive power with series capacitor compensation

compensator is connected in series with the line, the current rating of the capacitor must be equal to that of the transmission line. Also, its reactance shall match the system requirement. At light load, the series capacitor has little or no effect, and compensation increases with load current inductance. At higher load current, the compensating reactive power (I^2X_c) may equal the reactive power requirement (I^2X_l) of the line. A few of the 400 kV transmission lines that are compensated with fixed series capacitors in India [8] are tabulated in Table 6.2.

6.4 Variable reactive volt-ampere generators

As discussed, reactive power absorbed or generated by an inductance or capacitance is fixed. However, in an AC system, we need reactive compensation based on its demand. Mechanical switches to connect or disconnect an appropriate amount of inductance or capacitance are one of the early options for variable reactive power generation. As detailed earlier, with the advent of line-commuted thyristors,

Table 6.2 Few 400 kV transmission lines compensated with fixed series capacitors

Sl. No.	400 kV transmission line	Compensation
1	Purnea–Muzaffarpur (Bihar)	40% fixed, +15%/−5% dynamic
2	Ranchi (Jharkhand)–Sipat (Chhattisgarh)	40%
3	Seoni–Khandwa (Madhya Pradesh)	40%
4	Raipur–Raigarh (Chhattisgarh)	40% fixed, +15%/−5% dynamic
5	Lucknow–Gorakhpur (Uttar Pradesh)	30%
6	Karcham (hydroelectric plant)–Kala Amb (Himachal Pradesh)	40%

installations with thyristors started to realize fixed and variable reactive impedance. Accordingly, shunt-connected static var compensators (*SVC*) of thyristor-switched reactors (*TSR*), thyristor-controlled reactors (*TCR*), and thyristor-switched capacitors (*TSC*) became quite popular. It may be noted that these reactive power controllers have energy storage components such as a capacitor and an inductor. Also, its response is relatively slow due to the limited switching speed of the devices. Dynamic compensation based on converters and controllers is discussed in the next chapter.

6.5 Passive harmonic filters

We have seen from the earlier chapters that harmonics adversely affect the performance of most components, including the capacitor, the building block for power quality improvement. Also, a capacitor cannot improve power factor under non-sinusoidal supply conditions. However, capacitors are capable of operating up to 135% of their kVAr rating provided the current through them does not exceed 180% of their rated current and voltage does not exceed 110% of them rated rms voltage, including them harmonic content [6]. Hence, we continue to explore the capacitor in combination with other passive components for the elimination of harmonics. These electrical circuits, called passive filters, are used to filter out unwanted frequency components from the power supply system. These allow selected frequency components to pass through the filter circuit while blocking others from it. These circuits are made with the frequency-dependent impedance characteristic of passive components of circuit theory. Hence, passive filters are often characterized according to their frequency response, their impedance, and the number of reactive components [9,10]. As the harmonic current contains multiple frequency components in addition to the fundamental, a detailed study of the harmonic spectrum is desirable for selecting the rating of the capacitor and inductor [11–14]. Also, it provides reactive VAR to the system with loss of power, and hence a new term, quality factor Q, to express its efficiency is defined.

$$\text{Quality factor, } Q = \frac{\text{Reactive power of the filter}}{\text{Active power } P \text{ consumed at tunning frequency}}$$

The following passive filter topologies are generally used for filtering of harmonics.

6.5.1 Band-pass or Notch filter

It is the simplest circuit and the most common type of passive filter. It is designed to filter a single frequency (f) component from the harmonic spectrum. Figure 6.17 (a) shows its circuit diagram. It consists of a series circuit with a capacitor (C) and an inductor (L) with its own resistance (R). As the filter has two reactive components, it is a 2nd-order filter.

$$Z = R + j2\pi fL + \frac{1}{j2\pi fC} \tag{6.24}$$

At the tuned frequency f_o, the circuit offers minimum impedance (Z) governed by the resistance R of the circuit. Thus, it offers very little power loss, and at the tuned frequency, the filter allows f_o component to pass with minimum attenuation and blocks other frequency components.

At resonance, $X_L = X_C$ or $2\pi fL = \frac{1}{2\pi fC}$ and $Z = R$.

Therefore, tuned frequency, $f_o = \frac{1}{2\pi\sqrt{LC}}$ or angular frequency, $\omega_o = \frac{1}{\sqrt{LC}}$.

Impedance characteristics of the filter with capacitance of 12.5 F μH, inductance of 5 mH and resistance of 2 Ω, tuned for an angular speed of 4000 rad/sec, are drawn and shown as in Figure 6.17(b). It is seen that as the f increases to its tuned frequency f_o, Z decreases, and thereafter, Z increases. Thus, the circuit offers the

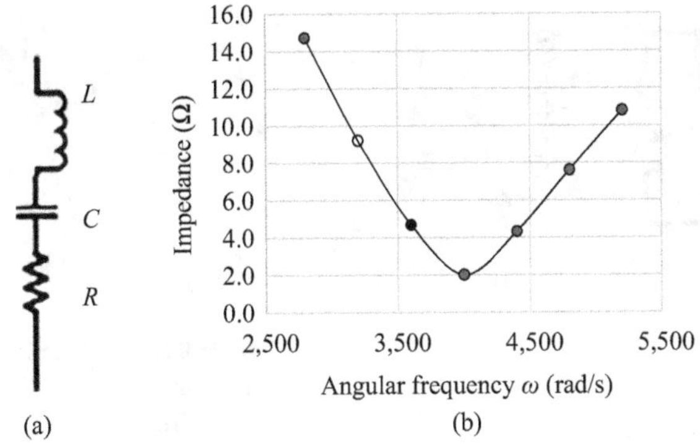

Figure 6.17 Band-pass filter. (a) Circuit diagram. (b) Filter impedance–frequency characteristic.

least Z or shows a notch at f_o. For filtering of multiple frequency components, several filter circuits are required so that each circuit filters one harmonic independently of the others.

6.5.2 High-pass filter

The filter circuit consists of a capacitor (C) in series with a parallel circuit of an inductor (L) and resistor (R), as shown in Figure 6.18(a). It is similar to a band-pass filter with a resistor parallel to the inductor. This filter is 2nd-order filter with a damping resistor R. The equivalent impedance of the filter is written as:

$$Z = \left(\frac{1}{R} + \frac{1}{j\omega L}\right)^{-1} + \frac{1}{j\omega C} \tag{6.25}$$

$$Z = \frac{R \times j\omega L(R - j\omega L)}{(R + j\omega L)(R - j\omega L)} + \frac{1}{j\omega C} = \frac{R^2 j\omega L + R\omega^2 L^2}{R^2 + \omega^2 L^2} + \frac{1}{j\omega C} \tag{6.26}$$

Separating real and imaginary parts

$$Z = \frac{R\omega^2 L^2}{R^2 + \omega^2 L^2} + j\frac{R^2\omega L}{R^2 + \omega^2 L^2} - \frac{1}{\omega C} \tag{6.27}$$

Impedance characteristic of the high-pass 60 Hz filter with inductance of 2.21 mH, resistance of 5 Ω, capacitance of 26.31 μF, fundamental frequency of 60 Hz, and tuned for a frequency of 660 Hz is shown in Figure 6.18(b). The computation of various parameters to derive the curve is shown as problem 7. The curve appears nearly flat towards the higher frequency side of the spectrum, allowing frequencies above the tuned frequency to pass. Thus, it is a better filter for higher frequencies or a targeted frequency and all above it. However, it generates

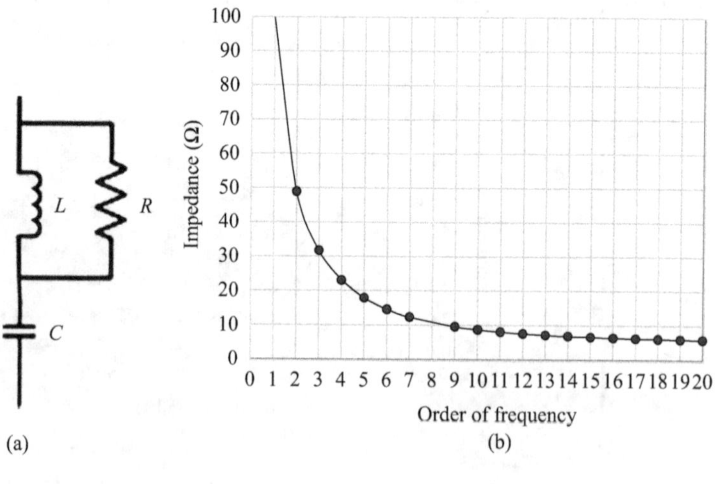

(a) (b)

Figure 6.18 High-pass filter. (a) Circuit diagram. (b) Filter impedance–frequency characteristic.

more I^2R or power loss. The resistor dampens the current waveform; the higher the value of R, the flatter the curve and the greater the loss.

6.5.3 C-type filter

The filter circuit consists of a capacitor C_2 in series with a parallel combination of a resistor (R) and inductor L-capacitor C_1, as shown in Figure 6.19(a). Thus, it has three frequency-dependent components, L, C_1, and C_2, and is known as a 3rd-order filter. A C-type filter is like a high-pass filter with an extra capacitor, C_2. The capacitor C_1 forms a resonant circuit at the fundamental frequency with inductor L that bypasses the resistor R. Thus, no fundamental frequency power loss through R. Also, capacitance C_1 and C_2 together form a resonance at a tuned frequency with inductance L, thus behaving like a high-pass filter. The impedance Z of the filter can be derived as:

$$Z = \left(\frac{1}{R} + \frac{1}{j\omega L - j\frac{1}{\omega C_1}}\right)^{-1} + \frac{1}{j\omega C_2} \tag{6.28}$$

$$j\omega L - j\frac{1}{\omega C_1} = \frac{j(\omega^2 LC_1 - 1)}{\omega C_1} \tag{6.29}$$

$$\frac{1}{R} + \frac{1}{j\omega L - j\frac{1}{\omega C_1}} = \frac{1}{R} + \frac{\omega C_1}{j(\omega^2 LC_1 - 1)} = \frac{j(\omega^2 LC_1 - 1) + \omega C_1 R}{j(\omega^2 LC_1 - 1)R} \tag{6.30}$$

$$Z = \frac{j(\omega^2 LC_1 - 1)R}{j(\omega^2 LC_1 - 1) + \omega C_1 R} - j\frac{1}{\omega C_2} \tag{6.31}$$

$$Z = \frac{j(\omega^2 LC_1 - 1)R\{\omega C_1 R - j(\omega^2 LC_1 - 1)\}}{\{\omega C_1 R + j(\omega^2 LC_1 - 1)\}\{\omega C_1 R - j(\omega^2 LC_1 - 1)\}} - j\frac{1}{\omega C_2} \tag{6.32}$$

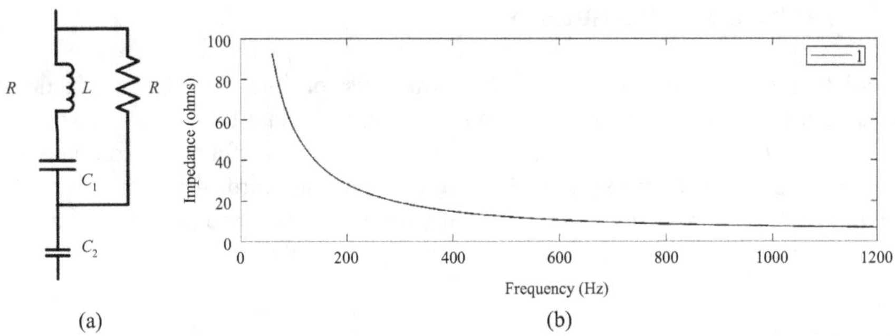

Figure 6.19 C-pass filter: (a) Filter circuit diagram. (b) Impedance-frequency characteristic.

Separating real and imaginary parts,

$$Z = \frac{(\omega^2 LC_1 - 1)^2 R}{(\omega^2 LC_1 - 1)^2 + (\omega C_1 R)^2} + j\left(\frac{\omega C_1 R^2(\omega^2 LC_1 - 1)}{(\omega^2 LC_1 - 1)^2 + (\omega C_1 R)^2} - \frac{1}{\omega C_2}\right) \tag{6.33}$$

The impedance characteristic of a C-pass filter is similar to that of a high-pass filter. To demonstrate it, identical parameters as detailed high-pass filter ($R = 5\ \Omega$, L = 2.21 mH with an equivalent capacitance $\frac{1}{C} = \frac{1}{C_1} + \frac{1}{C_2}$, ($C = 26.31\ \mu$F, $C_1 = 318\ \mu$F, $C_2 = 28.683\ \mu$F) is drawn using MATLAB® and is shown as Figure 6.19(b). The curve appears nearly flat towards the higher frequency side of the spectrum, attenuating a wide range of harmonic frequencies. At the fundamental frequency $X_L = X_c$, L–C path has zero impedance. Thus, no fundamental component power loss.

6.6 Conclusion

The chapter deliberated the importance and some of the techniques of controlling reactive power, voltage variation, and mitigation of harmonics in today's power generation, transmission, and customer's electrical energy dependence scenario. It focused on static var generation and a passive filter. Both of them depend on frequency and rely on energy storage components, inductance, and capacitance together with switching devices. Passive filters provide a low impedance path for the harmonic current to flow to its source, thereby blocking or suppressing specific harmonic frequencies from the system. They are tuned to a specific frequency or group of frequencies but are not suitable for varying load conditions. However, with the growing use of converter-based equipment and multiple harmonics in the system, these systems are ineffective under dynamic conditions. Chapter 7 details an insight into an emerging converter-based solution and active filter.

6.7 Problems and solutions

Problem 6.1 Consider an interconnected transmission line with 0.2 pu reactance connected between two areas with 1 pu. voltage. Calculate the power that can be transmitted through the line: (i) without compensation; (ii) if a shunt compensator is connected (a) at the mid-point of the line; (b) at one-third of the line from the sending end; (c) three compensations at regular intervals. Assume 1 pu voltage at the point of installation of the compensator (Figure 6.20).

Solution:

Case (i) Without compensation

$$P_{\max} = P = \frac{V_s V_r}{X_l} \sin\delta = \frac{1 \times 1}{0.2} = 5 \text{ pu}$$

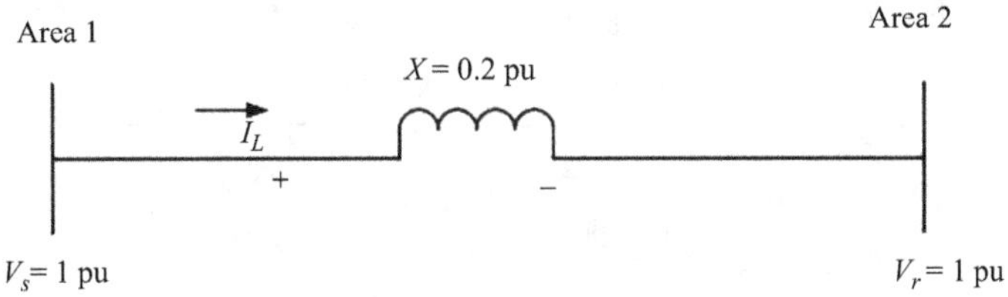

Figure 6.20 Equivalent circuit for problem 6.1 without compensation

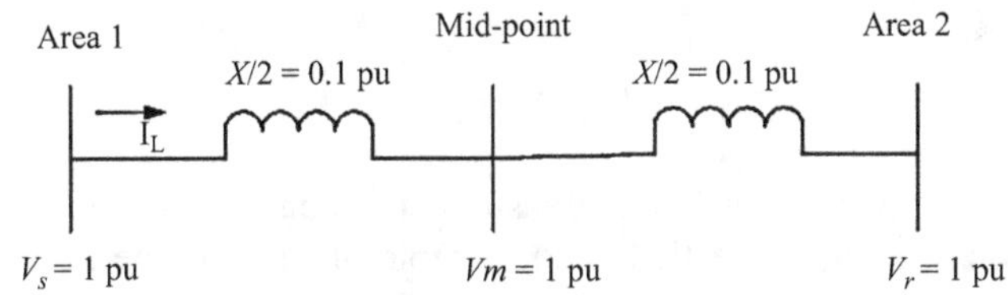

Figure 6.21 Equivalent circuit for problem 6.1 with mid-point compensation

Case ii (a) With mid-point compensation

Assume that a shunt compensator is connected at the mid-point of the transmission line to make voltage at this point same as sending-end voltage. The equivalent circuit is shown in Figure 6.21.

Power flow between the sending end and mid-point (ref Eq.6.12) is calculated as $\frac{2V^2}{X_l}\sin\frac{\delta}{2}$

$$P_{\max} = \frac{|V_s||V_m|}{\frac{X}{2}} = \frac{1 \times 1}{0.1} = 10\ \text{pu}$$

The power flow from mid-point to receiving end is also the same, as the total reactance of the transmission line is equally divided. Thus, the net maximum power that can be transmitted from one area to another is 10 *pu*, which indicates that mid-point compensation doubles the steady-state power transmitted without compensation.

Case ii (b) With compensation at one-third of the line

In this case, the shunt compensator is connected at one-third portion (marked 1) of the transmission line to make the voltage at this point (V_1) the same as the sending end voltage. The equivalent circuit is shown in Figure 6.22.

P_{max} at $\frac{X}{3}$ section ($P_{\max}$) is given by:

$$P_{\max} = \frac{|V_s||V_p|}{\frac{X}{3}} = \frac{1 \times 1}{0.066} = 15\ \text{pu}$$

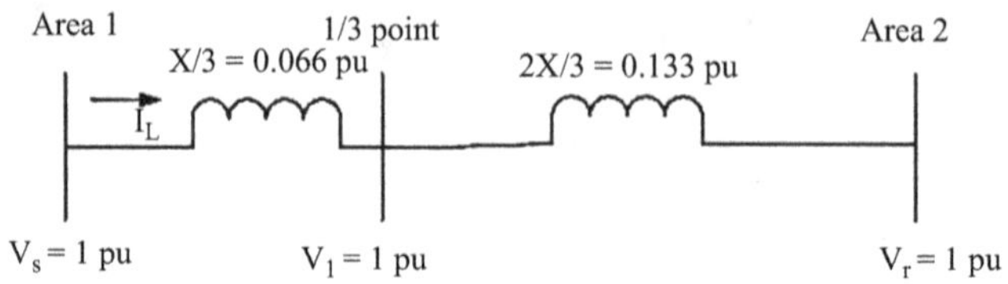

Figure 6.22 Equivalent circuit for problem 6.1 with compensation at 1/3 of the transmission line

$P_{\max}$ at $\frac{2X}{3}$ section is given by:

$$P_{\max} = \frac{|V_P||V_r|}{\frac{2X}{3}} = \frac{1 \times 1}{0.133} = 7.5 \text{ pu}$$

Thus, the net power that can be transmitted through the line is decided by the longer section. The longer section is not capable of carrying the P_{max} that can be handled by the shorter section. Thus, P_{max} that can be transmitted from the sending end to the receiving end is 7.5 *pu* only, and this power is greater than P_{max} without compensation but less than 10*pu* with mid-point compensation. Hence, the best location for shunt compensation placement is at the mid-point of the line.

Case (iii) three compensation elements at regular intervals

In this case, the shunt compensator is connected at regular intervals (marked 1,2,3) of the transmission line to make the voltage at these points (V_1, V_2, V_3) the same as the sending end voltage. The equivalent circuit is shown in Figure 6.23.

$P_{\max}$ at each section is given by:

$$P_{\max} = \frac{|V_S||V_1|}{\frac{X}{4}} = \frac{1 \times 1}{0.05} = 20 \text{ pu}$$

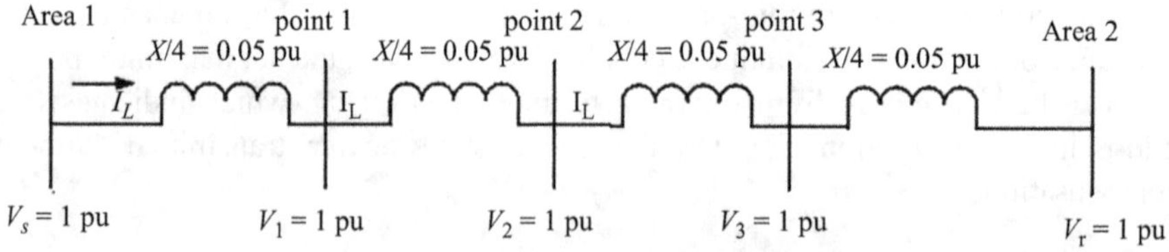

Figure 6.23 Equivalent circuit with compensation at regular intervals for problem 6.1

The same power can be transmitted in each section from the sending end to the receiving end as reactance in each section, and the voltage at each point is the same. Thus, with three compensations at regular intervals, the net power that can be transmitted through the line is 20 *pu* and the voltage at regular intervals will be the same as the sending-end voltage, resulting in a flat voltage profile over the entire transmission line.

Problem 6.2 Assume that there are parallel transmission lines connecting two areas with a bus voltage of 132 kV, as shown in Figure 6.24. The length of one of the lines is double that of the other, while the power-carrying capacity of this line is limited to 0.5 pu. It is found that in area two, there is a deficiency of 1 *pu* power. If there is surplus power in area 1, calculate how this power will be distributed between the two parallel lines (paths). Comment on the results and suggest a solution in case of any problem.

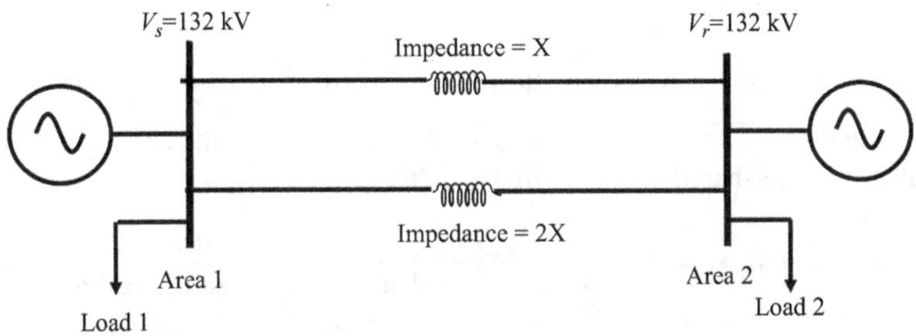

Figure 6.24 Two-machine model of a power system with parallel lines for Problem 6.2

Solution:
Assume the maximum power that can be transmitted from area 1 to 2 as 1 *pu*. $P_{\max} = 1$ pu. As $P_{\max}$ is inversely proportional to the impedance of the line, power flow through a long line (P_{max-l}) is given by:

$$P_{\max-l} = P_{\max}\frac{\text{Impedance of short line}}{\text{Sum of impedance of short and long line}} = 1 \times \frac{X}{3X}$$

$$= 0.33 \text{ pu}$$

Similarly, power through the short line (P_{max-s}) is given by:

$$P_{\max-s} = 1 \times \frac{2X}{3X} = 0.66 \text{ pu}$$

Discussion:
Line capacity is 0.5 pu. So, the short line is overloaded, and the long line is underutilized. If we connect an inductor of X in series with a short line, known as a compensator, as shown in Figure 6.25, power will be shared equally, which is within the limit as given below.

$$P_{\max-l} = P_{\max-s} = P_{\max}\frac{\text{Impedance of short line}}{\text{Sum of impedance of short and long line}} = 1 \times \frac{2X}{4X} = 0.5 \text{ pu}$$

Thus, using inductive series compensation, power sharing between parallel lines can be controlled.

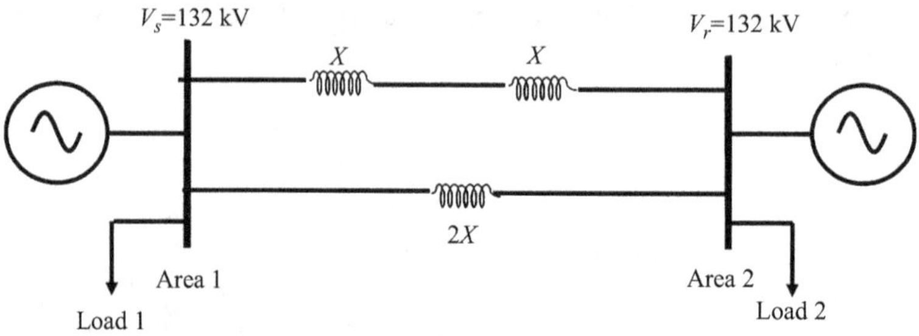

Figure 6.25 Model of the system with parallel lines with series inductive compensation

Alternate solution: As power flow is governed by $P = \frac{V_s V_r}{X_l - X_c.} \sin \delta$, with a series capacitor, the effective reactance of the long line is reduced to X, and hence the power will be shared equally (Figure 6.26).

$$P_{\max-l} = P_{\max-s} = P_{\max} \frac{\text{Impedance of short line}}{\text{Sum of impedance of short and long line}}$$

$$= 1 \times \frac{X}{2X} = 0.5 \text{ pu}$$

Thus, the power transmitted through the line can also be controlled using series capacitive compensation.

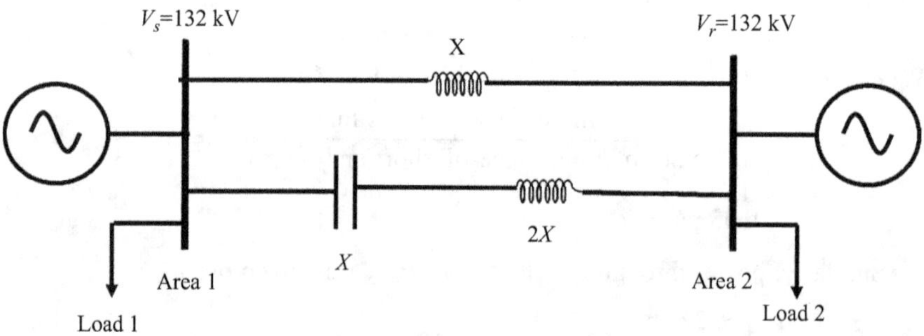

Figure 6.26 System with parallel lines with series capacitive compensation for Problem 6.2

Problem 6.3 An industrial consumer with a connected load of 10 MW at 0.8 pf receives power at 33 kV level. The distribution company charges the consumer for electricity based on energy used in kWh and maximum demand at Rs. 395/kVA, apart from other charges and duties on a monthly basis. The consumer plans to reduce the maximum demand by 15% by installing a capacitor at 11 kV level,

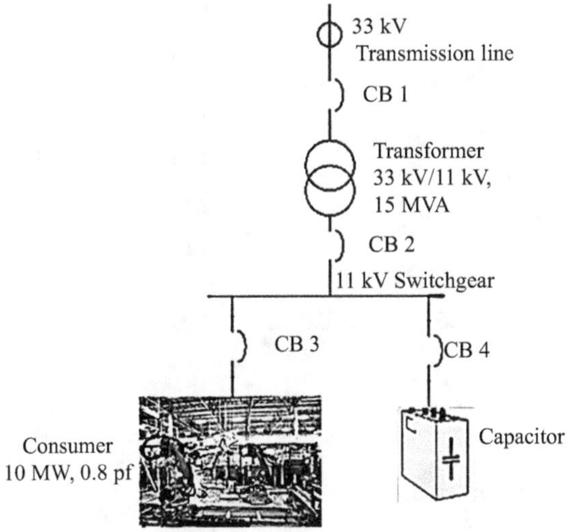

Figure 6.27 Single-line diagram of a receiving station of the industry. Problem 6.3.

keeping the same load. Figure 6.27 indicates the single-line diagram of its receiving station. Calculate the following:

1. Reduction in maximum demand, required capacitance compensation, and savings in the electricity charges. Represented by a phasor diagram.
2. Reduction in transmission losses, if any. Assume transmission line length as 50km and impedance as 0.5 + j0.4 Ω/km.
3. Capacity effect on the transformer and transmission lines.

Solution:

1. Reduction in maximum demand, required capacitor, saving in electricity charges and vector diagram (see Figure 6.28)

Power demand of the consumer (P) = 10 MW

Existing values of maximum demand and reactive volt-amperes are calculated as follows.

Power factor of the consumer, $\cos\phi_1 = 0.8$ lag. Therefore $\phi_1 = 36.87°$.

The maximum demand of the consumer (S_1) is calculated as:

$$S_1 = \frac{P}{\phi_1} = \frac{10}{0.8} = 12.5 \text{ MVA}$$

Reactive volt-ampere of the consumer (Q_1) is given by:

$Q_1 = 10 \tan 36.87 = 7.5$ MVAr

It is given that the consumer plans to reduce the maximum demand by 15% by installing capacitor compensation at the 11 kV level. The reactive volt ampere to be compensated using capacitor compensation is obtained from the targeted reactive volt ampere. Targeted values are calculated as follows.

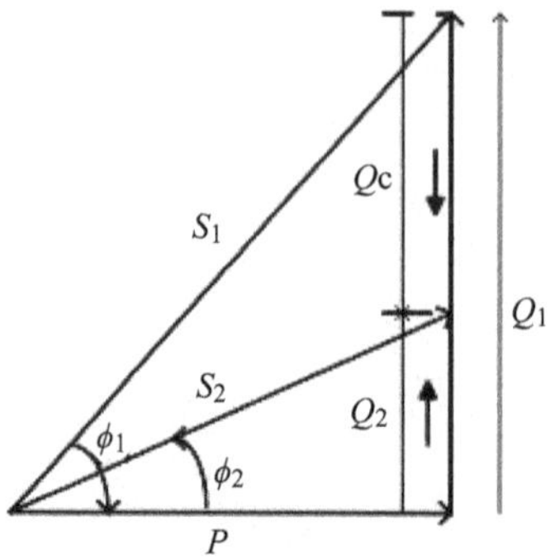

Figure 6.28 Vector diagram for Problem 6.3

Targeted maximum demand (S_2) is given by:

$$S_2 = 85\% \text{ of } S_1 = 12.5 \times 0.85 = 10.625 \text{ MVA}$$

Targeted power factor (PF_2) is calculated as:

$$PF_2 = \frac{P}{S_2} = \frac{10}{10.625} = 0.941 \text{ Therefore } \phi_2 = 19.75$$

Targeted reactive volt-ampere of the consumer (Q_2) is:

$$Q_2 = P \tan\phi_2 = 10 \tan 19.75 = 3.59 \text{ MVAr}$$

So, the required compensating capacitor reactive volt-ampere

$$Q_c = Q_1 - Q_2 = 7.5 - 3.59 = 3.91 \text{ MVAr}$$

Cost saving is calculated as follows:
Reduction in maximum demand is given by:

$$S_1 - S_2 = 12.5 - 10.625$$

$$= 1.875 \text{ MVA } (15\% \text{of consumer maximum demand})$$

Savings on the monthly bill for reduction on demand = $1.875 \times 10^3 \times 395$ = Rs. 7,40,625.

2. Reduction in transmission losses, if any. Assume transmission line length as 50 km and impedance as $0.5 + j0.4\ \Omega/\text{km}$.

Therefore resistance,

$$R = \frac{0.5\ \Omega}{\text{km}}$$

Length of line = 50 km

Total resistance of the line = $0.5 \times 50 = 25\ \Omega$

Loss is calculated using I^2R loss.

$$\text{Load current for 10 MW at 0.8 } PF(I_1) = \frac{S_1}{\sqrt{3} \times V_{L-L}} = \frac{12.5 \times 10^3}{\sqrt{3} \times 33} = 2187\ \text{A}$$

$$\text{Load current for 10 MW at 0.94 } PF(I_2) = \frac{S_2}{\sqrt{3} \times V_{L-L}} = \frac{10.625 \times 10^3}{\sqrt{3} \times 33} = 1859\ \text{A}$$

$$\text{Reduction in } I^2R \text{ loss} = 3 \times (I_1 - I_2)^2 \times 0.5 \times 50$$

$$\text{Reduction in } I^2R \text{ loss} = 3 \times (2187 - 1859)^2 \times 0.5 \times 50\ \text{W} = 8068.8\ \text{kW}$$

3. Capacity effect on transformer and transmission line

With the installation of capacitive compensation, maximum demand is reduced by 15% (1.875 MVA). This is now made available for other loads to transmit on the line. Thus, the transmission system (transformer and transmission line) can now take an additional load of 1.875 MVA. Also, transmission loss is reduced by 8068.8 kW, leading to a reduction in temperature rise of the transformer and line conductors, thereby slowing the ageing process of the insulating material. MVA.

Problem 6.4 A distribution company imposes a power factor penalty on non-residential/commercial/industrial consumers on low tension (*LT*) having a sanctioned load above 20 kW whenever the average power factor (*PF*) is less than 0.9. It also gives *PF* an incentive for *PF* more than 0.95 in steps of 0.01 up to 1.0, and at 0.97, 1%. An educational institute has a load of 20 kW at 0.8 *PF*. It pays electricity charges based on its maximum demand at a fixed rate, energy charges based on energy used in kWh, and other charges. It pays a penalty of 1% for a monthly electricity bill below 0.89 PF. It is proposed to install a capacitor to avail the *PF* incentives of 1% for a power factor above 0.97. Calculate the required capacitance value, savings in electricity charges, and reduction in maximum demand after the installation of the capacitor.

Solution:

Power factor of the consumer = 0.8 lag. Therefore $\phi_1 = 36.87°$.

It is given that a penalty of 1% is imposed on the monthly electricity bill for below 0.89 PF. Thus, an educational institution pays the penalty of 1% on the monthly electricity bill.

The maximum demand of the consumer (S_1) is given by:

$$S_1 = \frac{P}{\cos\phi_1} = \frac{20}{0.8} = 25\ \text{kVA}$$

Reactive VA of the consumer:

$$Q_1 = P \times \sin\phi = 20 \times \sin 36.87 = 12\ \text{kVar}$$

To get incentives of 1%, on the electricity bill, the power factor shall be above 0.97 or $\phi_2 = 14.07$.

Reactive VA of the consumer (Q_2) to get PF above 0.97 is:

$$Q_2 = P \times \sin\phi_2 = 20 \times \sin 14.07 = 4.86\ \text{kVar}$$

So, the required compensating capacitor reactive power (Q_c) is:

$$Q_c = Q_1 - Q_2 = 12 - 4.86 = 7.14\ \text{kVar}$$

Saving on monthly bill = 2% of electricity bill (1% penalty + 1% incentive)
The maximum demand of the consumer (S_2) is given by:

$$S_2 = \frac{P}{\cos\phi} = \frac{20}{0.97} = 20.62\ \text{kVA}$$

Reduction on maximum demand is:

$$S_1 - S_2 = 25 - 20.62 = 4.38\ \text{kVA}$$

Problem 6.5 A 400 kV single-circuit transmission line of length 400 km has $0.33343\ \angle 84.87^\circ\ \Omega$ series impedance and shunt admittance of $3.4725\times10^{-6}\angle 90^\circ\ \mho$ per km. Assume the total load on the line to be 125 MW at a *PF* of 1.0. Calculate the no-load receiving-end voltage when identical shunt reactors are connected at both ends of the line, compensating for 70% of the total shunt admittance of the line (Figure 6.29).

Solution:

$$V_r(\text{Line voltage}) = 400\ \text{kV} \therefore V_r(\text{phase voltage}) = \frac{400}{\sqrt{3}} = 231\angle 0^\circ\ \text{kV}$$

$$L = 0.33343\angle 84.87^\circ \quad C = 3.4725 \times 10^{-6}\ \text{Length of the line} = 400\ \text{km}$$

$$\text{Series impedance } Z = 400 \times 0.33343\angle 84.87^\circ\ \Omega = 133.372\angle 84.87^\circ\ \Omega$$

$$\text{Shunt admittance } Y = 400 \times 3.4725 \times 10^{-6}\angle 90^\circ\ \mho = 1.389 \times 10^{-3}\angle 90^\circ\ \mho$$

Load current (I_r) at 125 MW with *PF* 1.0 is calculated as:

$$I_r = \frac{MW \times 10^6}{\sqrt{3} \times V_{L-L} \times PF} = \frac{125 \times 10^6}{\sqrt{3} \times 400 \times 1} = 180.422\angle 0^\circ$$

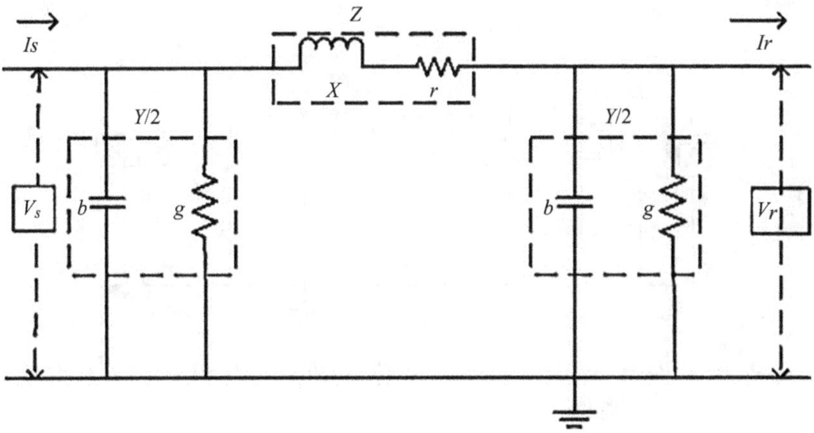

Figure 6.29 π model of medium transmission line for Problem 6.5

For a medium transmission line sending end voltage is calculated using ABCD constants and is given by:

$$V_s = A \times V_r + A \times I_r$$

where $A = \dfrac{Z \times Y}{2} + 1 \; B = Z$

$$A = \frac{ZY}{2} + 1 = 0.9077\angle 0.5227^\circ \text{ and } B = Z = 133.372\angle 84.87^\circ \ \Omega$$

$$V_s = AVr + BIr = 0.9077\angle 0.5227 \times 231000\angle 0 + 133.372\angle 84.87 \times 180.422\angle 0$$
$$V_s = 213.3967\angle 6.96^\circ \text{ kV}$$
$$V_s = AVr + BIr$$

At no load, $I_r = 0$, $V_r = \dfrac{V_s}{A} = \dfrac{213.3967\angle 6.96^\circ}{0.9077\angle 0.5227^\circ} = 235.096\angle 6.43$ kV.

Receiving end voltage when shunt reactors are connected at both ends of the line (Figure 6.30).

As shunt reactors are connected at both ends of the line, compensating for 70% of the total shunt admittance of the line, the modified shunt admittance of the line is given by:

$$Y = (1 - 0.7) \times 1.389 \times 10^{-3}\angle 90^\circ \text{ S} = 4.167 \times 10^{-4}\angle 90^\circ \ \mho$$

However, there is no change in the series inductance of the line, as the shunt reactor is connected in parallel with the lines. Hence, $Z = 133.372\angle 84.87^\circ \ \Omega$.

As a modification in Y, which affects the receiving end voltage, A is calculated as follows.

$$A = \frac{ZY}{2} + 1 = \frac{1}{2 \times 133.372}\angle 84.87 \times 4.167 \times 10 - 4\angle 90 + 1$$

$$= 1.002862\angle 1.5814$$

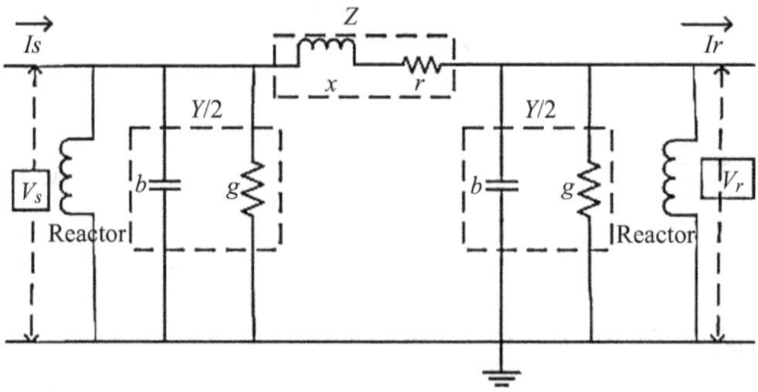

Figure 6.30 *π model of the transmission line with shunt reactor at the receiving and sending end*

And at no load, $I_r = 0$, $\therefore V_r = \frac{V_s}{A} = \frac{213.3967\angle 6.96}{1.002862\angle 1.5814} = 212.7877\angle 5.3986$ kV.

Thus, by adding a shunt reactor, the receiving end no-load terminal voltage changed from 235 kV to 213 kV.

Problem 6.6 Design a band-pass passive *R–L–C* filter with 5 Ω resistance and 12.5 μF capacitance to filter the 5th-order harmonic frequency (f) component from a 50 Hz supply. Show magnitude of impedance as a function of frequency (f) from 50 Hz to 550 Hz. Also, plot its characteristics.

Solution:

Given resistance $R = 5\ \Omega$, capacitance $C = 12.5\ \mu$F, fundamental frequency (f) = 50 Hz, tuned frequency of the filter, 5th-order harmonic, i.e., 250 Hz.

For bandpass filter, impedance $Z = R + j2\pi fL + \frac{1}{j2\pi fC}$ and at tuned frequency,

$$X_L = X_C \text{ or } 2\pi fL = \frac{1}{2\pi fC}$$

Therefore, at the tuned frequency, $f_o = \frac{1}{2\pi\sqrt{LC}}$; inductance $L = \frac{1}{C(2\pi f_o)^2}$

$$L = \frac{1}{C(2\pi f_o)^2} = \frac{1}{12.5 \times 10^{-6}(2\pi \times 250)^2} = 32.423 \text{ mH}$$

Tabulation of reactance and impedance is shown in Table 6.3.

Plot of impedance versus frequency characteristics is shown in Figure 6.31.

As inductive reactance is proportional to frequency and capacitive reactance is inversely proportional, net reactance below the tuning frequency is capacitive and above f_o is inductive. Also, at f_o, the filter circuit has an impedance of 5 Ω resistive. Hence, the filter circuit has a power loss given by I^2R.

Table 6.3 Frequency response of band-pass filter – problem 6

Frequency f, Hz	Inductive reactance $X_L = 2\pi fL$	Capacitive reactance $Xc = \frac{1}{2\pi fC}$	Impedance $Z\Omega$ $\sqrt{R^2 + (X_L - Xc)^2}$
50	10.19	254.65	244.51
100	20.37	127.32	107.07
150	30.56	84.88	54.55
200	40.74	63.66	23.46
250	50.93	50.93	5.00
300	61.12	42.44	19.33
350	71.30	36.38	35.28
400	81.49	31.83	49.91
450	91.67	28.29	63.58
500	101.86	25.46	76.56
550	112.05	23.15	89.04

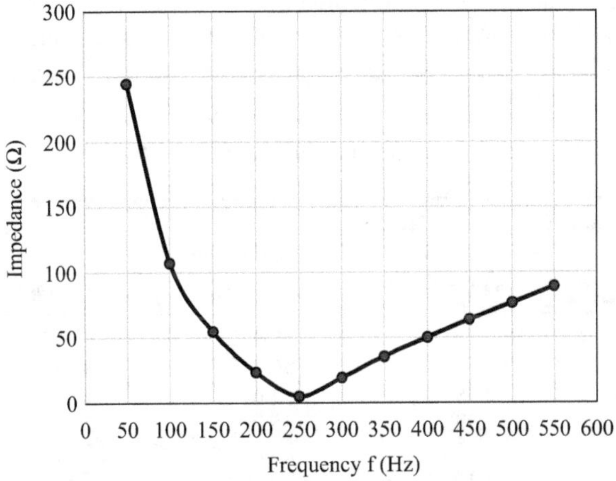

Figure 6.31 Impedance frequency characteristics of a band-pass filter – Problem 6.6

Problem 6.7 Design a high-pass passive filter for a 60 Hz supply for a 12-pulse rectifier, with 5 Ω resistance and 26.31 μF capacitance. Show the magnitude of impedance as a function of frequency (f) from 60 Hz to 1200 Hz. Also, plot its impedance characteristics up to the harmonic order of 20.

Solution:

Given resistance R = 5 Ω, capacitance C = 26.31 μF, fundamental frequency (f) = 60 Hz, 12-pulse rectifier. For a 12-pulse rectifier circuit, the characteristic frequency starts from 11. Therefore, for considering the tuned frequency of the filter, the 11th-order harmonic frequency = 660 Hz.

For bandpass filter, impedance

$$Z = \left(\frac{1}{R} + \frac{1}{j2\pi fL}\right)^{-1} + \frac{1}{j2\pi fC}$$

and at tuned frequency, $X_L = X_C$ or $2\pi fL = \frac{1}{2\pi fC}$

Therefore, at the tuned frequency, $f_o = \frac{1}{2\pi\sqrt{LC}}$; inductance $L = \frac{1}{C(2\pi f_o)^2}$

$$L = \frac{1}{C(2\pi f_o)^2} = \frac{1}{26.31 \times 10^{-6}(2\pi \times 660)^2} = 2.21 \text{ mH}$$

Tabulation of reactance and impedance is shown in Table 6.4.

The following inference may be drawn from the table and its characteristics (Figure 6.32). At a tuned frequency of 660 Hz, both inductive reactance and capacitive reactance are equal (9.165 Ω) as designed. Below tuning frequency f_o, capacitive reactance is more than inductive. Also, at f_o, the filter circuit has an impedance of 8.05 Ω, with 5 Ω resistance. The various values corresponding to tuned frequency of 660 Hz are highlighted in the Table 6.4. Hence filter circuit has a power loss given by I^2R. The filter has a flat characteristic at higher frequencies.

Table 6.4 Frequency response of high-pass filter – Problem 6.7

Order of harmonic	**Frequency *f*, Hz**	**Inductive reactance** $X_L = 2\pi fL$	**Capacitive reactance** $Xc = \frac{1}{2\pi fC}$	**Impedance (Ω)** $Z = \left(\frac{1}{R} + \frac{1}{j2\pi fL}\right)^{-1} + \frac{1}{j2\pi fC}$
1	60	0.833	100.820	100.01
2	120	1.666	50.410	48.91
3	180	2.499	33.607	31.62
4	240	3.333	25.205	22.95
5	240	3.333	25.205	22.95
6	300	4.166	20.164	17.82
7	360	4.999	16.803	14.52
8	420	5.832	14.403	12.28
9	540	7.498	11.202	9.54
10	600	8.332	10.082	8.69
11	**660**	**9.165**	**9.165**	**8.05**
12	720	9.998	8.402	7.55
13	780	10.831	7.755	7.16
14	840	11.664	7.201	6.85
15	900	12.497	6.721	6.60
16	960	13.330	6.301	6.40
17	1020	14.164	5.931	6.23
18	1080	14.997	5.601	6.09
19	1140	15.830	5.306	5.97
20	1200	16.663	5.041	5.87

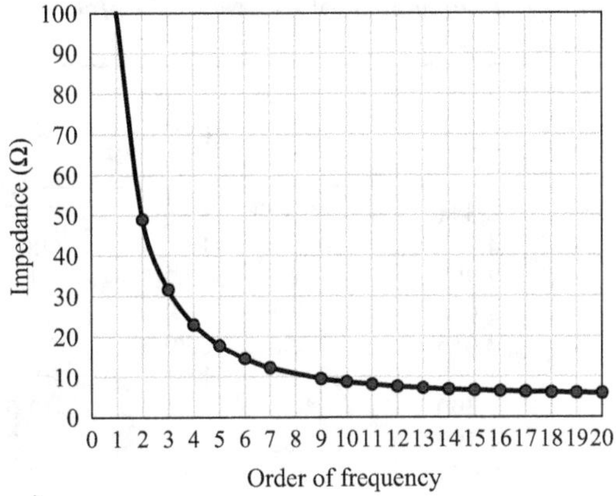

Figure 6.32 Impedance frequency characteristics of high-pass filter – Problem 6.7

Problem 6.8 Design a C-pass passive filter for 60 Hz supply tuned for 660 Hz, with 5 Ω resistance, 2.21 mH inductance, and 26.31 μF capacitance. Show the magnitude of impedance as a function of frequency (f) from 60 Hz to 1200 Hz. Also, plot its impedance characteristics up to the harmonic order of 20.

For the C-pass filter, impedance

$$Z = \frac{(\omega^2 LC_1 - 1)^2 R}{(\omega^2 LC_1 - 1)^2 + (\omega C_1 R)^2} + j\left(\frac{\omega C_1 R^2(\omega^2 LC_1 - 1)}{(\omega^2 LC_1 - 1)^2 + (\omega C_1 R)^2} - \frac{1}{\omega C_2}\right)$$

and at fundamental frequency, $X_L = X_{C_1}$ or $\omega L = \dfrac{1}{\omega C_1}$

Therefore, $C_1 = \frac{1}{\omega^2 L} = \frac{1}{(2\pi 60)^2 22.1 \times 10^{-6}} = 318\,\mu\text{F}$

capacitance $\frac{1}{C} = \frac{1}{C_1} + \frac{1}{C_2}$. Therefore, $C_2 = \frac{C \times C_1}{C_1 - C} = \frac{26.31 \times 318}{318 - 26.31} = 28.683\,\mu\text{F}$

Tabulation of computed reactance, impedance etc. with bold values at tuned frequency is shown in Table 6.5. Figure 6.33 shows the curve of impedance frequency characteristics.

The above curve appears nearly flat towards the higher frequency side of the spectrum, attenuating a wide range of harmonic frequencies. At the fundamental frequency $X_L = X_c$ or L–C path offers zero impedance. Thus, no fundamental frequency component power loss. Also, the filter circuit has a lower impedance (94.85 Ω) compared to the high-pass filter (100.01 Ω; Table 6.4) of identical resistance, inductance and equivalent capacitance. Hence, it is more suitable for an industrial environment, where higher-order harmonics are likely to be present.

Table 6.5 Harmonics and impedance of C-filter – Problem 6.8

Order of harmonics	**Inductive reactance** $X_L = \omega L$	$X_{C_1} = \frac{1}{2\omega C1}$	$X_C = \frac{1}{2\omega C}$	***Z* real part Ω**	***j* part of *Z* Ω**	**Magnitude, *Z* Ω**
1	0.833	8.341	100.820	3.46	–94.79	94.85
2	1.666	4.171	50.410	1.00	–48.24	48.25
3	2.499	2.780	33.607	0.02	–31.11	31.11
4	3.333	2.085	25.205	0.29	–21.95	21.95
5	3.333	2.085	25.205	0.29	–21.95	21.95
6	4.166	1.668	20.164	1.00	–16.50	16.53
7	4.999	1.390	16.803	1.71	–13.04	13.15
8	5.832	1.192	14.403	2.31	–10.72	10.97
9	7.498	0.927	11.202	3.17	–7.87	8.48
10	8.332	0.834	10.082	3.46	–6.94	7.75
11	**9.165**	0.758	**9.165**	**3.69**	**–6.21**	**7.23**
12	9.998	0.695	8.402	3.88	–5.62	6.83
13	10.831	0.642	7.755	4.03	–5.14	6.53
14	11.664	0.596	7.201	4.15	–4.73	6.29
15	12.497	0.556	6.721	4.25	–4.38	6.11
16	13.330	0.521	6.301	4.34	–4.09	5.96
17	14.164	0.491	5.931	4.41	–3.83	5.84
18	14.997	0.463	5.601	4.47	–3.60	5.74
19	15.830	0.439	5.306	4.52	–3.40	5.66
20	16.663	0.417	5.041	4.57	–3.22	5.59

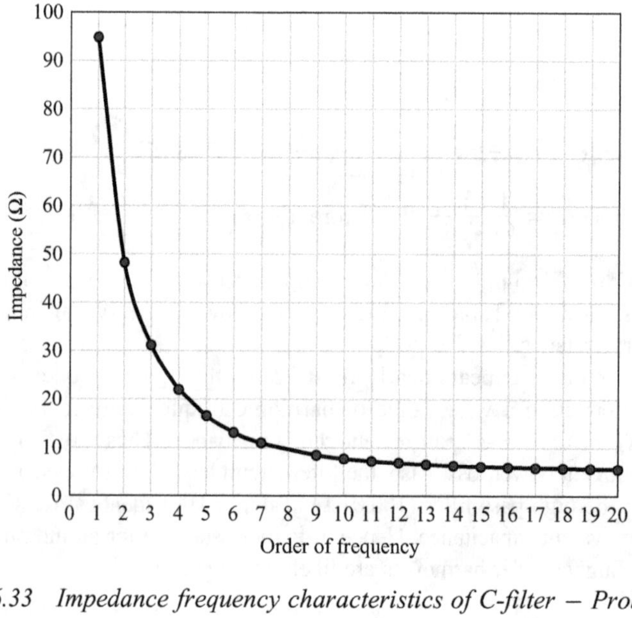

Figure 6.33 Impedance frequency characteristics of C-filter – Problem 6.8

References

[1] Technical background and recommendations for defence plans in the continental Europe synchronous area. European Network of Transmission System Operators for Electricity, ENTSO-E. https://eepublicdownloads.entsoe.eu/clean-documents/pre2015/publications/entsoe/RG_SOC_CE/RG_CE_ENTSO-E_Defence_Plan_final_2011_public.pdf1)

[2] Ministry of Power, India. Report on the Grid Disturbance on 30th July and 31st July 2012. https://www.cercind.gov.in/2012/orders/Final_Report_Grid_Disturbance.pdf

[3] D. P. Kothari, and I. J. Nagrath, *Power System Engineering*, Tata McGraw-Hill Publishing Company Limited, New Delhi India, Third edition, 2019.

[4] Narain G. Hingorani, and Laszlo Gyugyi, *Understanding FACTS Concepts and Technology of Flexible AC Transmission Systems*, IEEE Press Standard Publishers Distributors, First edition, 1999.

[5] Central Electricity Authority, Electricity Distribution Network Planning Criteria. https://cea.nic.in › notification › 2024/01 › Fina...

[6] IEEE Std 18-1992 - IEEE Standard for Shunt Power Capacitors.

[7] Central Electricity Authority, April 2021: Standard specifications and technical parameters for transformers and reactors (66 kV & above Voltage class) https://cea.nic.in › uploads › pse___td › 2022/01

[8] Report on Reactive Power Management and Voltage Control Ancillary Services (VCAS) in India, March 2021, Power System Operation Corporation Ltd.

[9] IS 15395 -2003 / IEC 61642- Industrial AC network affected by harmonics application of filters and shunt capacitors.

[10] IEEE Std 1531-2020, IEEE Guide for the Application and Specification of Harmonic Filters.

[11] Abdelhamid Hamadi, Salem Rahmani, and Kamal Al-Haddad "A Hybrid Passive Filter Configuration for VAR Control and Harmonic Compensation", *IEEE Transactions on Industrial Electronics*, Vol. 57, No. 7, 2010, pp. 2419–2434.

[12] Salem Rahmani, Abdelhamid Hamadi, Kamal Al-Haddad, and Louis A. Dessaint, "A Combination of Shunt Hybrid Power Filter and Thyristor-Controlled Reactor for Power Quality", *IEEE Transactions on Industrial Electronics*, Vol. 61, No. 5, 2014, pp. 2152–2164.

[13] Luis Morán, Cristóbal A. Albistur, and Rolando Burgos, "Multimega VAR Passive Filters for Mining Applications: Practical Limitations and Technical Considerations", *IEEE Transactions on Industry Applications*, Vol. 52, No. 6, 2016, pp. 5310–5317.

[14] Sina Shakeri, Saeid Esmaeili, and Mohammad Hossein Rezaeian Koochi, "Passive Harmonic Filter Design Considering Voltage Sag Performance - Applicable to Large Industries", *IEEE Transactions on Power Delivery*, Vol. 37, No. 3, 2022, pp. 1714–1722.

Chapter 7
Active power quality enhancement techniques

An uninterrupted electrical supply system demands dynamic control of reactive power and mitigation of multiple harmonics that are present in the power system in recent times apart from managing real power at any instant.

Goals of the chapter

The need for dynamic control of reactive power and elimination of power system harmonics is highlighted in previous chapters. This chapter focuses on recent techniques of converter-based solutions for dynamic control of voltage variation, reactive power, and mitigation of harmonics. A few problems and solutions are also included to make the concept of an active filter that complies with IEEE 519 and its optimization techniques.

7.1 Introduction

With the ongoing expansion and growth of the power system, including deregulation in many countries, transmission systems are being forced to work closer to their stability and thermal limits while maintaining the quality of power delivered to the consumers. This requires more efficient utilization and control of the system. As explained in the preceding chapter, with passive components, viz., capacitors, inductors (reactor) help in compensating reactive power or control of variation in system voltage as well as harmonics. One of the main drawbacks of passive compensation is that it is specific to an installation and the instant of operating conditions. Moreover, the number and size of passive filters depend on the harmonic order and the amount of harmonic power to be compensated, respectively. Some drawbacks of fixed or manually switched passive compensation are eliminated using thyristor-controlled compensators such as thyristor-controlled reactors (TCR), thyristor-controlled series capacitor (TCSC), etc., detailed in Chapter 6. Though it helps in improving the response time to some extent, such compensators inject harmonics into the system and are not suitable for varying load conditions. Later, with the advancement of power electronic circuits, it was realized that the shunt, series, and series-shunt compensation can be done without using an inductor (reactor) and or a capacitor. That is, it can be achieved using a converter-based flexible AC transmission (FACT) system [1], which uses devices with turn-off capability and operates on pulse width modulation (PWM) technique. Thus, harmonic injection and limitations on switching speed of thyristor-based compensators are mitigated. Of the converter type with a turn-off device, voltage source converter-based compensators [2] are used in FACTS applications. In fact, many other desirable functions of the passive compensators are incorporated in a single unit, as explained in the later sections. These compensators are suitable for dynamic compensation of reactive power and harmonics. The following section briefly details converter-based FACTS devices.

7.2 Converter-based FACTS devices

Converter-based compensators used to compensate for reactive power and harmonic power in power systems, which enhance the power quality of the system, are considered under FACTS. The main component of converter-based compensators is a DC to AC converter (also known as an inverter). Other components of the converter-based compensators and their working principles are explained in the next section. Unlike passive filters, the converter-based compensators are active compensators as they can inject current and/or voltage into the power system, which is independent of the system voltage or current, respectively. The high switching frequency devices used in such compensators help a dynamic compensation with fast response time. There are two types of DC to AC converters that can be used for FACTS applications, as given below; however, as of today, voltage source converters are more widely used:

(i) Voltage source converter (VSC)
(ii) Current source converter (CSC)

In the case of VSC, the DC source of the converter is a DC voltage source with a large value of parallel capacitor, which acts as a constant voltage source, whereas in the case of CSC, it is a controlled DC voltage source with a large value of series inductor, which acts as a constant current source. Thus, VSC maintains a constant voltage at the input side with a variable voltage at the output side of the converter, while CSC maintains a constant current at the input side with a variable current at the output side. The other components of both inverters, such as high frequency switches, gate driver circuits, filters, etc., are almost the same; however, they differ in their approach to the conversion process. Modern VSC and CSC use IGBT as switches, which helps with faster switching at higher volage level, resulting in better dynamic compensation. Selection of converters depends on the availability of DC power sources and switching devices, application requirements, system complexity, cost, etc.

From application point of view, there are three types of compensators: (a) shunt, (b) series and (c) series- shunt, which are used to improve power quality. In the case of a shunt compensator, a DC to AC converter, along with its additional components required to operate as compensator, is connected in parallel with the power system, while it is connected in series with the power system in the case of a series compensator. The shunt compensator directly controls current quality, which indirectly improves voltage quality, while series compensator directly controls voltage quality and hence the current quality of power systems. In case of series – shunt compensator, two converters are used where one acts as shunt compensator and other as series compensator so that both voltage and current quality can be controlled at the same time. The specialty of converted-based compensators is that a fast dynamic control of various parameters such as (i) reactive power/voltage in power system (ii) harmonics in voltage and current and (iii) power flow through the transmission line etc. can be achieved. Such transmission line compensators are also widely classified under FACTS devices.

A converter-based shunt compensator implemented with the objective of eliminating reactive power in power systems is called as static shunt compensator (STATCOM), whereas if its objective is mainly to eliminate harmonic power, it is termed a shunt active filter. Similarly, reactive power consumed by the transmission system can be directly controlled by converter-based series compensation, which is used to enhance the power transfer capability of the line. Such compensators are called static synchronous series compensator (SSSC), and if the objective of converter- based series compensation is to eliminate harmonics in voltage, it is termed as series active filter. STATCOM controls the voltage of the system by directly controlling current, while SSSC controls power by directly controlling the voltage drop across the transmission line. Both these activities can be implemented using the combination of shunt and series reactive power compensation known as unified power flow controller (UPFC), and if it is used to eliminate both voltage and current harmonics, it is called as series-shunt active filter. As this book focuses on power quality, shunt, series and series-shunt active filters based on voltage source converters are explained in the remaining sections.

In the case of shunt connected converter-based (active) compensator, the converter connected in parallel with power systems, as shown in Figure 7.1(a), injects compensating current (i_{ca}, i_{cb}, i_{cc}) into the system to control the magnitude and shape of source current (i_{sa}, i_{sb}, i_{sc}). In the case of series connected converter-based compensator, the magnitude and shape of PCC voltage at point '*x*' is modified at point 'y' (v_{sa}, v_{sb}, v_{sc} to v_{sa}', v_{sb}',v_{sc}') by injecting compensating voltage (v_{ca}, v_{cb},v_{cc}) using series connected converter as shown in Figure 7.1(b). In the case of series-shunt compensators, two converters are used as shown in Figure 7.1(c), one to control the magnitude and shape of voltage by injecting desired voltage in series using series connected converter, and the other, parallel connected converter, injects desired current into the system for enhancing mainly the current quality. The basic structure of the compensator is the same; however, the function of the compensator may vary depending on its application requirements.

In the case of FACTS devices, the availability of high-power switches is an important criterion that determines the cost of FACTS equipment. The number of series and parallel connections of devices for a particular application depends on the highest rating of the devices available and the voltage level of FACTS equipment. Switching frequency is limited by the power dissipation allowed by the switching devices. The filter required to make the output voltage free from higher-order harmonics is also an important factor while designing converter-based

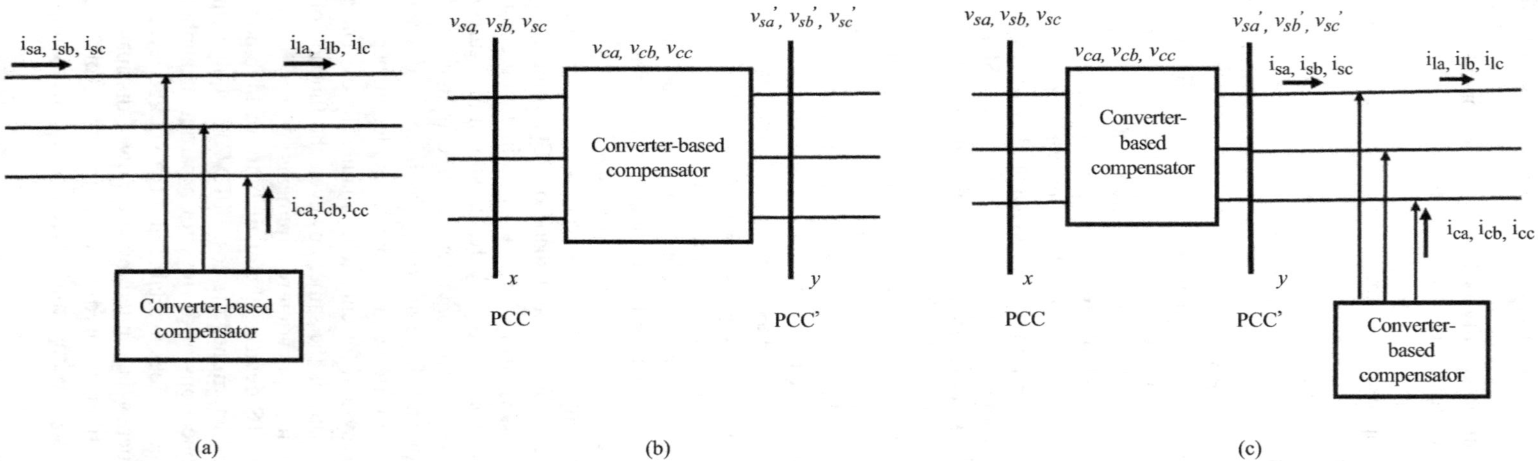

Figure 7.1 *Connection of various types of converter-based compensators. (a) Shunt compensator (b) Series compensator. (c) Series–shunt compensator.*

compensators. The filter design depends on the order of harmonics and their magnitude present in the output waveform. The selection of the PWM technique used for converter operation plays an important role in controlling the order of harmonics along with their magnitude present in its output waveforms. As VSC-based FACTS devices are widely used, the basic components of VSC-based compensators and their operations are explained in the next section.

7.3 Voltage source converter-based compensators

The selection of proper components and control strategies used are important factors that decide the efficiency of operation of compensators. The following points are discussed for the basic understanding of operation of converter-based compensators used for enhancement of power quality in power systems.

(i) Components of the voltage source converter
(ii) Concepts of real and imaginary instantaneous powers
(iii) Reference signal generation for VSC operation
(iv) Implementation of VSC-based compensator

7.3.1 Components of voltage source converter

The VSC-based compensators used in FACTS applications have major segments common for their function as a shunt, series or series-shunt compensator. However, the connection of each of these compensators to the grid varies depending on its application. The common segments of such compensators are given below and is shown in the block diagram in Figure 7.2.

1. DC source
2. DC–AC converter (Voltage source inverter)
3. Output filter
4. Interfacing unit

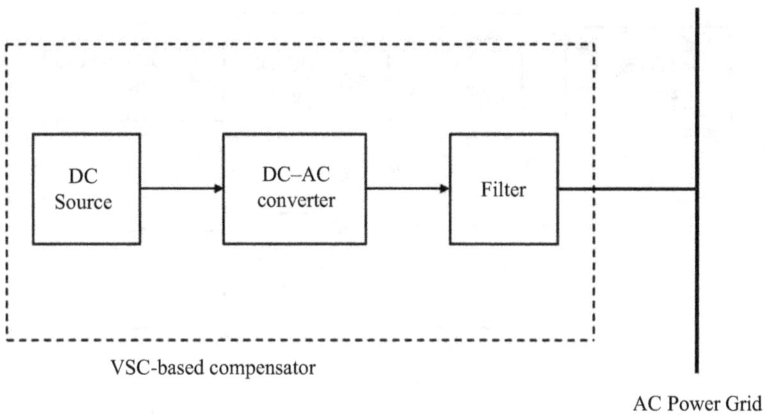

Figure 7.2 Block diagram of VSC-based compensator connected to grid

7.3.1.1 DC source

As shown in the block diagram in Figure 7.2, a DC voltage source acts as an input to the VSC. The DC voltage source can be a battery, a charged capacitor using rectified AC supply or a DC energy supply from any renewable energy sources such as solar PV or fuel cell stack or wind power, as in Figure 7.3(a), (b) and (c), respectively. A battery as DC source may not be an economical solution and hence not generally used for high power applications. Generally, in the case of VSC-based FACTS devices or compensator whose function is to inject reactive power or distortion power into the grid, DC source may be a capacitor that is charged from AC supply. If the compensator is required to inject active power into the grid, DC source needs to be a DC energy source such as any renewable energy sources. This is required in the case of compensator whose function is to compensate for imbalance in the system to make it a balanced system.

In case of a wind energy system used as DC source to VSC, a set of AC–DC and DC–DC converters is required to regulate the DC voltage, as shown in Figure 7.4(a), while in the case of solar energy as DC source, a DC–DC converter is used to regulate the DC voltage as shown in Figure 7.4(b).

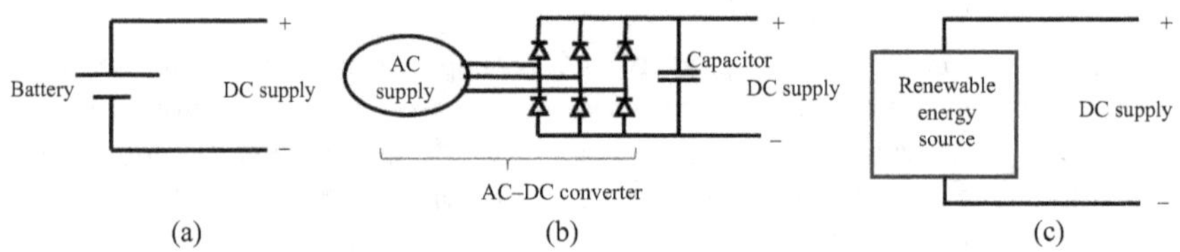

Figure 7.3 Various methods of generation of DC supply for VSC-based compensators. (a) Battery. (b) DC capacitor charged using AC–DC converter. (c) Renewable energy sources.

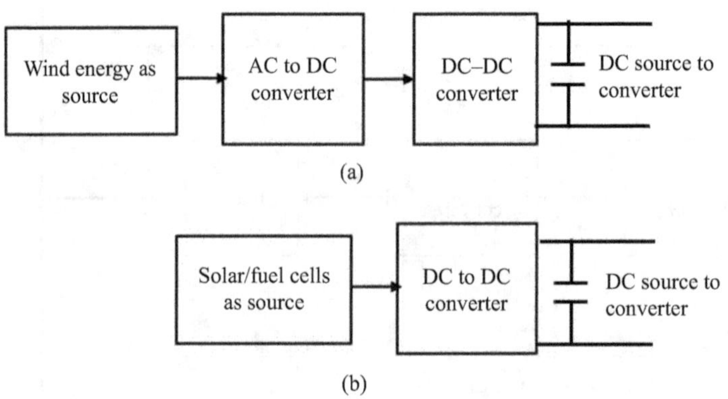

Figure 7.4 A typical block diagram of generation of DC supply. (a) From wind energy. (b) From solar/fuel cell energy.

Like a battery source, a DC supply generated from renewable energy sources can be given to the converter, which helps in supplying active power to the grid in addition to reactive power and distortion power. Let us analyse the working of VSC in the next section.

7.3.1.2 Voltage source converter

The schematic diagram of a three-phase voltage source DC to AC converter (a 3-ϕ inverter) is shown in Figure 7.5. These converters are categorized into two types based on control of their output parameters.

(i) Voltage-controlled VSC (In this type, the output voltage is directly controlled).
(ii) Current-controlled VSC (In this type, the output current is directly controlled).

By controlling the switching operations of DC to AC converters, either voltage or current at the output side of the converter can be controlled. There are various techniques, such as sine triangular PWM, hysteresis current control, etc., to decide switching operations, which are explained in later sections. Generally, in the case of voltage-controlled VSC operation, a reference signal proportional to desired output voltage is to be generated, while in the case of current-controlled VSC, a reference signal proportional to the desired output current is to be found out using any algorithm to generate appropriate pulses to the switches. However, regardless of whether voltage or current control is used, the physical layout and components of the converter [3] remain the same, as shown in Figure 7.5.

It consists of six power semiconductor switches (S_1, S_2, …., S_6). Typically, power devices used for FACTS applications are of high voltage and high-power rating, such as IGBTs. Though silicon-controlled rectifiers can support the desired voltage and power ratings, they need commutation circuits and large filter circuits and hence are not preferred over IGBTs. Now, let us analyse the difference between voltage-controlled VSC and current-controlled VSC.

The specialty of voltage-controlled VSC is that the output AC voltage magnitude and frequency can be easily controlled, though the input DC voltage is fixed.

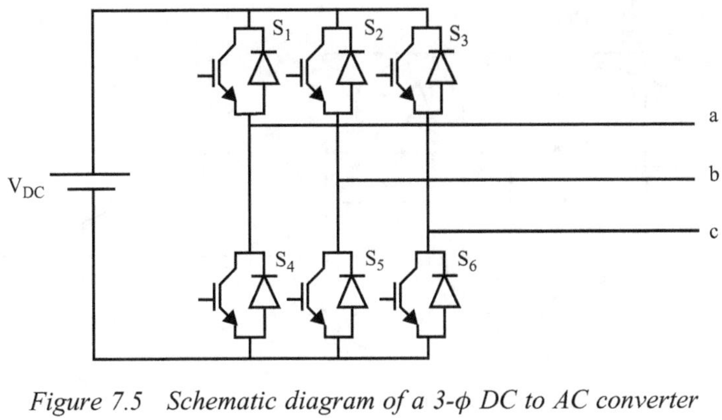

Figure 7.5 Schematic diagram of a 3-ϕ DC to AC converter

By giving proper pulses to the switches, it is possible to convert DC to AC at a desired voltage magnitude and frequency. The pulses to top switches (S_1, S_2, S_3) and bottom switches (S_4, S_5, S_6) are complementary to each other. The pulse generation technique (commonly known as pulse width modulation (PWM) technique) plays an important role in the performance of these converters.

There are various PWM techniques available for pulse generation of these switches. The commonly used sine-triangular PWM technique and hysteresis control technique for controlling VSC are explained below for understanding purposes. These techniques are generally used for the implementation of active filters, which are used to control voltage and/or current quality in power systems.

Sine triangular PWM technique

Sine triangular PWM (SPWM) technique is used to generate pulses to the switches of VSC to get a desired output at its terminals. To understand the working of a VSC using sine triangular PWM technique, assume a simple schematic diagram of a 3-ϕ, DC to AC voltage source converter connected to a resistive load as shown in Figure 7.6. For easy explanation, the supply voltage (V_{DC}) is divided into two halves ($V_{DC}/2$) and the centre points are grounded as shown in the figure. The load is also grounded so that each phase voltage is independently controlled.

Generation of pulses for VSC switches in the SPWM technique. Assume that the function of the VSC is to generate a desired voltage in magnitude and frequency at its output, which is termed as voltage controlled VSC. Waveforms of a VSC simulated in the MATLAB® platform using the SPWM technique are given in Figure 7.7. In this case, a reference sinusoidal or control signal, as in Figure 7.7(a), which is proportional and in phase with the desired phase ‘a’ voltage, is compared with a carrier frequency signal as in Figure 7.7(a) to generate pulses for the switches S_1 and S_4 connected at phase ‘a’. The frequency of reference sinusoidal signal needs to be selected the same as the desired frequency of output voltage, and the frequency of triangular waveform depends on switching frequency of the switches used for the purpose. The higher the switching frequency, lesser is the filter requirement. When the magnitude of the sine wave is greater than triangular waveform (such as from A to B), a high pulse is generated as shown in Figure 7.7(b),

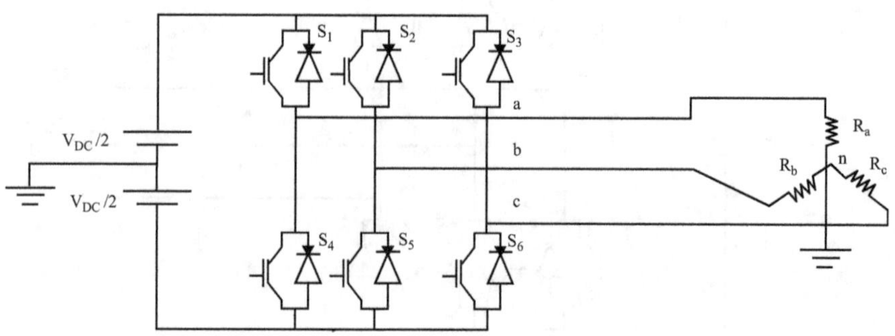

Figure 7.6 Schematic diagram of 3-ϕ DC to AC VSC connected to a resistive load

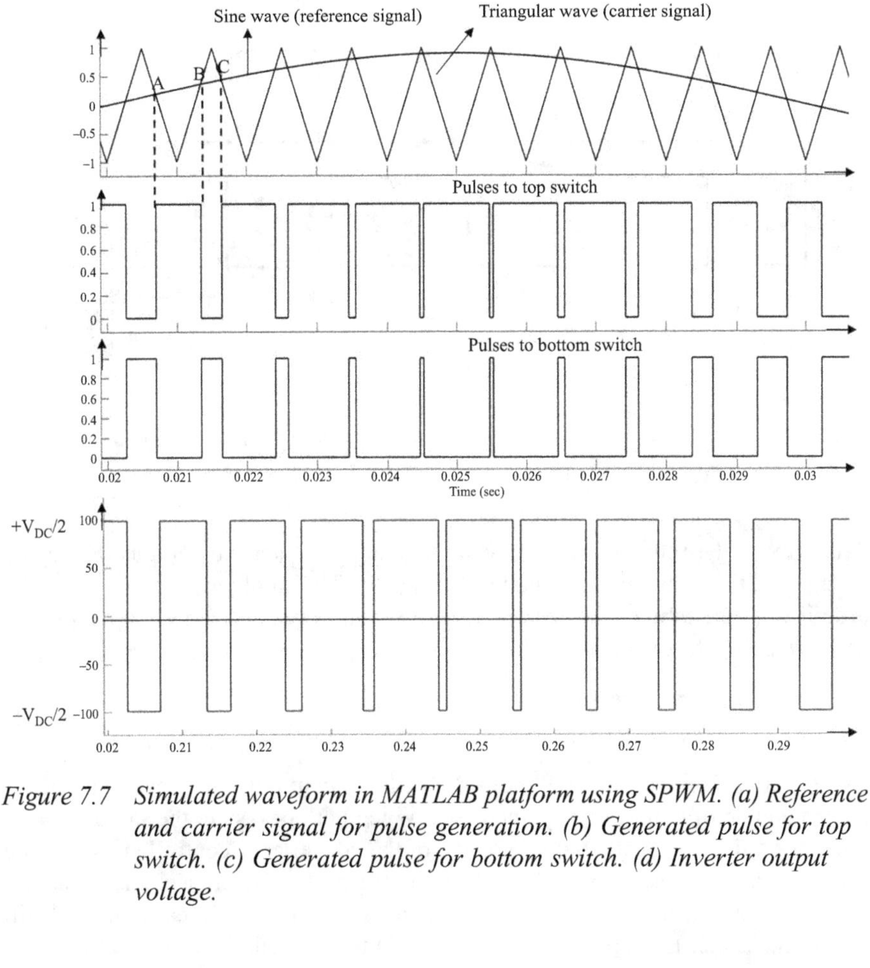

Figure 7.7 *Simulated waveform in MATLAB platform using SPWM. (a) Reference and carrier signal for pulse generation. (b) Generated pulse for top switch. (c) Generated pulse for bottom switch. (d) Inverter output voltage.*

and otherwise (such as from B to C) the pulse magnitude is made low or zero. This pulse is to be given to the top switch (S_1) of phase 'a' leg of the converter, while its complement, as in Figure 7.7(c), is to be given to bottom switch (S_4) of the same leg. When the top switch of 'phase a' is on, the output phase voltage v_{an} is $+V_{DC}/2$ and with the bottom switch on, it is $-V_{DC}/2$ as shown in Figure 7.7(d). The pulses of top and bottom switches are complementary to each other with a dead band between them. If dead band is not provided, the bottom and top switch turn on at the same time, and hence the circuit gets shorted. In simulations, VSC may work without a dead band; however it is necessary for hardware implementation. The pulses, including a dead band for a hardware implementation of VSC, is given in Figure 7.8. In a similar way, switching pulses to other phases can be generated using a sinusoidal reference signal proportional to its respective phase voltages.

Control of magnitude of output voltage of VSC in SPWM technique. It is known that the fundamental frequency of the output voltage depends on the

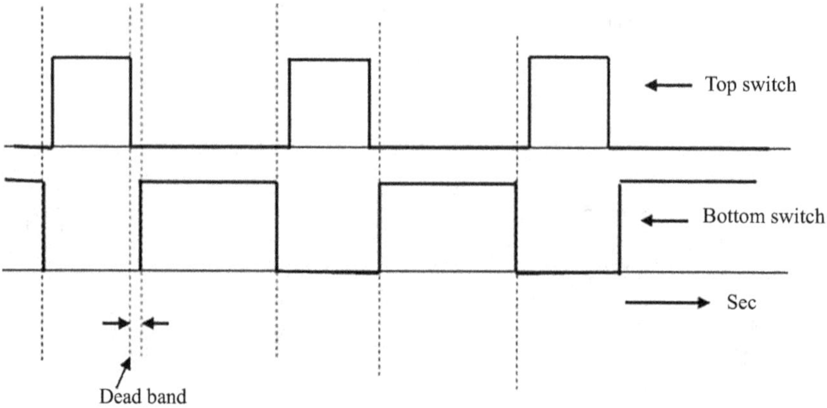

Figure 7.8 Pulses of the top and bottom switches for hardware implementation of VSC

frequency of the reference signal, while its magnitude depends on the DC supply voltage and the amplitude modulation index (M_a). The amplitude modulation index is defined as the ratio of the peak of control signal to peak of carrier signal used to generate pulses to switches.

$$M_a = \frac{\text{Peak of reference signal}}{\text{Peak of carrier signal}} \tag{7.1}$$

As M_a varies from zero to one, the width of the output pulse increases, and hence the magnitude of output voltage increases. When both peaks are the same $M_a = 1$. Thus, for a fixed V_{DC}, the output voltage of the converter is controlled by varying M_a. Generally, M_a is controlled by changing the peak of reference sinusoidal signal, as it is proportional to the desired voltage, while the peak of carrier signal will remain the same. It is clear from Figure 7.7(c) that the output voltage of the converter is in the form of pulses that vary from $+V_{DC}/2$ to $-V_{DC}/2$ in each half cycle. This indicates that in addition to the fundamental component, there are other higher-order harmonic component present in output voltage. It is seen that the peak amplitude of the fundamental component of the ac output in 'phase a' (V_{an1}) in each phase is given by:

$$V_{an1} = M_a \times \frac{V_{DC}}{2} \tag{7.2}$$

if centre point of DC input voltage is grounded as shown in Figure 7.6 and $v_{sine} \leq v_{tri}$.

The line to line rms voltage (V_{LL1}) at fundamental frequency is given by:

$$V_{LL1} = \left(\frac{\sqrt{3}}{2\sqrt{2}}\right) M_a \times V_{DC} \tag{7.3}$$

if centre point of DC input voltage is not grounded as in Figure 7.5 and $v_{sine} \leq v_{tri}$

Where, v_{sine} is the peak of reference sinusoidal signal

v_{tri} is the peak of the triangular carrier signal

From the above equations, it is clear that by controlling M_a, it is possible to control the magnitude of output voltage $(v_a,\ v_b,\ v_c)$ for a fixed value of V_{DC}.

VSC can also be used to generate non-sinusoidal voltage by modifying the reference sinusoidal signal to a non-sinusoidal signal as per the requirement. For example, when VSC is used as a series active filter to compensate for voltage harmonics, the reference signal will be non-sinusoidal.

Control of frequency of output voltage of VSC in SPWM technique. Now, let us find out how to control the frequency of output voltage of the converter. The fundamental frequency of output voltage depends on the frequency of the reference signal. So, the generation of this reference signal with proper magnitude and frequency is an important factor that controls the output voltage. However, in addition to the fundamental component of voltage, higher-order harmonics (unwanted) are also present due to the switching operation, which is controlled by controlling frequency modulation index (M_f). Generally, the frequency modulation index is defined as the ratio of frequency of triangular wave to sine wave.

$$M_f = \frac{\text{Frequency of carrier signal}}{\text{Frequency of reference signal}} \tag{7.4}$$

As shown in Figure 7.7(d), the output voltage varies from $+V_{DC}/2$ to $-V_{DC}/2$ with a frequency depending on the frequency of triangular wave. THD in output voltage is high, which is to be limited as per the stipulated standards. It is found that the higher order of harmonics present in the output voltage depends on M_f. In case of sine triangular PWM, harmonics in output voltage contains additional harmonics of order M_f, $2M_f$, $3M_f$ and so on, along with side bands around it. This general pattern is for M_a in the range of 0 to 1. So, by increasing the frequency of triangular waveform or M_f for a fixed reference signal frequency, the higher-order harmonics present in the output voltage are shifted away from the fundamental component, which is easy to filter out using passive filters. As M_f increases, THD decreases. However, there is a limitation on the switching frequency allowed for a particular switching device. So M_f value is to be selected properly based on the switch selected for the purpose.

Output voltage of VSC with and without a filter. A typical waveform of the output voltage of the inverter without filter along with its fundamental component is plotted using MATLAB is shown in Figure 7.9. The frequency of reference sine wave is assumed as 50 Hz. It is seen that the frequency of fundamental component of output voltage is same as the frequency of reference wave and in phase with each other.

It can be concluded that using SPWM technique:

(i) The magnitude of output voltage of inverter can be controlled by controlling M_a for fixed value of V_{DC}.
(ii) The output voltage tracks the shape of reference signal and is in phase with it.

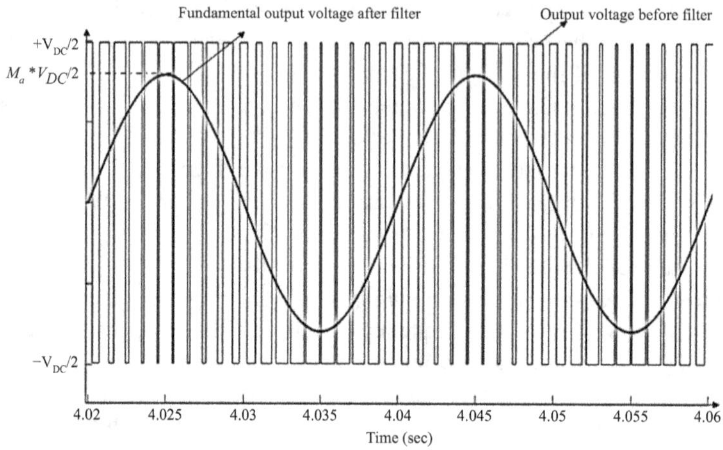

Figure 7.9 Typical voltage waveform of a DC–AC converter and its fundamental wave

(iii) The fundamental frequency of output voltage is same as frequency of reference signal.
(iv) The higher order of harmonics present in output voltage depends on frequency of carrier signal for a fixed reference signal.
(v) It can generate sinusoidal or non-sinusoidal voltage waveform by selecting proper reference signal.
(vi) The technique is applicable to the implementation of active filters under sinusoidal as well as non-sinusoidal conditions.

Hysteresis control technique

Generally, the hysteresis current control technique is used where output current of VSC needs to be controlled. There are various methods available to control output current of an inverter. However, a basic hysteresis current control technique used to track a desired output current of VSC, which acts as a current-controlled VSC, is explained in this section. In this case, a reference signal proportional to the desired output current of the converter of all three phases to be formed using any technique and one such reference signal in phase 'a' is as shown in Figure 7.10. Then, an upper band and a lower band signal for this reference signal to be generated by considering a hysteresis band, as shown in the figure. Switching frequency depends on the width of the band, and hence the width is to be decided based on the availability of switches. Assume that it is required to control the output current of the converter to a desired value. So, sense the actual current signal in phase 'a' and compare it with the respective reference signal. At the start, make the top switch on by giving a pulse to it, then the actual current increases from A to B. When it touches point B, make the pulse zero, which turns off the top switch. The complement of such a pulse generated with a dead band is to be given to the bottom switch on the same leg. Then the current starts decreasing from point B to C. When

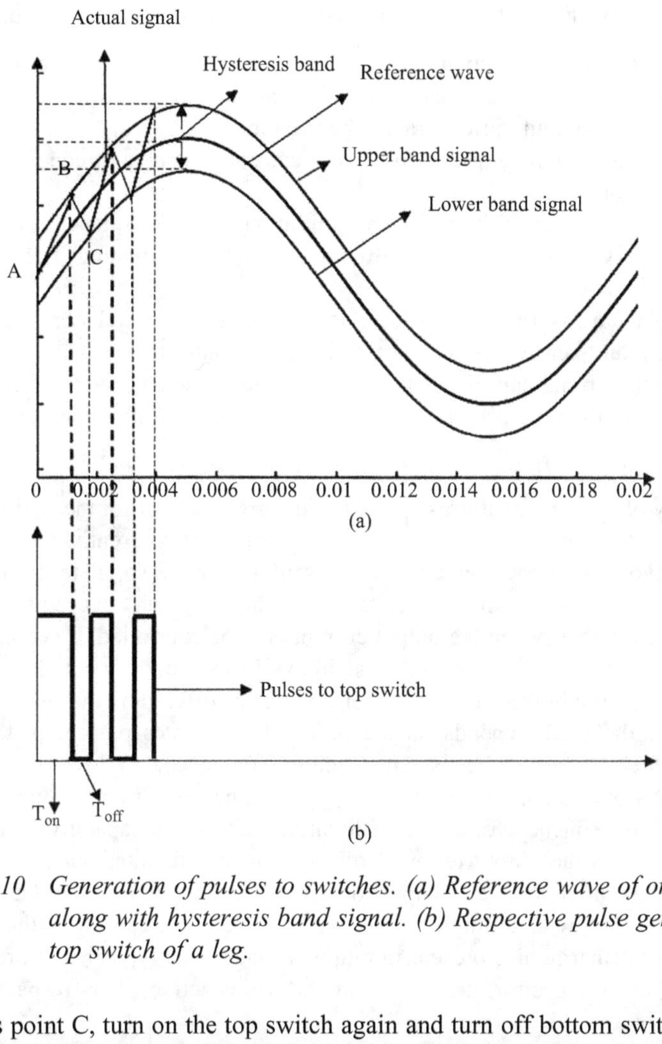

Figure 7.10 Generation of pulses to switches. (a) Reference wave of one phase along with hysteresis band signal. (b) Respective pulse generated of top switch of a leg.

it touches point C, turn on the top switch again and turn off bottom switch. Pulses for top switch, thus generated, is shown in Figure 7.10(b). Continue this switching pattern in other phases 'b' and 'c' so that the actual converter output current tracks the reference current in each phase.

In the hysteresis-controlled method too, the output voltage will vary from $+V_{DC}/2$ to $-V_{DC}/2$ as in the case of SPWM technique. However, the current output of the converter is independent of load connected and depends on the selected reference signal. It contains unwanted higher-order harmonic components in addition to fundamental component of current, which is the desired component. By decreasing the bandwidth, switching frequency can be increased, which shifts the higher order of harmonics present in output current further away from the frequency of fundamental component so that higher-order harmonic component can be easily removed using passive filter.

It can be concluded that using the hysteresis current control technique:

(i) The magnitude of output current of a converter can be controlled by controlling the reference signal to any desired value.
(ii) The actual output current tracks the reference signal.
(iii) The fundamental frequency of output current is same as frequency of reference signal.
(iv) The higher order of harmonics present in output current depends on width of hysteresis band. Higher width results in low switching frequency and vice versa.
(v) Output current of the converter is independent of load connected to the system and hence acts as controlled current source.
(vi) Output current can be sinusoidal or non-sinusoidal, which helps active compensators to work efficiently under non-sinusoidal conditions.

7.3.1.3 Output filter

The quality of output waveforms applied to the grid through compensating devices depends on the filter topology. A proper filter design [4] is essential to synchronize the output voltage of the converter to the grid as per the stipulated standards. A first-order passive filter, which consists of an inductor, is widely used at the output side of the converter, when the output current is to be controlled. It is connected in series with the converter output terminal so that high-frequency components in the current may be minimized (due to higher impedance offered by the filter inductor). The size of the filter depends on the order of harmonics present in the output current, which in turn depends on switching frequency. At low switching frequency, the size of the inductor will be bigger, and hence it is not recommended for low switching frequency conditions. Similarly, a first-order capacitive filter can be connected across the output terminal of the converter to filter out the harmonics present in the output voltage. A capacitor will absorb the higher-order harmonics as its impedance is low under high frequency. The size of the capacitor filter depends on the order of harmonics present in output voltage. The typical first order filters with inductor or capacitor, used in shunt and series active filters respectively are shown in Figure 7.11.

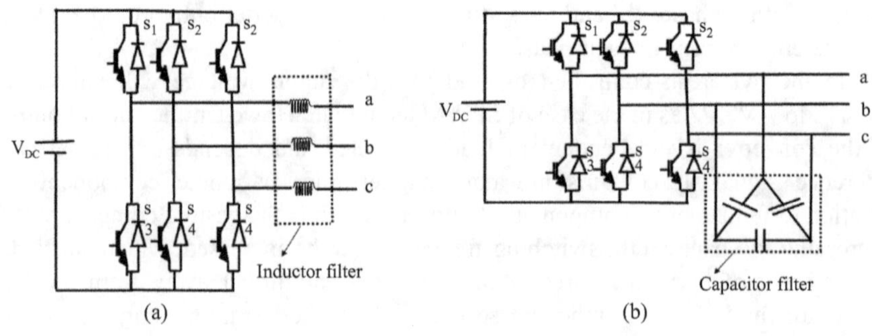

Figure 7.11 Inverters connection. (a) With inductive filter. (b) With capacitive filter.

Second order LC filters are widely used for voltage filtering, and third order filter is used for current control at low switching frequency situations. The design of simple L filter is explained below.

Inductive filter design

Assume the converter circuit of Figure 7.5 with an inductive filter connected to the grid [5] with a fundamental frequency of 50 Hz. In this case, the voltage at the input side of the inductor (filter) varies from zero to V_{DC} during the positive half cycle and $-V_{DC}$ to zero during the negative half cycle, as shown in Figure 7.12. It indicates that in addition to the fundamental component of sinusoidal waveform with 50 Hz, there are higher-order harmonics present in it, which are to be removed to connect it with the grid. It is known that the voltage at the output side of the inductor or at the grid side will be a sine wave with 50 Hz and appropriate magnitude and phase sequence. The magnitude and phase sequence of the converter are already controlled using an appropriate reference signal as explained in the earlier section. However, the design of the filter inductor to be done properly to remove higher-order harmonics from the output of the converter. Assume that an inductive filter as shown in Figure 7.12 (only one phase waveform is shown) is used as a filter for converter which operate at high switching frequency. The voltage at the input side of the filter is same as the VSC output, while the other side is the grid voltage, i.e., sinusoidal waveform as shown in the figure. The grid voltage is sinusoidal with a particular magnitude, frequency and phase sequence, while the VSC output is in the form of pulses as shown.

The input voltage of inductor is V_{DC}, the inductor current increases and the energy are stored in the inductor as shown in Figure 7.13(a). When V_{DC} is zero, the inductor energy will be released and current decreases as shown in Figure 7.13(b). Assume, i_i be the instantaneous current through the inductor and I_{iave} is the average current through it. Also, let ΔI_i be the peak-to-peak current through the inductor.

The maximum current through inductor (I_{imax}) is given by:

$$I_{imax} = I_{iave} + \Delta I_i/2 \tag{7.5}$$

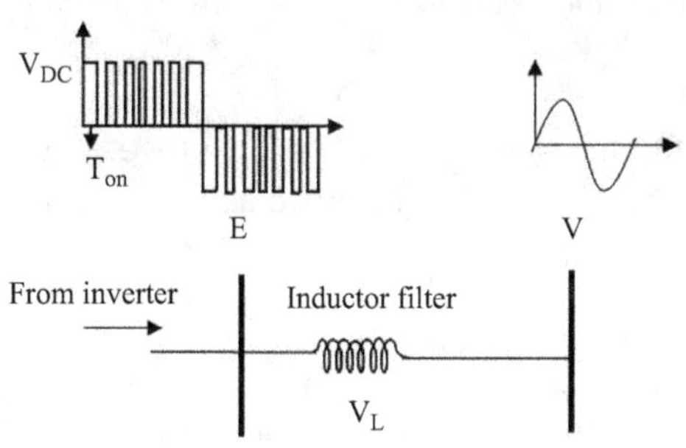

Figure 7.12 Filter inductor of the shunt active filter connected to the grid, along with its input and output terminal waveforms

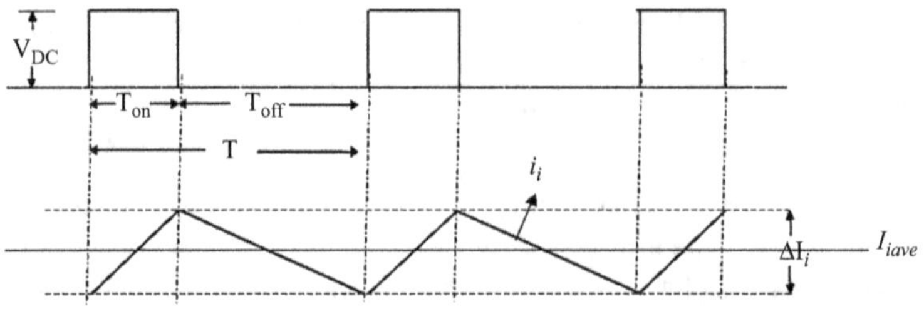

Figure 7.13 Waveforms of inverter terminal voltage and inductor current. (a) Voltage at the output terminals of the inverter. (b) The inductor current.

By Faradays laws of electromagnetic induction, voltage across the inductor, V_L is given by

$$V_L = L \times \frac{di_i}{dt} \tag{7.6}$$

Refer to Figure 7.12, assume E as average voltage at the output side of inductor filter (output of inverter terminal including filter) and V_{DC} as the DC bus voltage. During the ON time of the switch, voltage across the inductor is $E - V$. Hence, using (7.6), it can be written as:

$$E - V = L \times \Delta i_i / DT \tag{7.7}$$

where T is the switching period and is given by $T = T_{on} + T_{off}$ and duty cycle, $D = \frac{T_{on}}{T}$.

From (7.7), Δi_i is written as:

$$\Delta i_i = \frac{DT \times (E - V)}{L} \tag{7.8}$$

Assuming modulation index M_a, duty cycle at any time $(D(\omega t))$, which depends on M_a and instantaneous value of sinusoidal waveform is given by:

$$D(\omega t) = M_a \times Sin\ (\omega t) \tag{7.9}$$

Referring Figure 7.13, during on time $E = V_{DC}$ and $V = D \times V_{DC}$, $\frac{V}{E} = D$

Substituting (7.9) in (7.8), ΔI_i is modified as:

$$\Delta i_i = \frac{M_a \times Sin(\omega t)T\ (E - V)}{L} \tag{7.10}$$

Equation (7.10) can be modified as:

$$\Delta i_i = \frac{E \times M_a \times Sin(\omega t)T\ (1 - D)}{L} \tag{7.11}$$

Substituting (7.9) in (7.11),

$$\Delta i_i = \frac{V_{DC} \times M_a Sin(\omega t) \times T\ (1 - M_a Sin(\omega t))}{L} \tag{7.12}$$

The value of M_a at which the maximum ripple occurs is found out by making,

$$\frac{d(\Delta i_i)}{dt} = 0 \tag{7.13}$$

Simplify (7.12) as follows:

$$\Delta i_i = \frac{V_{DC} \times M_a \times T \times \left(\sin(\omega t) - M_a \times \sin^2(\omega t)\right)}{L} \tag{7.14}$$

$$\Delta i_i = \frac{V_{DC} \times M_a \times T}{L} \times \left(\sin\ (\omega t) - M_a \frac{1 - \cos(2\omega t)}{2}\right) \tag{7.15}$$

Differentiating (7.15) with respect to time:

$$\frac{V_{DC} \times M_a \times T}{L}\left(\omega\cos(\omega t) - M_a \frac{2\omega \times \sin(2\omega t)}{2}\right) = 0 \tag{7.16}$$

By rearranging (7.16)

$$\cos(\omega t) = M_a \sin(2\omega t) \tag{7.17}$$

$$\sin(\omega t) = \frac{1}{2M_a} \tag{7.18}$$

By substituting (7.18) in (7.12), maximum ripple current $\Delta i_{L\max}$ is given by:

$$\Delta i_{imax} = \frac{V_{DC} \times T}{4 \times L} \tag{7.19}$$

By re-arranging (7.19), the value of inductor for maximum ripple current is given by:

$$L = \frac{V_{DC} \times T}{4 \times \Delta i_{imax}} \tag{7.20}$$

Using the relation $T = \frac{1}{F_s}$, where F_s is the switching frequency, the value of filter inductor in terms of switching frequency is given by:

$$L = \frac{V_{DC}}{4 \times F_{s\times} \Delta i_{imax}} \tag{7.21}$$

So, using (7.21), the design of inductor for an inverter can be done by assuming $\Delta i_{L\max}$ to an acceptable value. Using only inductor as filter, the switching frequency should be too high to eliminate harmonics effectively. Various types of filters used in DC to AC converters such as first order (L filter), second order (LC filter) and third

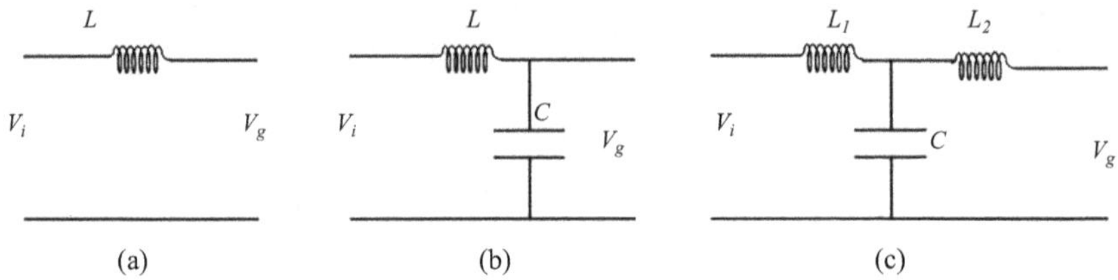

Figure 7.14 Filters used for interfacing VSC-based compensators with the grid. (a) L filter. (b) LC filter. (c) LCL filter.

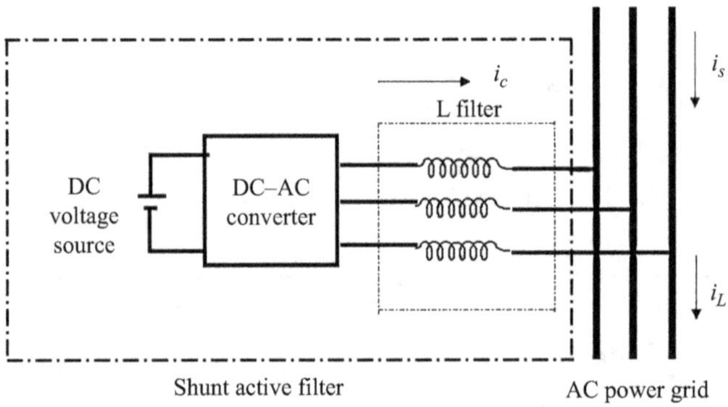

Figure 7.15 A VSC-based compensator connected to the grid as a shunt compensator – a shunt active filter

order (LCL Filter) are shown in Figure 7.14. Selection of proper based on requirement is very important for the implementation of efficient VSC-based compensators.

7.3.1.4 Interfacing unit

Perfect synchronization between converter output and the grid must be achieved when compensator is connected to the grid. The voltage, frequency and phase sequence must exactly match between converter output terminal and the grid. The converter is connected to the grid using a suitable filter, which eliminates the higher-order harmonic components in inverter output voltage and current waveforms. Based on the interface of VSC-based compensators to the power systems, it is classified as (i) shunt active filter, (ii) series active filter and (iii) shunt-series active filter. Integration of such filters to the grid is briefly explained here.

Integration of shunt active filter to grid

A shunt active filter helps in improving the current quality through the transmission line, which indirectly improves voltage quality too. A shunt active filter is used to control source current (i_s) to a desired value by injecting required compensation current (i_c) into the grid, which compensates for the unwanted component of load current (i_L) as shown in Figure 7.15. The shunt active filter is connected in shunt with the load so that the reactive component and distortion component of load current are supplied by the shunt active filter. In other words, the source current is controlled to meet just the active component of load current by injecting sufficient current into the grid using VSC-based compensators as a current-controlled voltage source inverter, which is explained later in this chapter.

By injecting the desired amount of current or reactive power into the system using a shunt active filter, the quality of current flowing through the transmission line can be controlled. Under sinusoidal supply voltage and linear load conditions, current i_c is sinusoidal in shape, whereas under sinusoidal supply and non-linear load conditions, it is non-sinusoidal for full compensation. With full compensation, the current i_s becomes sinusoidal and in phase with the supply voltage. This indicates that the shunt active filter works well under linear and non-linear load conditions.

Integration of a series active filter to the grid

Consider a 3-ϕ incoming feeder with a bus voltage v_{BUS} as shown in Figure 7.16. Assume that the voltage quality at the incoming point (v_{BUS}) is poor, and it is required to control the voltage to a desired value (v_{BUSmod}). In such cases, a series compensation can be used to improve voltage quality. A series active filter, which consists of VSC along with its filter, injects the required voltage v_c in series with

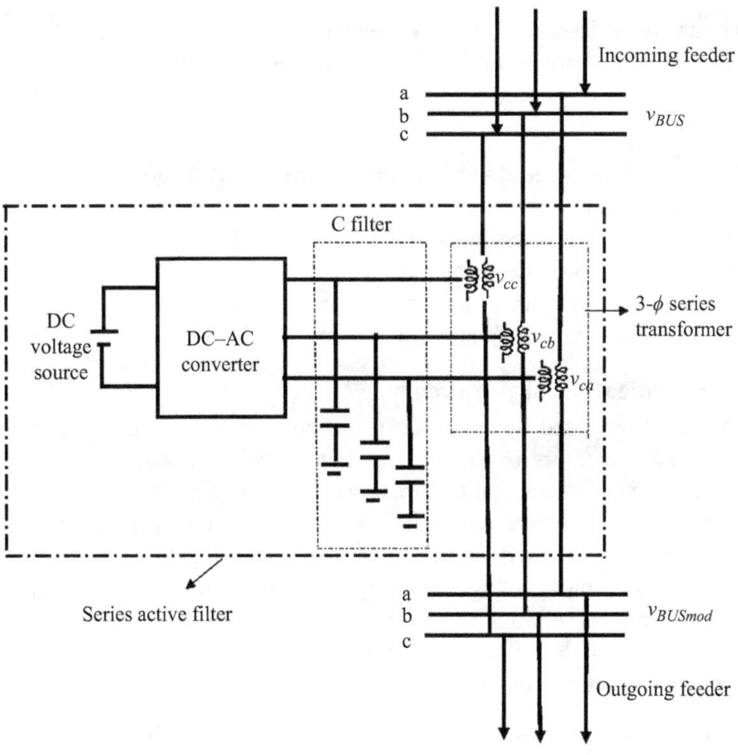

Figure 7.16 A VSC-based compensator connected to the grid as a series compensator – a series active filter

v_{BUS}, through a series transformer (generally 1:1 ratio) as shown in the figure. Thus, the bus voltage of the outgoing feeder is modified to $v_{BUS\text{mod}} = v_{BUS} \pm v_c$. The injected voltage, v_c may be sinusoidal or non-sinusoidal, depending on the voltage quality at the point of interest, v_{BUS}. This indicates that VSC acts as voltage controlled VSC, through which v_c can be controlled to improve voltage quality at a particular point. At the same time, a series active filter can also be used to control transmitted power through the line. Under this case, it indirectly controls the reactance of the transmission line, due to which the angle between the sending end and receiving end voltage is controlled, which results in active power flow control.

The performance of the compensators depends on the output current or voltage generated by the VSC, which, in turn, depends upon the technique used to generate the reference waveform for the converter operation. Generally, if VSC-based compensators operate as current controlled inverter, a current reference is generated using a proper technique to get a desired current at the output of VSC, whereas if VSC operates as a voltage-controlled inverter, a voltage reference waveform is required to get a desired voltage at the output of the inverter. The generation of the reference waveform is explained in the next section.

7.3.2 *Concepts of instantaneous real and imaginary powers*

The various methods based on the following theories are used to determine the useful and unwanted power in the system for power quality improvement. A few such methods [6–8] are listed below, and only p-q theory is explained in detail in this section.

(i) The instantaneous real and imaginary power (p-q) theory
(ii) Modified instantaneous reactive power theory
(iii) Synchronous reference frame theory
(iv) Modified synchronous frame theory
(v) The instantaneous active and reactive current (i_d-i_q) theory
(vi) Instantaneous power compensation using p-q-r theory

7.3.2.1 Application of p-q theory

The p-q theory proposed by Akagi *et al.* in the year 1983 is used for control of VSC-based compensators in the system. It is widely known as the generalized theory of the instantaneous real and imaginary power. p-q theory helps to find the reference current or voltage required for the implementation of converter-based compensators. As it is based on a time domain approach, it is applicable to sinusoidal and non-sinusoidal voltage and current. The advantage of p-q theory is that it works well with all conditions of power systems, i.e., when the system is balanced or unbalanced and with or without harmonics.

The following points are discussed in this section

(i) Transformation of voltage or current a-b-c frame to α-β-0 reference frame
(ii) Real and imaginary instantaneous power using α-β-0 reference frame
(iii) Classification of real and imaginary instantaneous power
(iv) Reactive and harmonic power compensation

Transformation of voltage or current in a-b-c frame to α-β-0 reference frame

The p-q theory is based on Clarke transformation, which transforms three-phase rotating instantaneous voltage or current in a-b-c frame into a stationary α-β-0 reference frame. The various axes in a-b-c frame are 120° displaced from each other, whereas the axes in α-β frame are orthogonal to each other as shown in Figure 7.17. 0-coordinate represents zero sequence component. The instantaneous voltage and current in a 3-ϕ system can be represented by instantaneous space vectors in a-b-c coordinates, i.e., a, b, and c axes are on the same plane separated by 120° as shown in Figure 7.17(a). These three vectors can be converted to three vectors in the stationary frame using the α-β-0 transformation as shown in Figure 7.17(b).

Using Clark's transformation principle, voltage and current in a-b-c coordinate (v_a, v_b, v_c and i_a, i_b, i_c, respectively) can be converted to α-β-0 reference frame (v_α, v_β, v_0 and i_α, i_β, i_0, respectively) as given in (7.22) and (7.23):

$$\begin{bmatrix} v_0 \\ v_\alpha \\ v_\beta \end{bmatrix} = \sqrt{\frac{2}{3}} \begin{bmatrix} \frac{1}{\sqrt{2}} & \frac{1}{\sqrt{2}} & \frac{1}{\sqrt{2}} \\ 1 & -\frac{1}{2} & -\frac{1}{2} \\ 0 & \frac{\sqrt{3}}{2} & \frac{-\sqrt{3}}{2} \end{bmatrix} \begin{bmatrix} v_a \\ v_b \\ v_c \end{bmatrix} \tag{7.22}$$

$$\begin{bmatrix} i_0 \\ i_\alpha \\ i_\beta \end{bmatrix} = \sqrt{\frac{2}{3}} \begin{bmatrix} \frac{1}{\sqrt{2}} & \frac{1}{\sqrt{2}} & \frac{1}{\sqrt{2}} \\ 1 & -\frac{1}{2} & -\frac{1}{2} \\ 0 & \frac{\sqrt{3}}{2} & \frac{-\sqrt{3}}{2} \end{bmatrix} \begin{bmatrix} i_a \\ i_b \\ i_c \end{bmatrix} \tag{7.23}$$

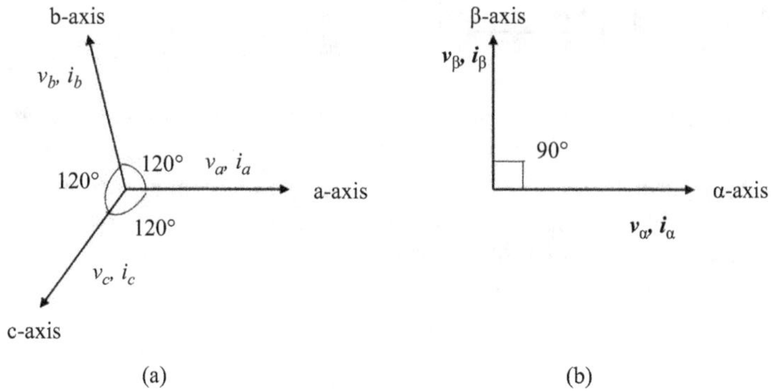

Figure 7.17 Conversion of a-b-c coordinate to α-β reference frames. (a) a-b-c reference frame. (b) α-β coordinate system.

The α-β-0 reference frame of voltage and current simplifies the calculation of instantaneous active and reactive power powers under any situation of power system, such as sinusoidal or non-sinusoidal, balanced or unbalanced conditions, with neutral or without neutral conductor, etc. For many applications, it is required to convert the α-β-0 reference frame of voltage and current back to its a-b-c frame for actual implementation of the system. From α-β-0 values, it is possible to calculate corresponding a-b-c values using the inverse transform. For example, the instantaneous values of current in a-b-c frame are calculated using α-β-0 values by applying inverse transformation in (7.24) as follows:

$$\begin{bmatrix} i_a \\ i_b \\ i_c \end{bmatrix} = \sqrt{\frac{2}{3}} \begin{bmatrix} \frac{1}{\sqrt{2}} & \frac{1}{\sqrt{2}} & \frac{1}{\sqrt{2}} \\ 1 & -\frac{1}{2} & -\frac{1}{2} \\ 0 & \frac{\sqrt{3}}{2} & \frac{-\sqrt{3}}{2} \end{bmatrix}^{-1} \begin{bmatrix} i_0 \\ i_\alpha \\ i_\beta \end{bmatrix} \tag{7.24}$$

Similarly, voltage in α-β-0frame can be transformed into the corresponding a-b-c frame. One of the advantages of this transformation is that the zero sequence reference frame (v_0 and i_0) can be eliminated if there is no zero-sequence component present in the system. Generally, in the transmission system, zero sequence current is not present due to the absence of a neutral conductor, and hence v_0 and i_0 can be neglected.

Thus, the equations (7.22) and (7.23) are modified as:

$$\begin{bmatrix} v_\alpha \\ v_\beta \end{bmatrix} = \sqrt{\frac{2}{3}} \begin{bmatrix} 1 & \frac{-1}{2} & \frac{-1}{2} \\ 0 & \frac{\sqrt{3}}{2} & \frac{-\sqrt{3}}{2} \end{bmatrix} \begin{bmatrix} v_a \\ v_b \\ v_c \end{bmatrix} \tag{7.25}$$

$$\begin{bmatrix} i_\alpha \\ i_\beta \end{bmatrix} = \sqrt{\frac{2}{3}} \begin{bmatrix} 1 & \frac{-1}{2} & \frac{-1}{2} \\ 0 & \frac{\sqrt{3}}{2} & \frac{-\sqrt{3}}{2} \end{bmatrix} \begin{bmatrix} i_a \\ i_b \\ i_c \end{bmatrix} \tag{7.26}$$

Real and imaginary instantaneous power using α-β-0 reference frame

Using the α-β transformation of instantaneous voltage and current from a-b-c frame, it is possible to find out the instantaneous active and reactive power, i.e., p and q, in the system in a simple way. Assume v and i are the voltage and current vectors, respectively, at any instant in the α-β frame (refer to Figure 7.17(b)), and these vectors in the α-β axis can be represented as:

$$v = v_\alpha + jv_\beta \text{ and } i = i_\alpha + ji_\beta \tag{7.27}$$

The complex power is given by:

$$s = v \times i^* = (v_\alpha + jv_\beta) \times (i_\alpha - ji_\beta)$$
$$= (v_\alpha i_\alpha + v_\beta i_\beta) + j(v_\beta i_\alpha - v_\alpha i_\beta) = p + jq \quad (7.28)$$

By separating the real and imaginary parts of (7.28), the instantaneous real and reactive power in α-β reference frame is given by:

$$p = v_\propto \cdot i_\alpha + v_\beta \cdot i_\beta \quad (7.29)$$

$$q = v_\beta \cdot i_\alpha - v_\alpha \cdot i_\beta \quad (7.30)$$

By combining (7.29) and (7.30), it can be written in Matrix form as:

$$\begin{bmatrix} p \\ q \end{bmatrix} = \begin{bmatrix} v_\propto & v_\beta \\ v_\beta & -v_\propto \end{bmatrix} \begin{bmatrix} i_\alpha \\ i_\beta \end{bmatrix} \quad (7.31)$$

Classification of real and imaginary instantaneous power

p and q, calculated using p-q theory, have the following main components.

(i) Instantaneous zero-sequence power (p_0)
(ii) Instantaneous real power (p)
(iii) Instantaneous imaginary power (q)

Each of the above power can be again divided into the mean value and alternating value of the respective power.

Instantaneous zero-sequence power (p_0). Instantaneous zero sequence power indicates power transferred or exchanged between the source and load due to zero sequence voltage and current. It contains two components of power (a) mean value of zero sequence power ($\overline{p_0}$) which is power transferred due to a similar order of zero sequence component present in voltage and current (b) alternating value of zero sequence power ($\tilde{p}_0$) which is power exchanged due to the dissimilar order of the zero-sequence component present in voltage and current.

The total zero sequence power is given by:

$$p_0 = v_0 \cdot i_0 = \overline{p_0} + \tilde{p}_0 \quad (7.32)$$

The zero-sequence power needs to be eliminated from the power system to have an efficient power system. In the absence of a neutral conductor, such power will not flow through the system.

Instantaneous real power (p). Instantaneous real power also consists of two components: (a) the mean value of instantaneous real power ($\overline{p}$) which is the real power transferred from source to load and useful power due to a similar order of harmonic components of voltage and current, (b) alternating value of real power ($\tilde{p}$) which is the real power exchanged between source and load due to higher-order harmonic components present in power, or may be due to the dissimilar order of harmonics of voltage and current. The component of power in (a) is required for the operation of the equipment and is to be supplied by the source. However, the power given in (b) is to be removed from the supply system to have a healthy system. The instantaneous real power in terms of $\alpha-\beta$ reference frame is written as:

$$p = v_{\alpha} \cdot i_{\alpha} + v_{\beta} \cdot i_{\beta} = \overline{p} + \tilde{p} \tag{7.33}$$

$\overline{p}$ is equivalent to the conventional average 3-ϕ power $P = 3 \times V \times I \times \cos\phi$

Instantaneous imaginary power (q). The instantaneous imaginary power in terms of $\alpha-\beta$ reference frame is written as:

$$q = v_{\beta} \cdot i_{\alpha} - v_{\alpha} \cdot i_{\beta} = \overline{q} + \tilde{q} \tag{7.34}$$

As in the case of real power, the imaginary power, q, also has two components (a) mean value of imaginary power ($\overline{q}$), which indicates the reactive power flow due to a similar order of harmonic components of voltage and current (b) alternating value of imaginary power ($\tilde{q}$), which is the same as reactive power flow due to the dissimilar order of the harmonic component of voltage and current. $\overline{q}$ is equivalent to conventional reactive power $Q = 3 \times V \times I \times \sin\phi$

By combining (7.32), (7.33) and (7.34), the power matrix including zero sequence power is written as:

$$\begin{bmatrix} p_0 \\ p \\ q \end{bmatrix} = \begin{bmatrix} v_0 & 0 & 0 \\ 0 & v_{\alpha} & v_{\beta} \\ 0 & v_{\beta} & -v_{\alpha} \end{bmatrix} \begin{bmatrix} i_0 \\ i_{\alpha} \\ i_{\beta} \end{bmatrix} \tag{7.35}$$

Using active filters, it is possible to minimize system current by eliminating the unwanted power component p_0, $\tilde{p}$ q from the supply system. In other words, power quality and power factor in a power system can be improved using active filters.

Reactive and harmonic power compensation

With the understanding of useful and unwanted power in the systems, the reference compensating current or/and voltage required for the operation of VSI to improve power quality can be easily decided. Assume a sinusoidal supply system connected to a load, which is a combination of linear and non-linear loads for analysis. Basic understanding of the removal of unwanted power for power quality improvement for (i) shunt compensation, (ii) series compensation and (iii) series-shunt compensation is explained in this section. Since harmonic compensation is included, active filters are used in the remaining section for explanation. However, the basic principle is applicable to other compensators too.

Power quality improvement using shunt active filter. Consider a line diagram of 3-ϕ sinusoidal supply system with loads (combination of linear and non-linear loads) at a receiving end bus through a transmission line with impedance, z as shown in Figure 7.18(a). As there is no compensation in the system, the load current i_{La}, i_{Lb}, i_{Lc} is same as the supply source current i_{sa}, i_{sb}, i_{sc}.

Without shunt compensation, the power balance equation is $p_s = p_L$ and $q_s = q_L$ as shown in Figure 7.18(a). As the load is a combination of linear and

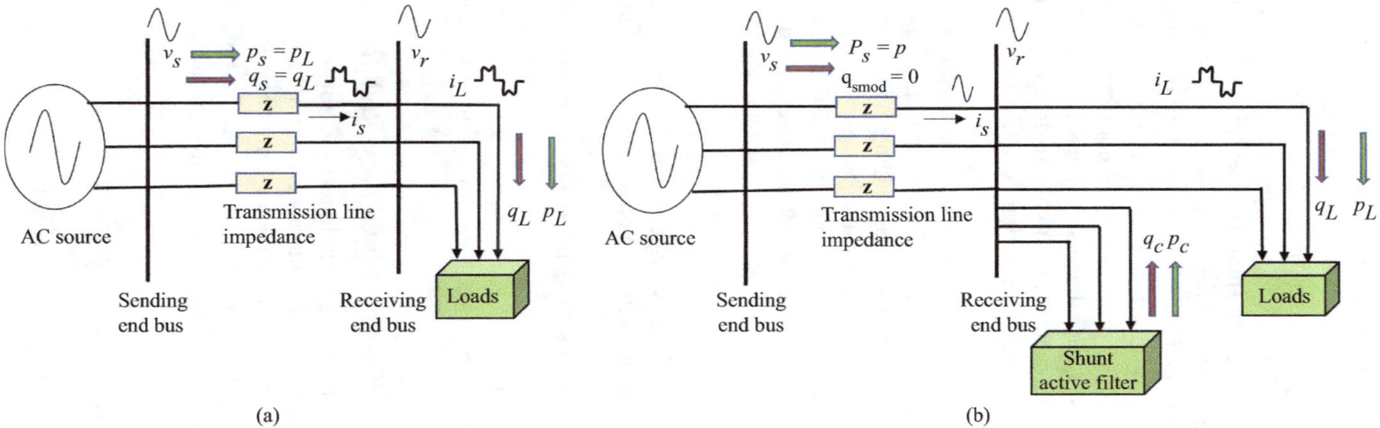

Figure 7.18 Line diagram of a simple 3-ϕ supply system. (a) Without shunt compensation. (b) With shunt compensation.

non-linear loads, i_s lags v_r and distorted as shown in the figure. Here, the power quality is poor due to reactive and harmonic power which is termed as unwanted power. As explained in earlier sections, alternating power ($\tilde{p}$) from p_L and complete reactive power q_L to be removed from the system, so that is becomes sinusoidal and in phase with the load voltage. Thus, the power quality can be improved. This is achieved using a shunt compensator [9] connected across the load as shown in Figure 7.18(b). The shunt compensator supplies the reactive power and harmonic power demanded by the load by injecting a compensating power $q_c = \tilde{p} + q$ into the system resulting in improvement in power quality. Under this condition, following parameters get modified (i) p_s becomes $\overline{p}$ (ii) q_s is reduced to zero (iii) i_s is modified to a sinusoidal current in phase with v_r (iv) the magnitude of i_s becomes minimum (v) voltage drop across transmission line impedance z reduces and hence load voltage quality is improved (vi) power factor is improved to unity.

Power quality improvement using series compensation. Assume that the receiving end voltage is deviated from its ideal one in its magnitude and shape as shown in Figure 7.19. Also, for better understanding, assume that linear loads are connected to such a system. Under such situation, as voltage quality is poor, the current becomes non-sinusoidal though linear loads are connected to the system as shown in Figure 7.19(a) and thus, both the voltage and current quality become poor. Using series compensation, the non-sinusoidal voltage at receiving end bus v_r can be modified to a sinusoidal voltage (v_{rmod}) and thus the load current and source current becomes sinusoidal as shown in Figure 7.19(b). However, if non-linear loads are connected to the system, voltage can be made sinusoidal using series compensation, but load current will be non-sinusoidal due to non-linearity of the load.

Thus, in series compensation, the following operations are possible: (i) to shape the non-sinusoidal voltage to sinusoidal, (ii) to increase or decrease the magnitude of v_r, (iii) to control the phase shift between v_r and v_{rmod}, which is generally used to control active power flow through the line (for example, in the case of SSSC).

7.3.3 Generation of reference signal for active filters

Depending on the application and technique used for switching, a reference current or voltage signal is to be generated for implementation of current-controlled or voltage-controlled VSC- based compensators, respectively. The performance of the compensators depends on the method of generation of the reference compensation signal, implementation of the algorithm, design of the inverter used for injecting the compensation voltage or current into the system, and the technique by which the inverter injects the appropriate compensating signal. Generally, reference compensating current (i_c*) and reference compensating voltage (v_c*) are to be calculated for the operation of shunt and series active filters, respectively. Whereas, both i_c* and v_c* are required for series-shunt active filters.

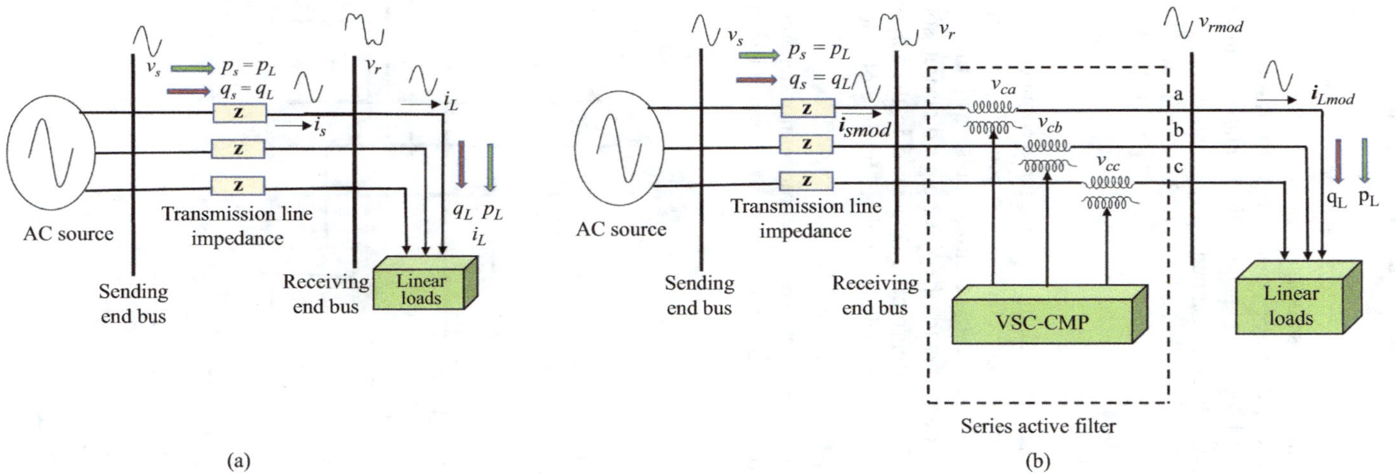

Figure 7.19 Line diagram of a simple 3-ϕ supply system. (a) Without series compensation. (b) With series compensation.

7.3.3.1 Generation of reference compensating current for shunt active filter

Generally, p-q theory is used to generate a reference compensating current for a shunt active filter. A shunt compensator directly injects the compensating current required to eliminate unwanted source power from the system. Consider the line diagram given in Figure 7.18(a), by sensing voltage and current in a-b-c rotating frame, active and reactive power consumed by the load can easily be calculated using the stationary α-β reference frame using (7.31) as follows:

$$\begin{bmatrix} p \\ q \end{bmatrix} = \begin{bmatrix} v_{s\alpha} & v_{s\beta} \\ v_{s\beta} & -v_{s\alpha} \end{bmatrix} \begin{bmatrix} i_{L\alpha} \\ i_{L\beta} \end{bmatrix} \tag{7.36}$$

Now consider Figure 7.20, where a shunt active filter is connected across a load. To implement a shunt active filter, the reference compensating current in each phase (i_{ca}*, i_{cb}*, i_{cc}*) is to be calculated, and then any pulse width modulation (PWM) technique is used to generate pulses for the VSC to get a desired source current, i_s.

The reference compensating current is calculated in the following ways: (a) direct method of calculation for i_c* and (b) indirect method of calculation for i_c*.

*Direct method of calculation for i_c**

The compensating current to be injected by the compensator depends on the portion of load power that is to be removed from the system or transmission lines. In other

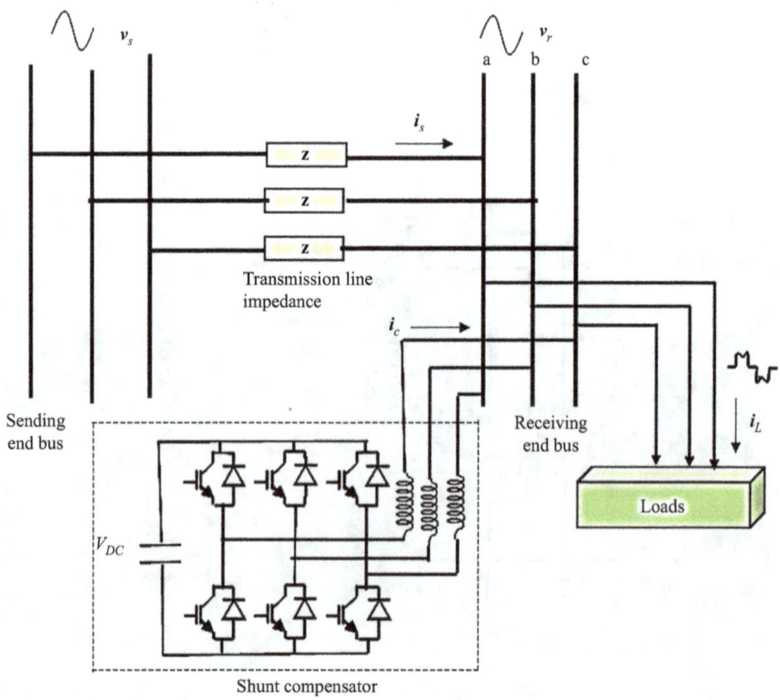

Figure 7.20 Schematic diagram of a shunt active filter connected in power system

words, the injected compensating current needs to supply the unwanted power from the source to the load. Using this concept and taking the inverse of (7.36), the reference compensating current, $i_c{*}$ in α-β frame, is calculated by putting $p = -\tilde{p}$ and $q = -\bar{q} - \tilde{q}$ as follows:

$$\begin{bmatrix} i_{c\alpha}* \\ i_{c\beta}* \end{bmatrix} = \begin{bmatrix} v_{s\alpha} & v_{s\beta} \\ v_{s\beta} & -v_{s\alpha} \end{bmatrix}^{-1} \begin{bmatrix} -\tilde{p} \\ -\bar{q} - \tilde{q} \end{bmatrix} \tag{7.37}$$

The reference compensating current in a-b-c coordinates is required to be generated for actual implementation of the shunt active filter. This can be obtained by the inverse transform of (7.23) and is given by:

$$\begin{bmatrix} i_{ca}* \\ i_{cb}* \\ i_{cc}* \end{bmatrix} = \sqrt{\frac{2}{3}} \begin{bmatrix} \frac{1}{\sqrt{2}} & \frac{1}{\sqrt{2}} & \frac{1}{\sqrt{2}} \\ 1 & \frac{-1}{2} & \frac{-1}{2} \\ 0 & \frac{\sqrt{3}}{2} & \frac{-\sqrt{3}}{2} \end{bmatrix}^{-1} \begin{bmatrix} i_{c\alpha}* \\ i_{c\beta}* \end{bmatrix} \tag{7.38}$$

This indicates that the shunt compensator should inject a current equal to the reference compensating current during its real time implementation.

Indirect method of calculation for $i_c{*}$

In this case, the reference source current ($i_s{*}$) is calculated using p-q theory, and then $i_c{*}$ is calculated. This is based on the principle that after compensation, the source current should meet only the useful power demanded by the load, (7.37) can be modified for reference source current in the α-β frame as:

$$\begin{bmatrix} i_{s\alpha}* \\ i_{s\beta}* \end{bmatrix} = \begin{bmatrix} v_{s\alpha} & v_{s\beta} \\ v_{s\beta} & -v_{s\alpha} \end{bmatrix}^{-1} \begin{bmatrix} \bar{p} \\ 0 \end{bmatrix} \tag{7.39}$$

The reference compensating current in the α-β frame is given by:

$$\begin{bmatrix} i_{c\alpha}* \\ i_{c\beta}* \end{bmatrix} = \begin{bmatrix} i_{L\alpha} \\ i_{L\beta} \end{bmatrix} - \begin{bmatrix} i_{s\alpha}* \\ i_{s\beta}* \end{bmatrix} \tag{7.40}$$

The compensating current in a, b and c phases can be found in a similar way as (7.38). The indirect method is easy compared to the direct method under non-linear load conditions. The next step is real-time implementation of the shunt active filter, which is explained below.

The equivalent circuit of a shunt active compensator connected to a power system is shown in Figure 7.21. VSC acts as a controlled current source, as shown in the figure, when it operates as a shunt compensator.

A shunt compensator can be installed by consumers or the utility. If all the consumers connect shunt compensators at their respective ends, it is possible to make power factor of unity with the best power quality or voltage quality.

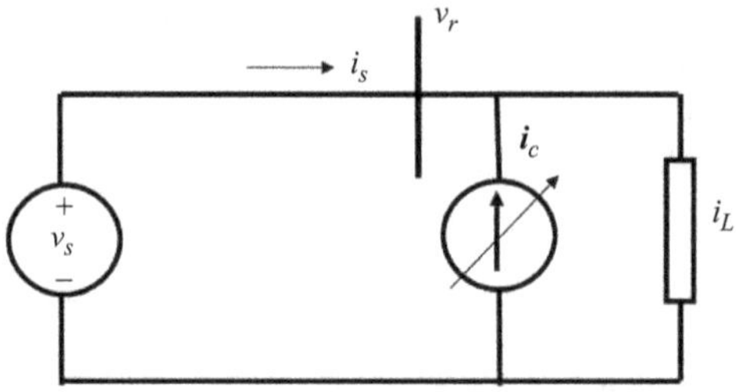

Figure 7.21 Equivalent circuit of the system with a shunt active filter

However, it may not be feasible for a low-power consumer to connect a shunt compensator at the consumer's end. In such cases, a penalty can be levied on the consumer based on the reactive and harmonic power drawn by the consumer. Then, the utility may install a shunt compensator at the PCC so that the power quality to other consumers can be improved. As it is the duty of utility to supply ideal power to its consumers, the implementation of shunt compensator may be the best solution to satisfy the consumers with a good quality power supply.

7.3.3.2 Generation of reference compensating voltage for series active filter

Another solution to improve voltage quality is to install a VSC-based series compensator, as shown in Figure 7.22, which helps in direct control of voltage quality. For the implementation of a series compensator, a reference voltage compensation (v_c*) is to be generated. Using v_c*, pulses to the switches are to be generated using any PWM technique. The generation of compensating voltage for a series compensator is simple compared to reference current generation for shunt compensation. Sense the voltage at any point (v_L) and subtract the desired voltage v_s* from it to get the reference compensation voltage. The compensating reference voltage of series active filter in α-β frame is given by:

$$\begin{bmatrix} v_{c\alpha}* \\ v_{c\beta}* \end{bmatrix} = \begin{bmatrix} v_{L\alpha} \\ v_{L\beta} \end{bmatrix} - \begin{bmatrix} v_{s\alpha}* \\ v_{s\beta}* \end{bmatrix} \tag{7.41}$$

To generate pulses for switches in respective phases, the compensating voltage in α-β frame is transformed to a-b-c frame as in the case of shunt active filter. After the generation of reference compensating voltage v_c* in a-b-c frame a suitable PWM technique is used to turn on and off switches of VSC to inject v_{ca}, v_{cb}, v_{cc}, into the system across secondary winding of series transformer connected in series with the receiving end voltage bus voltage v_r, so that the bus voltage is modified to $v_{rmodified}$. A capacitor filter is connected across VSC to filter higher-order harmonic components of voltage present at the output terminal of the series compensator.

VSC acts as a voltage-controlled voltage source inverter and dynamically controls the voltage at PCC to a desired value, unlike a series passive filter. The equivalent circuit of series compensated network is shown in Figure 7.23. VSC acts

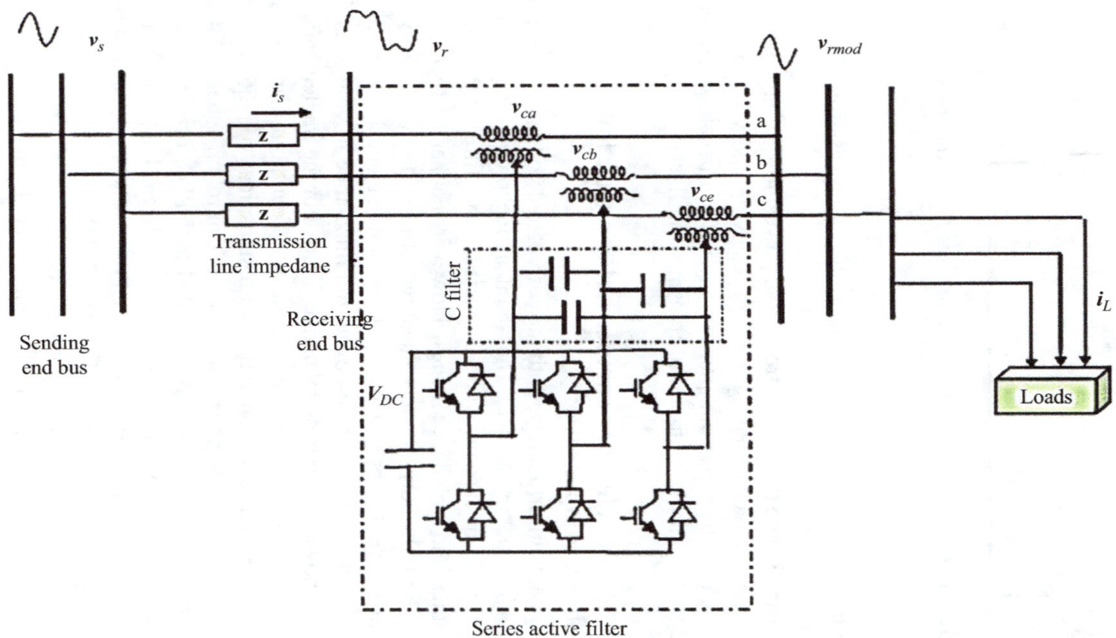

Figure 7.22   Schematic diagram of VSC- based series compensator

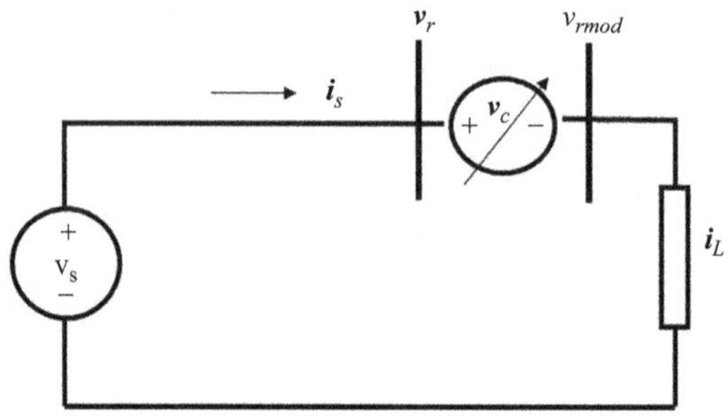

Figure 7.23 Equivalent circuit of the system connected with a series active filter

as a controlled voltage source (v_c) in series active filter compensation, as shown in the figure.

A series filter can also be used to control active power flow through the line by controlling phase angle between sending end and receiving end voltage, which is known as static synchronous series compensation (SSSC). In such cases, the best location for SSSC will be at the midpoint of transmission lines.

7.3.3.3 Generation of reference compensating signal for series-shunt compensation

In today's power system, unfortunately, both the current and voltage waveforms are polluted by harmonics and imbalance. The shunt active provides a solution to harmonics and imbalance in current, while the series active filter is the answer to the voltage harmonics and imbalance issues. However, neither of them alone is sufficient to control both the current and voltage waveform quality. A combination of series and shunt active filters can be used for simultaneously compensating harmonics power as well as the reactive power and imbalance in voltage and current. Such filters are known as series-shunt active filters. For the implementation of a series-shunt active filter, both i_c* and v_C* need to be found out using the above methods. Thus, a unity power factor operation with sinusoidal voltage and current is possible with a series-shunt combination. Such a controller, when used under sinusoidal voltage and current conditions, is called as UPFC, or if it is used for shaping voltage and current, which happens under non-linear load conditions, it is called as a unified power quality conditioner (UPQC). The basic structure of a UPQC and UPFC is similar and shown in Figure 7.24.

From Figure 7.24, it is clear that i_s is controlled by i_c (i.e., shunt part of the compensator) and v_L (same as v_r shown in the figure) is controlled by v_c (i.e., series component of the compensator). Out of the two converters, one acts as a voltage-controlled VSC while the other acts as a current-controlled VSC. In this case too, the reference voltage and current for implementation of the series and shunt part of the compensator can be determined using any of the suitable techniques explained in the above section. Subsequently, the converters can be suitably operated to inject

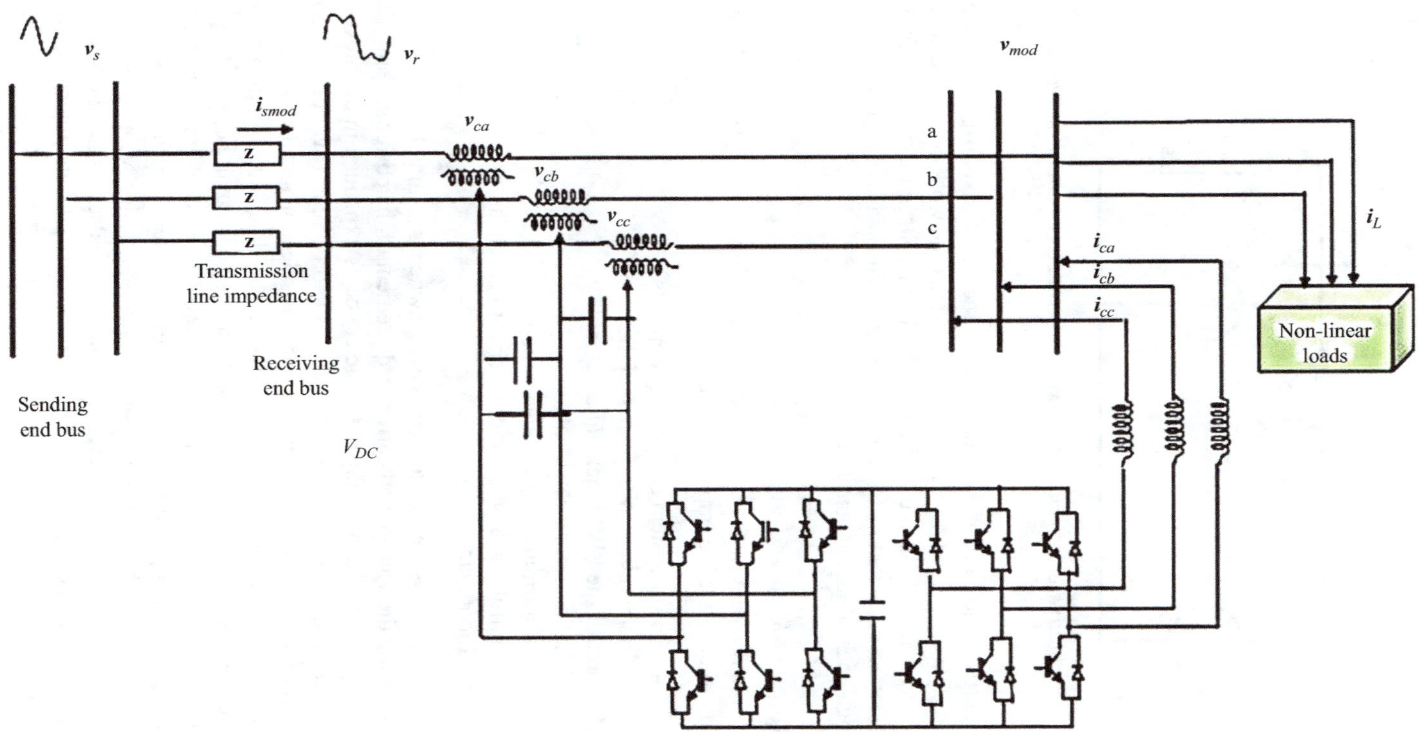

Figure 7.24   A series-shunt active filter placed in a power system network

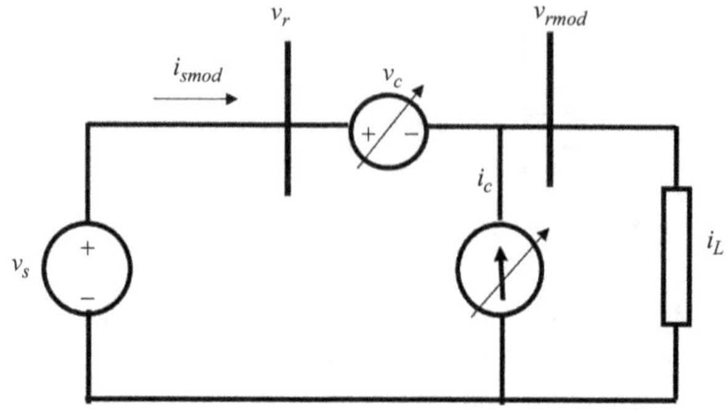

Figure 7.25 Equivalent circuit of the system with series-shunt active filter

the desired current and voltage into the grid to meet the desired power quality requirements. The equivalent circuit of a power system network connected with a series-shunt active filter is shown in Figure 7.25. Both voltage (v_r) and current (i_s) can be modified to v_{rmod} and i_{smod}, respectively, to a sinusoidal voltage and current by controlling series voltage v_c and i_c, respectively.

7.3.4 Implementation of active filters

Based on the requirement of the compensators, the following process is to be carried out before their effective implementation:

(i) Study of the power system and its requirements
(ii) Selection of components with their specification
(iii) Identification of the PWM technique
(iv) Method of finding reference signals
(v) Design of an algorithm for pulse generation
(vi) Design of controllers and additional circuits such as protection circuits, phase-locked loop, etc.

DC source with its specification, rating of switches available and number required for meeting the voltage and current requirement of the power system, filter type and its design can be done using the basic concepts explained in the previous sections. The design of protection circuits, such as short circuit (over current protection) and over voltage protection (Snubber circuits selection and design), are two basic protection needed for high-power switches to be done while implementing high-power active filters. Implementation can be analog or digital; however, digital implementation is preferred over analog because of its ease of implementation. Nowadays, a digital signal processor is used for efficient real-time implementation of the algorithm. Once the circuit is made as per the design values, sense the voltage and current, generate a reference signal and pulses using the selected algorithm and give the pulses properly to the switches for real-time implementation as depicted in Figure 7.26. As voltage and current is varying instantaneous, a phase

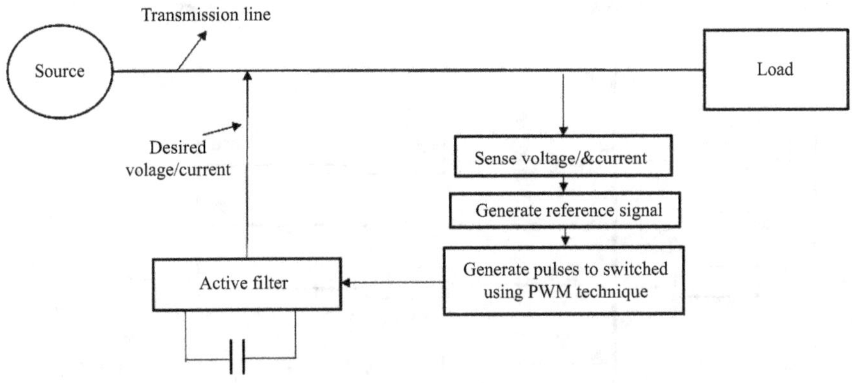

Figure 7.26 A typical process for implementation of algorithm for active filters

locked loop (PLL) is to be implemented to make frequency of the injected voltage and current match the supply frequency of the system. The theory behind PLL is out of scope of this book. PLL makes sure that the compensating voltage and current are injected at the same instant of desired point. In other words, it makes sure the frequency of the output signal tracks the frequency of the input signal. The switching frequency in compensating voltage or current depend on the selection of the control algorithm and its implementation. Generally, active filters are controlled through two control loops (i) for DC bus voltage control (ii) out voltage or current control. Many literatures [10–12] are available to select proper controllers.

7.4 Optimal compensation under non-sinusoidal supply voltage conditions

Consider a sample system with three PCC points as in Figure 7.27. Due to non-linear loads connected at PCC-2, assume that the voltage at PCC-2 becomes non-sinusoidal as shown in the diagram due to the non-sinusoidal drop across transmission line impedance 'z', though the voltage at PCC-1 is sinusoidal. As PCC-3 is connected to PCC-2 voltage at PCC-3 is the same as PCC-2, assuming the voltage drop across impedance between PCC-2 and 3 is negligible. Under such conditions, a non-linear load at PCC-3, worsens the voltage and current quality of the system. In such cases, an optimal control will give the best solution for the implementation of VSC-based compensators.

Necessity of optimal compensation

Refer to Figure 7.27. Assume the source voltage (v_s) and current (i_s) are non-sinusoidal as in Figure 7.28(a) and (b), respectively. Any attempt to improve current quality using a shunt active filter results in harmonic free current, as in Figure 7.28(c), or i_s with THD = 0, but the resultant power factor power is low as voltage contains harmonics. On the other hand, making the current waveform

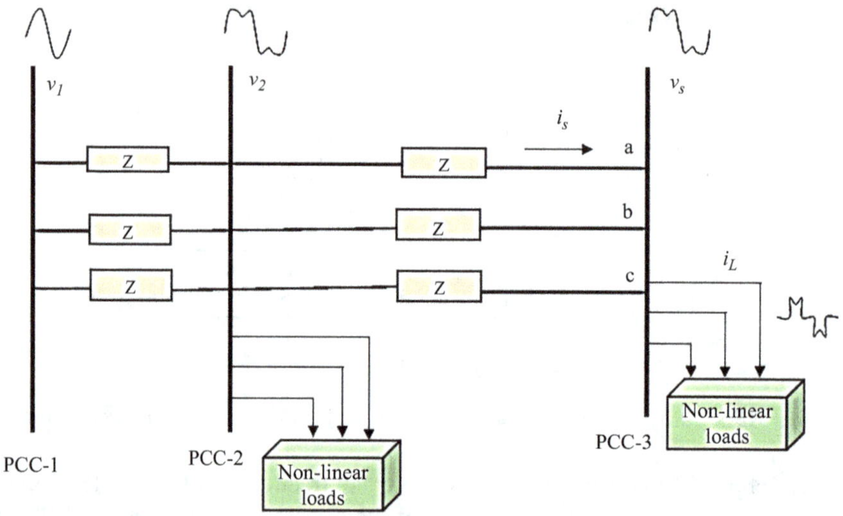

Figure 7.27 Generation of non-sinusoidal voltage and current in power system

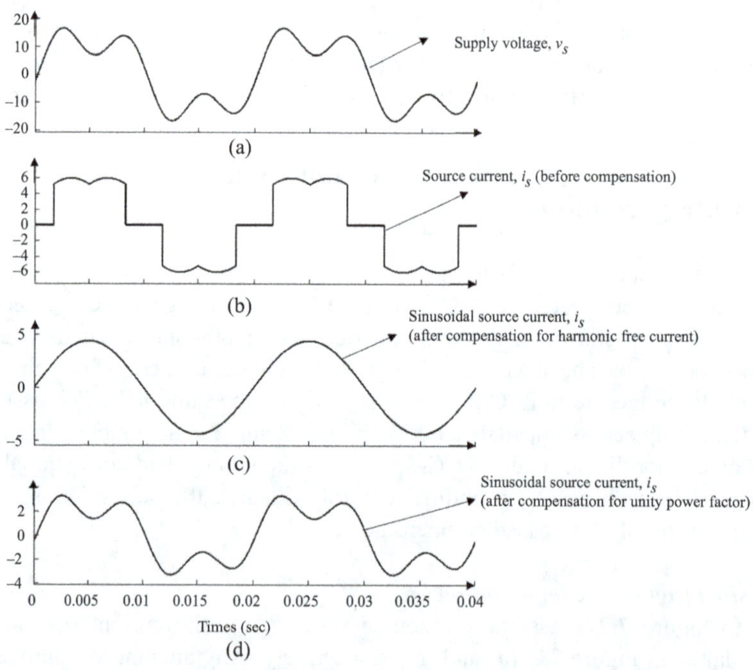

Figure 7.28 Representation to show 'need for optimization' of PF and THD in power system. (a) supply voltage, (b) source current before compensation, (c) source current after compensation for harmonics free current, and (d) source current after compensation for unity power factor.

similar in shape to the supply voltage and in phase with it, as in Figure 7.28(d), results in a unity power factor; however, THD in current is high. Thus, here is a trade-off between power factor and THD. Trying to achieve a high power factor using active filters results in high THD, while low THD results in a poor power factor. This trade-off between the power factor and THD must be optimized to achieve the best compromise. Similarly, trying to shape the voltage to a sinusoidal one using a series active filter results in poor power factor, while trying to achieve a high-power factor results in high THD. Ideally, THD should be zero and the power factor one. It is not possible to achieve this ideal condition when both supply voltage and load current are non-sinusoidal. So, an optimal solution [13–19] is necessary to get the best solution to make THD and power factor within the limit specified by stipulated standards. Using a proper algorithm, it is possible to control the shape and phase of supply current and/or voltage by properly injecting the current and/or voltage using suitable active filters to meet certain objectives with proper constraints. An algorithm to generate such a current or voltage is explained in the next sections. The specialty of the algorithm is that it can be used to generate a reference signal for shunt, series and series-shunt filters by suitably changing the control parameters.

7.4.1 Optimal shaping and shifting of current using shunt compensator

Now, assume that the objective of implementation of shunt compensator is to maximize power factor and limit its THD within the limit specified by power quality standards. This will help in efficient design of active filters. Consider a system with a sending voltage as v_s at P_{CC}-3, which contains a set of harmonic components, h_1 to h_m, that produce load current (i_L) of identical frequencies (represented by "i") and another set of components, h_2 to h_n that do not result in the corresponding load current components (represented by "j"). Further, let the load current also contain a set of current components, h_3 to h_d, having no corresponding frequency components in the supply voltage (represented by "k"), due to its non-linearity. Also, $i \neq j \neq k$ and the summation does not proceed with integer steps in any set of harmonic components. The corresponding expressions for voltage and current may be written as:

$$v_s = \sum_{i=h_1}^{h_m} \sqrt{2} \times V_{si} \sin(i\omega t + \theta_i) + \sum_{j=h_2}^{h_n} \sqrt{2} \times V_{sj} \sin(j\omega t + \theta_j) \quad (7.42)$$

$$i_L = \sum_{i=h_1}^{h_m} \sqrt{2} \times I_{Li} \sin(i\omega t + \theta_i - \phi_i) + \sum_{k=h_3}^{h_d} \sqrt{2} \times I_{Lk} \sin(k\omega t - \phi_k) \quad (7.43)$$

where, V_s and I_L represents rms value of the supply voltage and load current, respectively.

i, j and k are the order of the harmonic component, θ_i – arbitrary angle of ith component and θ_j – angle of jth component of supply voltage with respect to a reference axis, ϕ_i – arbitrary angle of ith component and ϕ_k– angle of kth component of load current with respect to a reference axis.

For unity power factor, a shunt active filter should inject a compensating current into the system to achieve the requirement of maximum power factor while meeting the limit of THD in current. It is known that to make the power factor unity, the following conditions must be met:

(i) Order of harmonic component present in current and voltage should be same to have shape of current similar to voltage, (ii) currents drawn must be in phase with the source voltage, i.e., $\phi = 0$ for every harmonic component of current and (iii) the ratio of current to voltage (admittance ratio, Y_n) for each harmonic component should be equal. However, in this case, current THD may not be within an acceptable limit, which depends on voltage THD. Under this situation, the current THD is the same as the voltage THD. By adjusting the admittance ratios, THD in current can be improved; however, this may result in reduction in power factor. For example, assume that the supply voltage contains 1, 3, 5 order of harmonic component and load current contain 1, 3, 5,7, 9 order of harmonic component. Using shunt active filter, unity power factor can be achieved by meeting the following conditions:

(i) Inject the magnitude of 7th and 9th order of harmonic component of currents present in the load current, into the system so that 1, 3, 5 order of harmonic component of current are present in the supply current, which are similar order of harmonic component of voltage.
(ii) Make the phase angle between each order of harmonic component of voltage and current (1, 3, 5 order of harmonic component) to zero.
(iii) Maintain the admittance for each harmonic component equal,

i.e.,

$$\frac{I_1}{V_1} = \frac{I_3}{V_3} = \frac{I_5}{V_5} = Y_1 = Y_3 = Y_5.$$

If all the above conditions are met, the shape of current will be similar to the shape of voltage, and they will be in phase. Thus, the power factor will be unity. By properly selecting Y_h for each order of harmonic component, the shunt active filter helps in maximizing the power factor while meeting current THD limit. It is very clear that, by controlling the ratio (Y_h), it is possible to control the shape of current and hence maximize the power factor while limiting THD. This principle is used in this section to optimize the power factor. Thus, for achieving an optimum power factor, an expression for the desired source current (i_s*) is written as:

$$i_s^* = \sum\nolimits_{i=h_1}^{h_m} \sqrt{2}\times I_{si}\sin\left(i\omega t + \theta_i\right) + \sum\nolimits_{j=h_2}^{h_n} \sqrt{2}\times I_{sj}\sin\left(j\omega t + \theta_j\right) \tag{7.44}$$

i_s^* in terms of the control variable is obtained as:

$$i_s^* = \sum\nolimits_{i=h_1}^{h_m} \sqrt{2}\times Y_i \times V_{si}\sin\left(i\omega t + \theta_i\right) + \sum\nolimits_{j=h_2}^{h_n} \sqrt{2} \times Y_j \times V_{sj}\sin\left(j\omega t + \theta_j\right) \tag{7.45}$$

where, $I_{sh} = Y_h \times V_{sh}$ and Y_h is the control variable defined as the ratio of hth order current and voltage harmonic components, identified as h^{th} order admittance of the shunt active filter, called admittance factor. The crux of the proposed algorithm is that by controlling Y_h, both the power factor and the THD can be controlled and optimized. The most appropriate value of Y_h, is computed using an optimization technique described in the following section. The reference compensating current (i_c*) for the implementation of shunt active filter can be found using (7.45). A similar principle can be used to shape and shift voltage waveform using series active filters.

7.4.2 *Optimal shaping and shifting of the voltage waveform using series compensator*

A similar principle is used in the case of series compensator too to find out the reference compensating voltage. For achieving an optimum power factor, an expression for the desired load voltage (v_L*) is obtained using (7.42) and (7.43), by setting the order of harmonics and phase angle present in the supply voltage to be the same as that of the load current. Accordingly, the expression for the desired load voltage after compensation is:

$$v_s^* = \sum\nolimits_{i=h_1}^{h_m} \sqrt{2}\times V_{si}\sin\left(i\omega t + \theta_i - \phi_i\right) + \sum\nolimits_{k=h_2}^{h_d} \sqrt{2}\times V_{sk}\sin\left(k\omega t - \phi_k\right) \tag{7.46}$$

By substituting $V_{sh} = Z_h \times I_h$, where Z_h is the control variable and known as the hth order impedance factor, reference source voltage is given by:

$$v_s^* = \sum\nolimits_{i=h_1}^{h_m} \sqrt{2}\times Z_{i\times} I_{si}\sin\left(i\omega t + \theta_i - \phi_i\right) + \sum\nolimits_{k=h_2}^{h_d} \sqrt{2}\times Z_{k\times} I_{sk}\sin\left(k\omega t - \phi_k\right) \tag{7.47}$$

Using series active filter Z_h, can be controlled, using which both the power factor and voltage THD can be controlled and optimized. The most appropriate value of Z_h is computed by using an optimization technique explained in a later section. Now, let us analyse how to obtain the optimal value of Y_h and Z_h for optimal control of the shunt and series active filter, respectively.

7.4.3 *An algorithm for optimal control of active filters*

There are various non-linear optimization techniques available in the literature to find the reference compensating current or voltage for compensators. However, in this book, a Lagrange-multiplier technique is used to optimize the non-linear equation and explain it in a simple way. Optimization problems generally have three components: (i) objective function, (ii) constraints and (iii) control variables. As per this technique, an augmented Lagrange function can be written as:

$$LF = OF + \lambda \times EC + \mu \times IC \tag{7.48}$$

where OF, *EC* and *IC* represent the objective function, the equality constraint and the inequality constraint, respectively, which are to be written in terms of control variables mentioned in the earlier section. The control variables, λ and μ are unknown quantities. The necessary condition for constrained local minima of *LF* is that its derivatives with respect to its control variables should be zero. By applying these conditions, control variables and reference compensating current and voltage are found out.

7.4.3.1 Application of Langrange optimization technique for shunt active filters

Consider PCC-2 and PCC-3 in Figure 7.27, where the supply voltage v_s and supply current i_s are non-sinusoidal. Assume that a consumer needs to implement a shunt active filter as shown in Figure 7.29 with the objective of minimizing the source current i_s while meeting THD within the limit specified by any stipulated standard on power quality. While doing so, it should also maintain a power balance between P_s and P_D. From the previous section, it is clear that the control parameter is the admittance factor (Y_h) corresponding to each harmonic component present in the voltage.

Now, develop the Langrange function for shunt active filter by deriving equations for OF, EC and IC in terms of control variables. The following derivations are explained below:

(i) Derivation for the objective function (OF_{shunt})
(ii) Derivation for equality constraints (EC_{shunt})
(iii) Derivation for inequality constraints (IC_{shunt})
(iv) Development of the Langrange function (LF_{shunt}) and its solution

Derivation for objective function (OF_{shunt})

By maximizing power factor at the supply end, it is possible to achieve minimum transmission line current. As power factor (PF_s) is inversely proportional to reactive volt-amp (S_s), as discussed in Chapter 5, the objective function may be selected as the

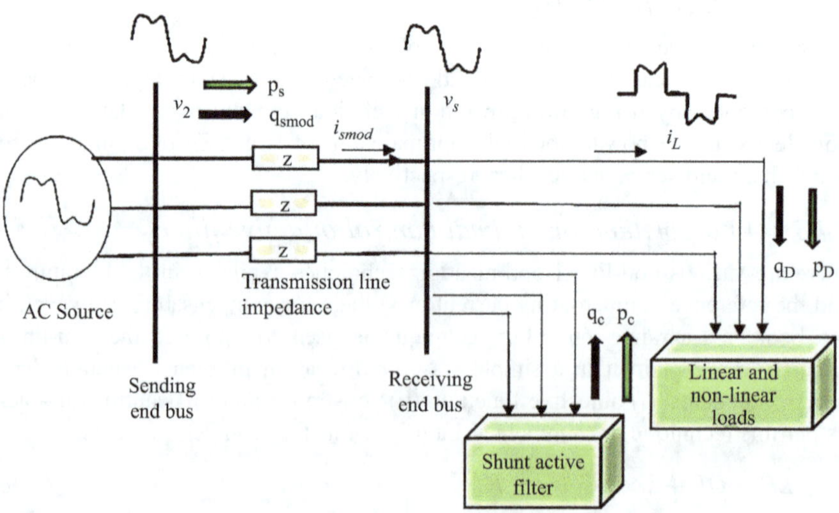

Figure 7.29 Single line diagram of the system with shunt active filter under non-sinusoidal supply voltage

minimization of S_s, which results in maximum power factor and hence minimum source current. To define the objective function, consider S_s after compensation. Assume that under this situation, the shunt active filter injects the required harmonic component of load current into the system so that a similar order harmonic component is present in voltage and load current. The load, along with shunt active filter, acts as a linear load to system under this condition. So, the objective function is constructed as follows:

$$OF_{shunt} = S_s^2 = \sum_{i=1}^{h} V_{si}^2 \sum_{i=1}^{h} I_{si}^2 \tag{7.49}$$

where, harmonics h_m and h_n have been combined and denoted by 'h' for convenience. The equation assumes the order of harmonic components present in the current is same as in the voltage. Now using the relation of current with voltage and control variable, Y_h., each harmonic component of the source current, I_{sh} can write:

$$I_{sh} = Y_h \times V_{sh} \tag{7.50}$$

The objective function is written in terms of the control variable as:

$$OF_{shunt} = S_S^2 = \sum_{i=1}^{h} V_{si}^2 \times \sum_{i=1}^{h} Y_i^2 \times V_{si}^2 \tag{7.51}$$

Derivation for equality constraints (EC_{shunt})

Equality constraints depend on the conditions to be met while controlling the shunt active filter. The power balance equation is taken as an equality constraint in this example. At PCC active power supplied by the source (P_s) is equal to useful power (P_D) demanded by the load after compensation. This indicates that the source should supply only reactive power, and the reactive or distortion power demanded by the load needs to be supplied by the shunt compensator. It is to be noted that the shunt compensator cannot directly control the voltage at the PCC. As the supply voltage is non-sinusoidal, $\overline{p}$ include power due to similar order harmonic components of voltage and current. Assume that the shunt active filer compensates for q and $\tilde{p}$ of the system. In such a case, the displacement angle between the supply voltage and current is zero, and the shape of voltage and current is the same. Hence, we can write the power equation in terms of the control variable as:

$$P_s = \sum_{i=1}^{h} V_{si} \times I_{si} = \sum_{i-1}^{h} V_{si}^2 \times Y_i \tag{7.52}$$

Thus, assuming P_s as the power from the source after compensation, using the power balance equation, the equality constraint (EC) as a function of Y_h, is written as:

$$EC_{shunt} = \sum_{i=1}^{h} V_{si}^2 \times Y_i - P_D = 0 \tag{7.53}$$

P_D may be calculated using the p-q theory by sensing load voltage and current.

Derivation for inequality constraints (IC_{shunt})

In this case, only one inequality constraint is taken into account for simplicity, which is THD limit in source current. However, literature is available with multiple inequality constraints, which are out of the scope of this book. Let the THD in source current be required to be limited to I_{sTHD_limit}. Then, the inequality constraint (IC_{shunt}) is taken as:

$$\frac{\sum_{i=2}^{h} I_{si}^2}{I_{s1}^2} \le I_{sTHD-limit}^2 \tag{7.54}$$

EC is written as a function of Y_h as below:

$$\frac{\sum_{i=2}^{h} Y_i^2 \times V_{si}^2}{Y_1^2 \times V_{s1}^2} \le I_{sTHD-limit}^2 \tag{7.55}$$

Now, using above (7.55), IC_{shunt} can be written in terms of control parameters as:

$$\sum_{i=2}^{h} Y_i^2 \times V_{si}^2 - I_{sTHD-limit}^2 \times Y_1^2 \times V_{s1}^2 \le 0 \tag{7.56}$$

Lagrange function (LF) for shunt active filter

After obtaining the objective function, equality and inequality functions in terms of the control variable, the Lagrange function is formed and solved to get the value of the control variable, which results in minimum reactive volt ampere in the system using an active filter. So, the objective is to minimize S_s given by (7.51) subject to the equality constraint, $EC_{shunt} = 0$, as defined by (7.53), and the inequality constraint, $IE_{shunt} \le 0$, as defined by (7.56). The augmented function (LF_{shunt}) is given by:

$$\begin{aligned} LF_{shunt} = {} & \sum_{i=1}^{h} V_{si}^2 \times \sum_{i=1}^{h} Y_i^2 \times V_{si}^2 + \lambda \times \left[\sum_{i=1}^{h} V_{si}^2 \times Y_i - P_D \right] \\ & + \mu \left[\sum_{i=2}^{h} Y_i^2 \times V_{si}^2 - I_{sTHD-limit}^2 \times Y_1^2 \times V_{s1}^2 \right] \end{aligned} \tag{7.57}$$

Here, unknowns are Y_h, λ and μ. By applying necessary conditions for local minima, the unknown variables are obtained. For the minimum value of Ss, the partial differentiation of Langrange function with respect to the variables needs to be zero. Accordingly, the equation is written as:

$$\frac{\partial LF_{shunt}}{\partial Y_i} = 0 \text{ i varies from 1 to h} \tag{7.58}$$

$$\text{Also } \frac{\partial LF_{shunt}}{\partial \lambda} = EC_{shunt} = 0 \text{ and } \frac{\partial LF_{shunt}}{\partial \mu} = IC_{shunt} \le 0 \tag{7.59}$$

By solving the above equations, all unknowns Y_1 to Y_n and λ and μ can be found. Using these values $i_s^* = \sum_{i=1}^{h} \sqrt{2} \times Y_i \times V_{,si} \sin(i\omega t + \theta_i)$.

By solving (7.58) and (7.59), Y_h is obtained for an optimum power factor within an acceptable THD of current. $\mu u = 0$ and $\mu > 0$ are the Kuhn-Tucker conditions to be satisfied at a relative minimum of the objective function.

Implementation process

The supply voltage and load current are sensed using a voltage sensor and a current sensor, respectively, and a Discrete Fourier Transform (DFT) is performed on the sensed supply voltage and load current [20–22]. Thus, the rms value of each harmonic component of voltage and current in each phase is calculated. Power demanded by the load is also calculated using sensed voltage and current. Then apply the above algorithm and find 'Y_n ', and the original time domain signal of the desired source current is obtained by performing inverse DFT on the frequency domain values. i_c* is obtained by subtracting the desired source current from the load current. This reference current is used to operate the inverter of shunt active filter to compensate for reactive power and harmonics. After the implementation of shunt active filter, the source current is in phase with the supply voltage, and its shape is nearly sinusoidal with THD within the specified limit. This results in minimum source current and maximum system power factor with current THD within the limit specified by the stipulated standard while meeting the power balance equation, which improves the efficiency of the power system.

7.4.3.2 Application of Langrange optimization technique for series active filters

As discussed earlier, instead of installing a shunt active filter by consumers, using a series active filter too, the desired compensation can be done. In this case, instead of shaping supply currents as in the case of a shunt active filter, shifting and shaping of voltage at PCC is controlled to meet minimum reactive volt amp from the system while maintaining the power balance equation as equality constraints and voltage THD limit as inequality constraints (Figure 7.30).

Consider that using a series active filter, it is required to implement a control to optimize system power factor while meeting equality constraints, same as in the case of shunt active filter, and in-equality constraints as THD limit on voltage. In this case, the control variable is the ratio of same order of harmonic component of voltage to current, known as impedance factor (Z_h), which can be controlled by injecting a desired amount of harmonic voltage in series with the line. So, as in the case of the shunt active filter, OF, EC and IC are to be framed in terms of Z_h.

Objective function (OF_{series}) for series active filter

For a given active power, the power factor can be improved by minimizing S_s, the total apparent power. Thus, the objective function OF_{series} is constructed as follows:

$$OF_{series} = S_s^2 = \sum_{i=1}^{h} V_{si}^2 \sum_{i=1}^{h} I_{si}^2 \tag{7.60}$$

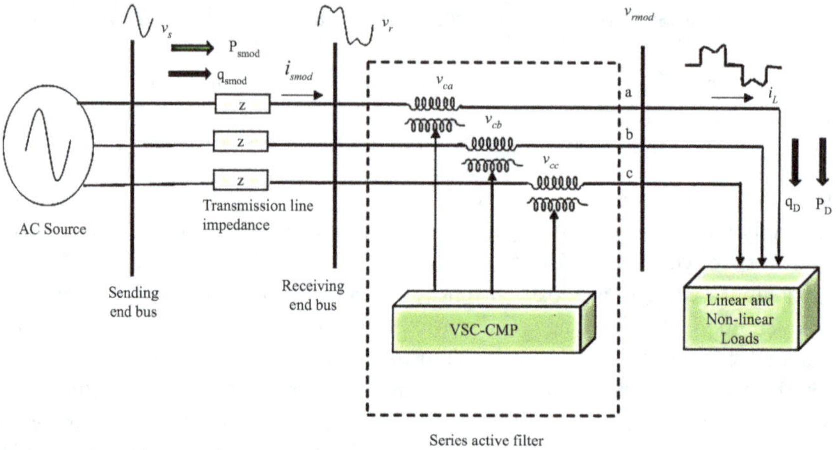

Figure 7.30 A line diagram of a system with a series active filter under a non-sinusoidal supply voltage

Substituting $V_{sh} = Z_h \times I_{sh}$ in (7.60), OF_{series} is written in terms of control variables, Z_h as:

$$OF_{series} = S_S^2 = \sum_{i=1}^{h} Z_i^2 \times I_{si}^2 \times \sum_{i=1}^{h} I_{si}^2 \tag{7.61}$$

Equality constraints

When the displacement angle between the voltage and current is zero, and the order of harmonics present in them are same, the mean value of the instantaneous real power in terms of control variables, Z_h is given by:

$$P_s = \sum_{i=1}^{h} V_{si} \times I_{si} = \sum_{i=1}^{h} Z_i \times I_{si}^2 \tag{7.62}$$

Thus, assuming P_s as the power from the source after compensation and using the power balance equation, the equality constraint (EC_{series}) for the series active filter as a function of Z_h, is written as:

$$EC_{series} = \sum_{i=1}^{h} Z_i \times I_{si}^2 - P_D = 0 \tag{7.63}$$

Inequality constraints (IC_{series}) for series active filter

Suppose it is required that the voltage THD be limited to V_{THD}, then the inequality constraint (IC_{series}) is calculated as:

$$\frac{\sum_{i=2}^{h} V_{si}^2}{V_{s1}^2} \le V_{sTHD-limit}^2 \tag{7.64}$$

$$\frac{\sum_{i=2}^{h} Z_i^2 \times I_{si}^2}{Z_1^2 \times I_{s1}^2} \leq V_{sTHD-limit}^2 \tag{7.65}$$

Now, using above equations, IC_{series} can be written in terms of control parameters as:

$$\sum_{i=2}^{h} Z_i^2 \times I_{si}^2 - V_{sTHD-limit}^2 \times Z_1^2 \times I_{s1}^2 \leq 0 \tag{7.66}$$

Lagrange function (LF_{series}) for series active filter

Similar to shunt active filter, the augmented function (LF_{series}) is given by:

$$LF_{series} = \sum_{i=1}^{h} I_{si}^2 \times \sum_{i=1}^{h} Z_i^2 \times I_{si}^2 + \lambda \times \left[\sum_{i=1}^{h} I_{si}^2 \times Z_i - P_D\right] + \mu\left[\sum_{i=2}^{h} Z_i^2 \times I_{si}^2 - V_{sTHD-limit}^2 \times Z_1^2 \times I_{s1}^2\right] \tag{7.67}$$

For the minimum value of Ss, the partial differentiation of Langrange function with respect to these control variables needs to be zero. Accordingly, the equation is written as:

$$\frac{\partial LF_{series}}{\partial Z_i} = 0 \; i \; varies \; from \; 1 \; to \; h \tag{7.68}$$

$$\text{Also } \frac{\partial LF_{series}}{\partial \lambda} = EC_{series} = 0 \text{ and } \frac{\partial LF_{series}}{\partial \mu} = IC_{series} \leq 0 \tag{7.69}$$

By solving the above equations, all unknowns Z_1 to Z_n and λ and μ can be found. By substituting these values of Z in (7.47), the reference source voltage, v_s^* is determined and subtracted from the load voltage to get reference compensating voltage, v_c^* can be obtained, which is to be injected by VSC to control the supply voltage to loads.

7.4.3.3 Application of Langrange optimization technique series-shunt active filters

In case of shunt active filter current quality in the transmission line is controlled by generating a reference compensating current i_c^* using the algorithm for maximizing the power factor while meeting the power balance equation and current THD within the limit. In this case, the voltage at PCC cannot be controlled using a shunt active filter. Similarly, in the case of series active filter, the voltage quality at PCC is controlled by generating a reference voltage v_c^* using a similar algorithm for maximizing the power factor while meeting the power balance equation and limiting voltage THD. However, a series active filter cannot control the current quality flowing through the non-linear loads. To achieve control of both voltage and current quality at the same time, a series-shunt active filter can be used (Figure 7.31). The reference compensation voltage (v_s^*) and current (i_s^*) can be computed using the

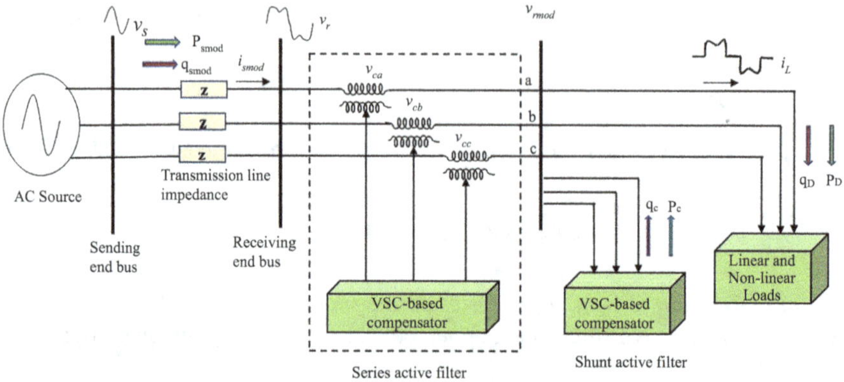

Figure 7.31 Layout of compensated power system with series-shunt active filter under non-sinusoidal supply voltage

same algorithm. Shunt portion of series-shunt filter controls Y_h, and the series portion of the filter controls Z_h.

It is possible to make both v_{mod} and i_{smod} near sinusoidal, which satisfies power quality standards while meeting all constraints using optimal control of series-shunt active filters.

7.5 Conclusion

This chapter briefly discussed VSC based reactive power compensators commonly referred to as STATCOM. Owing to the high switching frequency of VSC devices, STATCOM offers rapid response and shorter settling time, making it highly effective for dynamic reactive power support. Consequently, it serves as a reliable solution for mitigating the grid disturbance[1,2] outlined in Chapter 6. In addition to reactive power compensation, the chapter also elaborated active filters for mitigating voltage and current harmonics, meeting the requirements of IEEE 519 standard. An algorithm for optimal control of active filters to mitigate the trade-off between power factor correction and harmonic reduction under nonsinusoidal supply voltage condition was also covered.

7.6 Problems and solutions

Problem 7.1 A DC to AC converter is designed to generate a three-phase volage with a frequency of 50 Hz. A SPWM technique is used for its control implementation. The VA rating of the converter is 1 kVA. Assume switching frequency as 10 kHz and V_{DC} as 700 V. Calculate (i) the RMS value of output phase voltage of the converter with

[1]ENTSO-E report dated 26.10.2010 prepared by the subgroup "system protection and dynamics" under - regional group continental Europe

[2]Report on the Indian grid disturbances on 30th July and 31st July 2012

$M_a = 0.8$. Verify the results using simulation. (ii) What should be the magnitude and frequency of the reference sinusoidal wave used in the SPWM technique to get the desired magnitude and frequency? (iii) What is the value of the maximum resistance that can be connected to this converter? (iv) What will be V/f ratio if $M_a = 0.8$? (v) If V/f is to be maintained constant at 2 for the speed control of an induction motor application, suggest a suitable solution. Justify using simulation results.

Given data:
The circuit diagram is same as the figure given in Figure 7.6.
VA rating of the converter = 1 kVA
$\frac{V_{DC}}{2} = 350\ V$; Switching frequency = 10 kHz

Solution:

(i) RMS value of output phase voltage of the inverter with $M_a = 0.8$

$$V_{phmax} = M_a \frac{V_{dc}}{2} = 0.8 \times \frac{700}{2} = 280 \text{ V}$$

$$V_{phRMS} = \frac{280}{\sqrt{2}} = 197.98 \text{ V}$$

$$V_{L-LRMS} = \sqrt{3} \times V_{phRMS} = 342.93 \text{ V}$$

Simulation results:

The circuit diagram given in Figure 7.6 is simulated in open loop with $\frac{V_{DC}}{2} = 350$ V and $R = 10\ \Omega$ by adding a filter capacitor at the output. The simulated output voltage waveform is given below (Figure 7.32). It is seen that the RMS value of phase voltage is 196.1 V, nearly equal to the calculated value. It is clear from the waveform peak of the phase voltage that $\cong$ 280 V and frequency = 50 Hz.

(ii) What will be the magnitude and frequency of the reference sinusoidal wave used in SPWM technique to get the desired magnitude and frequency?

In the SPWM technique peak magnitude of the reference sine wave should be 280 V, and the frequency should be 50 Hz.

(iii) What is the value of the maximum resistance that can be connected to this converter?

As the VA rating is 1 kVA, the maximum load current (I_L) per phase is given by:

$$I_L = \frac{kVA \times 10^3}{3 \times V} = \frac{1000}{3 \times 197.98} = 1.68 \text{ A}$$

Minimum value of resistor (R) that can be connected per phase to the converter is:

$$R = \frac{V_{phRMS}}{I} = \frac{197.98}{1.68} = 117.845\ \Omega$$

(iv) What will be V/f ratio if $M_a = 0.8$?

$$V/f = 197.98/50 = 3.96$$

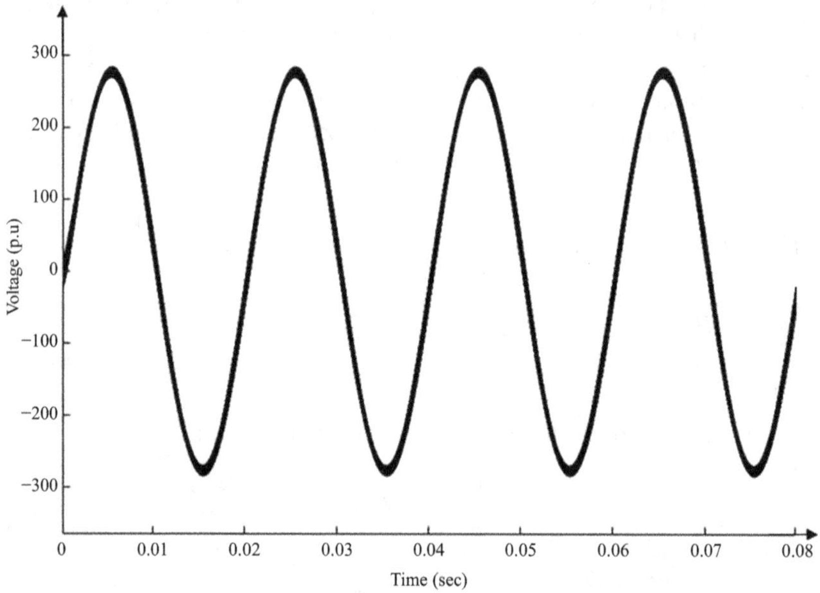

Figure 7.32 Output voltage of the converter specified in Problem 7.1 simulated using MATLAB

(v) If V/f is to be maintained constant at 2 for a speed control of an induction motor application, suggest a suitable solution.

V/f of a converter can be controlled by varying (i) $M_{a,}$ i.e., by controlling the magnitude of the peak value of reference signal (ii) $M_{f,}$ i.e., by controlling the frequency of the reference signal.

If frequency is to be maintained at 50 Hz, the peak magnitude of refence signal should be 141.42 V and to achieve this value $M_a = 0.404$. If the voltage is to be maintained at 197.98 V, the frequency of reference signal is to be 100 Hz. Simulated waveforms in each case are shown in Figure 7.33.

The conclusion is that using an inverter, V/f ratio can be maintained constant by adjusting both frequency and voltage magnitude.

Problem 7.2 Consider a 3-ϕ, 50 Hz, 440 V system with a resistive load of 10 Ω per phase (Figure 7.34). Convert three-phase voltage and current to the α-β co-ordinate system. Calculate (i) P and Q using conventional method (ii) verify P and Q obtained in (i) using the p-q theory.

Solution:

Let $V_{sL-L-RMS}$, RMS value of line voltage

$V_{sph-RMS}$, RMS value of phase voltage,

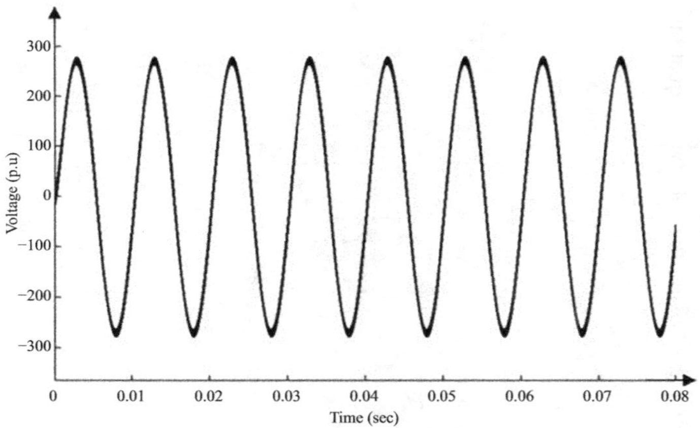

(a)

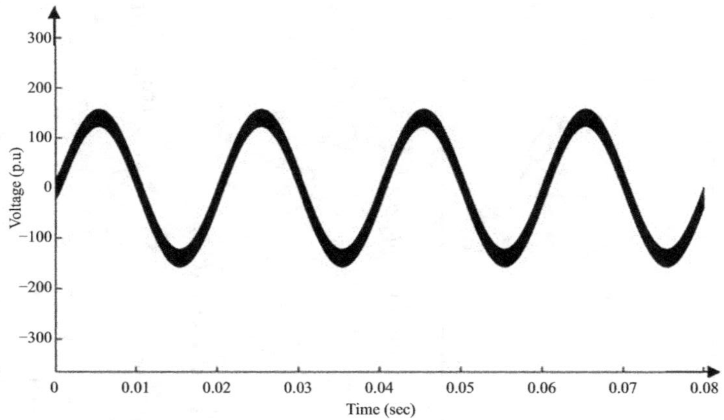

(b)

Figure 7.33 Simulated waveforms for problem 7.1 for V/f = 2. (a) With frequency control f = 100. (b) With voltage magnitude control M_a = 0.404.

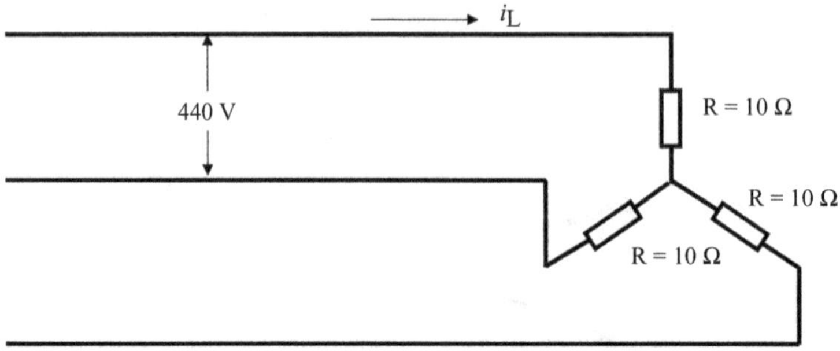

Figure 7.34 Circuit diagram of Problem 7.2

V_{sm}, maximum value of phase voltage, similar notation for load current (I_L) too.

(i) P and Q using the conventional method

$$V_{sL-L-RMS} = 440\text{ V} \quad V_{sPh-RMS} = 254.03\text{ V} \quad V_{sm} = 359.25\text{ V}I_{Lph-RMS}$$

$$= 25.4\text{ A}$$

$$I_{Lph-m} = 35.925\text{ A}\cos\phi = 1$$

Conventionally, P and Q are calculated using the following equation:

$$\text{Activepower, } p = 3V_{sPh-RMS} \times I_{LPh-RMS} \times \cos = 3 \times 254.03 \times 25.4 \times 1$$

$$= 19357.09\text{ W.}$$

Reactive power $Q = 0$ as the load is resistive.

(ii) Verify P and Q using p-q theory
α-β transformation of supply voltage v_s is given by:

$$\begin{bmatrix} v_{s\alpha} \\ v_{s\beta} \end{bmatrix} = \sqrt{2/3}\begin{bmatrix} 1 & -1/2 & -1/2 \\ 0 & \sqrt{3}/2 & -\sqrt{3}/2 \end{bmatrix}\begin{bmatrix} v_{sa} \\ v_{sb} \\ v_{sc} \end{bmatrix}$$

where, v_{sa}, v_{sb}, v_{sc} are the instantaneous voltage in respective lines and is given by:

$$v_{sa} = V_{sm}\sin(\omega t)$$

$$v_{sb} = V_{sm}\sin(\omega t - 120°)$$

$$v_{sc} = V_{sm}\sin(\omega t - 240°)$$

By substituting the instantaneous values of voltages, v_s in α-β frame is re-written as:

$$\begin{bmatrix} v_{s\alpha} \\ v_{s\beta} \end{bmatrix} = \sqrt{2/3}\begin{bmatrix} 1 & -1/2 & -1/2 \\ 0 & \sqrt{3}/2 & -\sqrt{3}/2 \end{bmatrix}\begin{bmatrix} V_{sm}\sin(\omega t) \\ V_{sm}\sin(\omega t - 120^\circ) \\ V_{sm}\sin(\omega t - 240^\circ) \end{bmatrix}$$

By simplifying the above equation using formula for *sin(A-B)*, it can be written as:

$$\begin{bmatrix} v_{s\alpha} \\ v_{s\beta} \end{bmatrix} = [1.225]\begin{bmatrix} V_{sm}\sin(\omega t) \\ -V_{sm}\cos(\omega t) \end{bmatrix} = \begin{bmatrix} 1.225 \times V_{sm}\sin(\omega t) \\ 1.225 \times V_{sm}\sin(\omega t - 90^\circ) \end{bmatrix}$$

Substituting the value of $V_{sm} = 359.25\ V$, we can write,

$$\begin{bmatrix} v_{s\alpha} \\ v_{s\beta} \end{bmatrix} = \begin{bmatrix} 440.081 \times \sin(\omega t) \\ 440.081 \times \sin(\omega t - 90^\circ) \end{bmatrix}$$

The peak value of $v_{s\alpha}$ and $v_{s\beta}$ are the same and $v_{s\beta}$ lags $v_{s\alpha}$ by 90°. Similarly, as the load is resistive, using $I_{Lph-m} = 35.925\ A$, i_L in α-β frame is re-written as:

$$\begin{bmatrix} i_{L\alpha} \\ i_{L\beta} \end{bmatrix} = \begin{bmatrix} 1.225 \times I_{Lm}\sin(\omega t) \\ 1.225 \times I_{Lm}\sin(\omega t - 90^\circ) \end{bmatrix} = \begin{bmatrix} 44.008 \times \sin(\omega t) \\ 44.008 \times \sin(\omega t - 90^\circ) \end{bmatrix}$$

The current peak values of $i_{L\alpha}$ and $i_{L\beta}$ are the same but $i_{L\beta}$ lags $i_{L\alpha}$ by 90°. The instantaneous value of p and q is given by:

$$\begin{bmatrix} p \\ q \end{bmatrix} = \begin{bmatrix} v_{s\alpha} & v_{s\beta} \\ -v_{s\beta} & v_{s\alpha} \end{bmatrix}\begin{bmatrix} i_{l\alpha} \\ i_{l\beta} \end{bmatrix}$$

$$= \begin{bmatrix} 440.081 \times \sin(\omega t) & 440.081 \times \sin(\omega t - 90^\circ) \\ -440.081 \times \sin(\omega t - 90^\circ) & 440.081 \times \sin(\omega t) \end{bmatrix}\begin{bmatrix} 44.008 \times \sin(\omega t) \\ 44.008 \times \sin(\omega t - 90^\circ) \end{bmatrix}$$

$$\begin{bmatrix} p \\ q \end{bmatrix} = \begin{bmatrix} 19367.15 \\ 0 \end{bmatrix}$$

In the above equation $p = 19367.15$ W indicates, $\overline{p}$ which is the mean value of useful power is transferred from source to load, and q indicates, $\overline{q}$ which is mean value of reactive power (= zero). It is seen that power calculation using the conventional method is nearly equal to the power calculated using p-q theory. Accuracy may be increased by taking 4 decimal points in the calculations.

Problem 7.3 Consider a 3-ϕ, 50 Hz, 440 V system supplying a resistive-inductive load with a resistance of 10 Ω and inductive reactance of 10 Ω per phase. Using p-q theory, (i) calculate p and q consumed by the load, (ii) determine the reference source current required to supply only active power of the load, (iii) if the reactive power is supplied by a shunt active filter, obtain the reference current for the inverter of the shunt compensator.

Solution:

$$V_{sL-L-RMS} = 440\text{V}\, V_{sPh-RMS} = 254.03\ \text{V}\ V_{sm} = 359.25\ \text{V}$$

$$Z = 10 + j10\ \Omega$$

$$I_{Lph-RMS} = \frac{V_{sPh-RMS}}{Z} = \frac{254.03}{10 + J10} = 17.96\angle - 45^{\circ}$$

$$I_{Lm} = 25.399\ \text{A}\ I_{Lrms} = 17.959\ \text{A}\ \cos\phi = 0.7071$$

The above calculations are verified using MATLAB simulations, and the waveforms of voltage and current are shown in Figure 7.35. It is observed that the calculated values matching with simulation results.

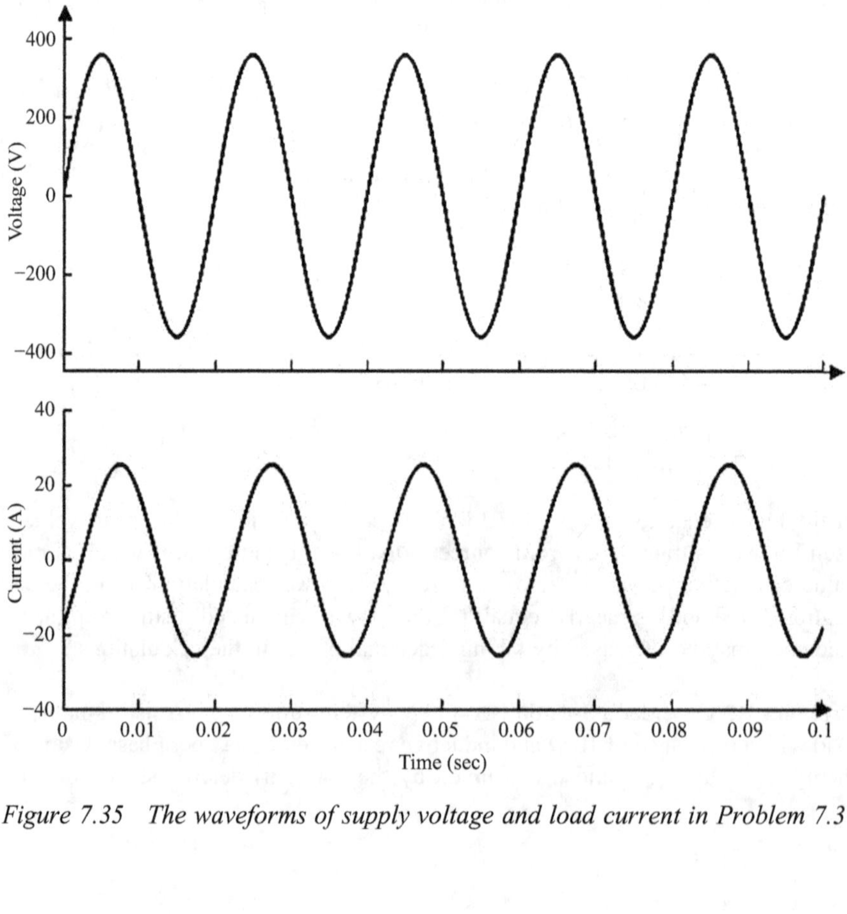

Figure 7.35 The waveforms of supply voltage and load current in Problem 7.3

Calculation of p and q using the conventional method,

$$\text{Activepower, } p = 3V_{sPh-RMS} \times I_{sPh-RMS} \times \cos$$

$$= 3 \times 254.03 \times 17.96 \times \cos 45 = 9678.267\text{ W}$$

$$\text{Rectivepower, } q = 3V_{sPh-RMS} \times I_{sPh-RMS} \times \sin$$

$$= 3 \times 254.03 \times 17.96 \times \sin 45 = 9678.267\text{ W}$$

(i) Calculate p and q consumed by the load

As supply voltage is same as in the case of problem 7.2, α-β transformation of supply voltage v_s is is given by:

$$\begin{bmatrix} v_{s\alpha} \\ v_{s\beta} \end{bmatrix} = \begin{bmatrix} 1.225 \times V_{sm}\sin(\omega t) \\ 1.225 \times V_{sm}\sin(\omega t - 90°) \end{bmatrix} = \begin{bmatrix} 440.081 \times \sin(\omega t) \\ 440.081 \times \sin(\omega t - 90°) \end{bmatrix}$$

However, as the load is inductive, the current in each phase lags the voltage by the power factor angle, load current is given by

$$i_{La} = I_{Lm}\sin(\omega t - 45)$$

$$i_{Lb} = I_{Lm}\sin(\omega t - 165°)$$

$$i_{Lc} = I_{Lm}\sin(\omega t - 285°)$$

So, $i_{L\alpha}$ and $i_{L\beta}$ are calculated as:

$$\begin{bmatrix} i_{L\alpha} \\ i_{L\beta} \end{bmatrix} = \sqrt{2/3}\begin{bmatrix} 1 & -1/2 & -1/2 \\ 0 & \sqrt{3}/2 & -\sqrt{3}/2 \end{bmatrix}\begin{bmatrix} I_{Lm}\sin(\omega t - 45) \\ I_{Lm}\sin(\omega t - 145°) \\ I_{Lm}\sin(\omega t - 285°) \end{bmatrix}$$

$$\begin{bmatrix} i_{L\alpha} \\ i_{L\beta} \end{bmatrix} = \begin{bmatrix} 0.865I_{Lm}\sin(\omega t) - 0.865I_{Lm}\cos(\omega t) \\ -0.865I_{Lm}\sin(\omega t) - 0.865I_{Lm}\cos(\omega t) \end{bmatrix}$$

$$\begin{bmatrix} i_{L\alpha} \\ i_{L\beta} \end{bmatrix} = \begin{bmatrix} 0.865I_{Lm}\sin(\omega t) - 0.865I_{Lm}\cos(\omega t) \\ 0.865I_{Lm}\sin(\omega t - 90) + 0.865I_{Lm}\cos(\omega t - 90) \end{bmatrix}$$

$$= \begin{bmatrix} 21.97\sin(\omega t) - 21.97\cos(\omega t) \\ 21.97\sin(\omega t - 90) + 21.97\cos(\omega t - 90) \end{bmatrix}$$

Load current peak values of $i_{L\alpha}$ and $i_{L\beta}$ are same but $i_{L\beta}$ lags $i_{L\alpha}$ by 90°

$$\begin{bmatrix} p \\ q \end{bmatrix} = \begin{bmatrix} v_{s\alpha} & v_{s\beta} \\ -v_{s\beta} & v_{s\alpha} \end{bmatrix}\begin{bmatrix} i_{l\alpha} \\ i_{l\beta} \end{bmatrix} = \begin{bmatrix} 440.081\sin(\omega t)\ 440.081\sin(\omega t - 90°) \\ -440.081\sin(\omega t - 90°)\ 440.081\sin(\omega t) \end{bmatrix}$$

$$\begin{bmatrix} 21.97\sin(\omega t) - 21.97\cos(\omega t) \\ 21.97\sin(\omega t - 90) + 21.97\cos(\omega t - 90) \end{bmatrix}$$

Simplifying and solving the equation, we can write:

$$\begin{bmatrix} p \\ q \end{bmatrix} = \begin{bmatrix} 9668.58 \\ -9668.58 \end{bmatrix}$$

In the above equation, p indicates $\overline{p}$, i.e., mean value of useful power, which is transferred from source to load, and q indicates $\overline{q}$, i.e., mean value of reactive power exchanged between source and load. Negative signs indicate inductive reactive power. The p and q calculated using the conventional method and the p-q theory are nearly the same. The minor difference may be due to the accuracy in decimal points taken in the two cases.

(ii) Determine the reference source current required to supply only the active power of the load

If the source needs to supply only the active power, the line current should be in phase with the supply voltage. This can be achieved by connecting a shunt active filter, which compensates for the reactive power consumed by the load. Using p-q theory, the reference source current (in α-β frame) for supplying only active power can be calculated using eqn. 7.39 as follows:

$$\begin{bmatrix} i_{s\alpha}* \\ i_{s\beta}* \end{bmatrix} = \begin{bmatrix} v_{s\alpha} & v_{s\beta} \\ -v_{s\beta} & v_{s\alpha} \end{bmatrix}^{-1} \begin{bmatrix} 9668.58 \\ 0 \end{bmatrix}$$

The inverse matrix is given by:

$$\begin{bmatrix} v_{s\alpha} & v_{s\beta} \\ -v_{s\beta} & v_{s\alpha} \end{bmatrix}^{-1} = \frac{1}{(1.225\times V_{sm})^2} \begin{bmatrix} 1.225\times V_{sm}\sin(\omega t) & 1.225\times V_{sm}\cos(\omega t) \\ -1.225\times V_{sm}\cos(\omega t) & 1.225\times V_{sm}\sin(\omega t) \end{bmatrix}$$

$$= \begin{bmatrix} 2.272\times 10^{-3}\sin(\omega t) & 2.272\times 10^{-3}\cos(\omega t) \\ 2.272\times 10^{-3}\cos(\omega t) & 2.272\times 10^{-3}\sin(\omega t) \end{bmatrix}$$

Substituting the value of the inverse matrix, the desired source current for supplying only active power can be found and is given by:

$$\begin{bmatrix} i_{s\alpha}* \\ i_{s\beta}* \end{bmatrix} = \begin{bmatrix} 2.272\times 10^{-3}\sin(\omega t) & 2.272\times 10^{-3}\cos(\omega t) \\ 2.272\times 10^{-3}\cos(\omega t) & 2.272\times 10^{-3}\sin(\omega t) \end{bmatrix} \begin{bmatrix} 9668.58 \\ 0 \end{bmatrix}$$

Solving the above equation

$$\begin{bmatrix} i_{s\alpha}* \\ i_{s\beta}* \end{bmatrix} = \begin{bmatrix} 21.967\sin(\omega t) \\ 21.967\cos(\omega t) \end{bmatrix}$$

Reference source current in a-b-c frame is calculated using (7.24).

Inverse of the conversion matrix in (7.24) is given by:

$$\sqrt{\frac{2}{3}} \begin{bmatrix} 1 & \frac{-1}{2} & \frac{-1}{2} \\ 0 & \frac{\sqrt{3}}{2} & \frac{-\sqrt{3}}{2} \end{bmatrix}^{-1} = \begin{bmatrix} 0.577 & 0.817 & 0 \\ 0.577 & -0.4085 & 0.7072 \\ 0.577 & -0.4085 & -0.7072 \end{bmatrix}$$

Substituting the value of the inverse matrix,

$$\begin{bmatrix} i_{sa}* \\ i_{sb}* \\ i_{sc}* \end{bmatrix} = \begin{bmatrix} 0.577 & 0.817 & 0 \\ 0.577 & -0.4085 & 0.7072 \\ 0.577 & -0.4085 & -0.7072 \end{bmatrix} \begin{bmatrix} 0 \\ -21.96\cos(\omega t) \\ -21.96\sin(\omega t) \end{bmatrix}$$

Solving the above equation, Reference source current in a-b-c frame is given by:

$$\begin{bmatrix} i_{sa}* \\ i_{sb}* \\ i_{sc}* \end{bmatrix} = \begin{bmatrix} 17.94\sin(\omega t) \\ 17.94\sin(\omega t - 120) \\ 17.94\sin(\omega t - 240) \end{bmatrix}$$

This indicates that the reference source current in each phase is in phase with the supply voltage, resulting in only active power transfer from source to load.

(iii) If the reactive power is supplied by a shunt active filter, get the reference compensating current for the inverter of the shunt compensator

As reactive power q is supplied by the shunt active filter, the reference compensating current is found out using following equation:

$$\begin{bmatrix} i_{c\alpha}* \\ i_{c\beta}* \end{bmatrix} = \begin{bmatrix} v_{s\alpha} & v_{s\beta} \\ -v_{s\beta} & v_{s\alpha} \end{bmatrix}^{-1} \begin{bmatrix} 0 \\ -9668.58 \end{bmatrix}$$

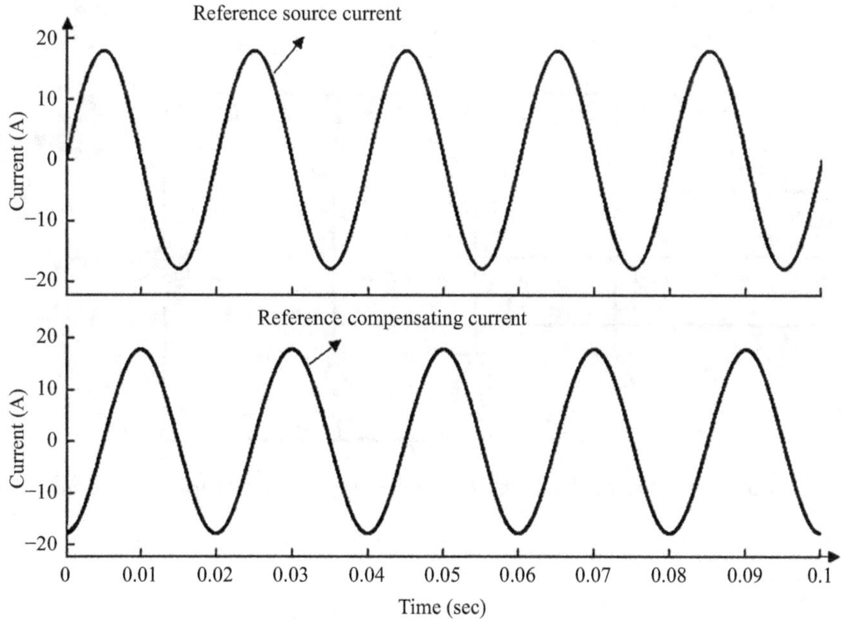

Figure 7.36 Waveforms of reference source current and compensating current in Problem 7.3

Solving the above equation as in case(ii),

$$\begin{bmatrix} i_{c\alpha}* \\ i_{c\beta}* \end{bmatrix} = \begin{bmatrix} 2.272\times10^{-3}\sin(\omega t) & 2.272\times10^{-3}\cos(\omega t) \\ 2.272\times10^{-3}\cos(\omega t) & 2.272\times10^{-3}\sin(\omega t) \end{bmatrix} \begin{bmatrix} 0 \\ -9668.58 \end{bmatrix}$$

$$= \begin{bmatrix} -21.96\cos(\omega t) \\ -21.96\sin(\omega t) \end{bmatrix}$$

Using (7.24) i_c in abc frame is calculated as:

$$\begin{bmatrix} i_{ca}* \\ i_{cb}* \\ i_{cc}* \end{bmatrix} = \begin{bmatrix} 17.94\sin(\omega t - 90) \\ 17.94\sin(\omega t - 210) \\ 17.94\sin(\omega t - 330) \end{bmatrix}$$

The above equation indicates the reference compensating current to be injected by the shunt active filter, which is 120° with respect to each other and lagging by 90° with respect to the respective phase voltage. In other words, reactive power demanded by the load is supplied by the compensator, and only active power is supplied by the source. The reference source current obtained in part (ii) and reference compensating current obtained in part (iii) are shown in Figure 7.36.

Problem 7.4 In Problem 7.2, if the load is replaced by a 3-ϕ diode rectifier with a load resistance of 18.055 Ω, resulting in the same active power as in Problem 7.2, i.e., 19357 W, obtain the source and compensating current using p-q theory (Figure 7.37).

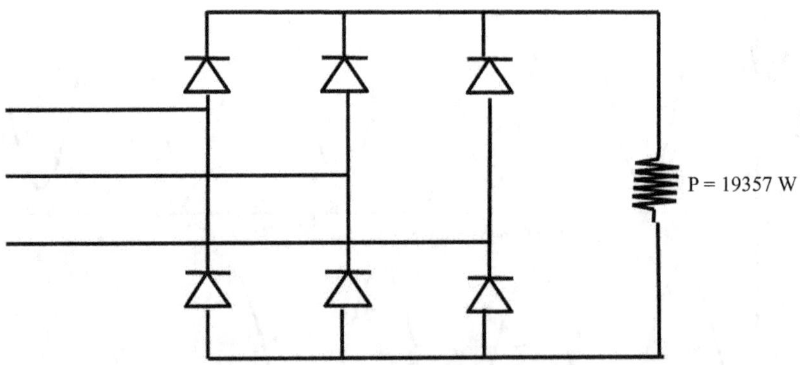

Figure 7.37 The circuit diagram of a diode rectifier circuit used in Problem 7.4

Solution:

$$p_{output} = 19357\ W Loadresistance,\ R_L = 18.055\ \Omega$$

As the supply voltage is sinusoidal, with $V_{sL-L} = 440\ V$

Phase voltage is given by:

$$V_{sph} = \frac{V_{sL-L}}{\sqrt{3}} = \frac{440}{\sqrt{3}} = 254.034\ V$$

Average load current I_L is calculated as:

$$I_L = \sqrt{\frac{P}{R}} = \sqrt{\frac{19357}{18.055}} = 32.743\ A$$

Reference source current generation using p-q theory

$$\begin{bmatrix} i_{s\alpha}* \\ i_{s\beta}* \end{bmatrix} = \begin{bmatrix} v_{s\alpha} & v_{s\beta} \\ -v_{s\beta} & v_{s\alpha} \end{bmatrix}^{-1} \begin{bmatrix} \overline{p} \\ 0 \end{bmatrix}$$

Substituting the values and $\overline{p} = 19357\ W$, we can find out the desired source current that supplies only useful power.

$$\begin{bmatrix} i_{s\alpha}* \\ i_{s\beta}* \end{bmatrix} = \frac{1}{(1.225 \times V_{sm})^2} \begin{bmatrix} 1.225\times V_{sm}\sin(\omega t) & 1.225 \times V_{sm}\cos(\omega t) \\ -1.225 \times V_{sm}\cos(\omega t) & 1.225\times V_{sm}\sin(\omega t) \end{bmatrix} \begin{bmatrix} 19{,}357 \\ 0 \end{bmatrix}$$

$$= \begin{bmatrix} 43.984\sin(\omega t) \\ -43.984\cos(\omega t) \end{bmatrix}$$

Line current in a, b, c phases is calculated using the inverse transform

$$\begin{bmatrix} i^*_{sa} \\ i^*_{sb} \\ i^*_{sc} \end{bmatrix} = \begin{bmatrix} 0.577 & 0.817 & 0 \\ 0.577 & -0.4085 & 0.7072 \\ 0.577 & -0.4085 & -0.7072 \end{bmatrix} \begin{bmatrix} 0 \\ 43.984\sin(\omega t) \\ -43.984\cos(\omega t) \end{bmatrix}$$

By comparing it with the equation for $\sin(\omega t)$, $\sin(\omega t - 120)$, $\sin(\omega t - 240)$ we get the following solution

$$\begin{bmatrix} i^*_{sa} \\ i^*_{sb} \\ i^*_{sc} \end{bmatrix} = \begin{bmatrix} 35.934\sin(\omega t) \\ 35.934\sin(\omega t - 120) \\ 35.934\sin(\omega t - 240) \end{bmatrix}$$

Using p-q theory, it is seen that the RMS value of the desired source current is 25.409 A. With this value the current power transferred is $p = 3 \times 254.03 \times 25.409 = 19{,}363$ W which is nearly equal to the output power. It is seen that supply voltage and desired source current are in phase and sinusoidal, and hence power factor is unity. This can be achieved by using shunt active filter which supply unwanted power (distortion power in this case) to the load.

The reference compensating current is calculated as follows:

As the load is a six-pulse converter, in addition to the fundamental component of current, higher-order harmonic component present in it is 5, 7, 11, 13, ... Also, as the fundamental component of current is in phase with the supply voltage in the case of six pulse converter with resistive load, reactive power q consumed by the load will be zero. However, $\tilde{p}$, an unwanted power is present in the input due to distortion in current. So, the instantaneous value of load current is given by:

$$\begin{aligned} i_L &= 35.934\sin(\omega t) + 35.934/5\sin(5\omega t) + 35.934/7\sin(7\omega t) \\ &\quad + 35.934/11\sin(11\omega t) + 35.934/13\sin(5\omega t) + \cdots \\ &= 35.934\sin(\omega t) + 7.187\sin(5\omega t) + 5.133\sin(7\omega t) + 3.267\sin(11\omega t) \\ &\quad + 2.764\sin(5\omega t) + \cdots \end{aligned}$$

The calculation of reference compensating current from load current using p-q theory is complicated compared to reference source current generation. So, the reference source current is obtained by subtracting the reference source current from the load current. Compensating current in phase 'a' is given by:

$$\begin{aligned} i^*_{ca} &= i_L - i^*_{sa} \\ &= 7.187\sin(5\omega t) + 5.133\sin(7\omega t) + 3.267\sin(11\omega t) + 2.764\sin(5\omega t) + \cdots \end{aligned}$$

The waveforms of phase 'a' supply voltage, load current, the reference source current and reference compensating current obtained using p-q theory, plotted using MATLAB simulation, are shown in Figure 7.38. It is seen that the reference

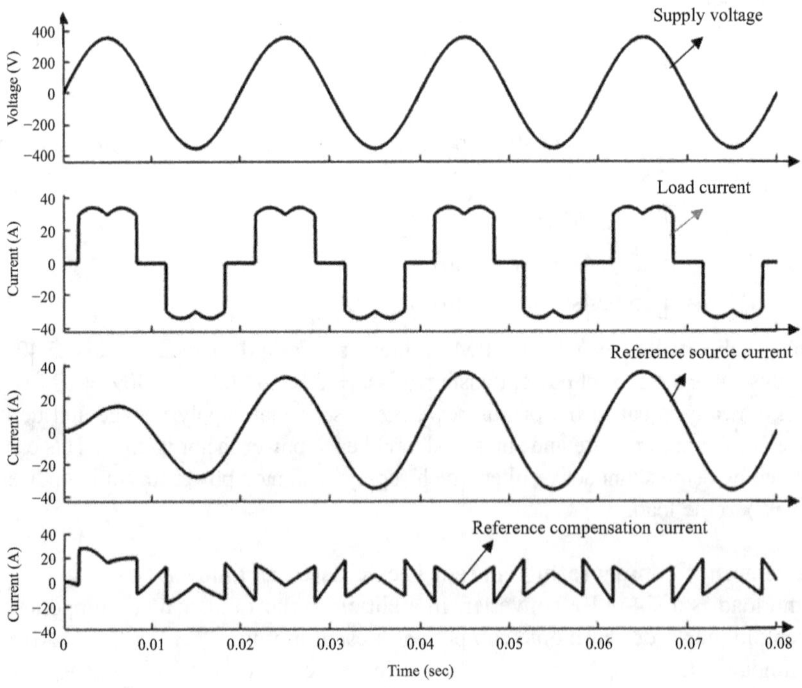

Figure 7.38 Waveform of phase voltage, load current and reference source current to provide only active power compensation of the load

source current generated using p-q theory shows a sinusoidal waveform in phase with the supply voltage, and reference compensating current supplies all unwanted harmonic components of load current. In other words, $\tilde{p}$ and $\tilde{q}$ present in p and q respectively, is supplied by the shunt active filter so that only active power exists in supply side.

References

[1] Diego Soto, and Tim C. Green, "A Comparison of High-Power Converter Topologies for the Implementation of FACTS Controllers", *IEEE Transactions on Industrial Electronics*, Vol. 49, No. 5, 2002, pp. 1072–1080.

[2] Bo Jie, Takao Tsuji, and Kenko Uchida, "Coordinated Voltage Control by Inverters and FACTS Devices in Distribution System", 2018 *China International Conference on Electricity Distribution (CICED)*, Tianjin, China, 17–19 September 2018.

[3] Ned Mohan, Tore M. Undeland, and William P. Robbins, *Power Electronics: Converters, Applications and Design*, Second edition, 1995, John Wiley & Sons (Asia) Pte Ltd.

[4] Khaled H. Ahmed, Stephen J. Finney and Barry W. Williams, "Passive Filter Design for Three Phase Inverter Interfacing in Distributed Generation", *2007 Compatibility in Power Electronics*, 2007, pp. 1–9.

[5] Design Guide: TIDM-HV-1PH-DCAC Grid Connected Inverter Reference Design, Texas Instruments.

[6] Vasco Soares, Pedro Verdelho, and Gil Marques, "Active Power Filter Control Circuit based on the instantaneous Active and Reactive Current i_{d}-i_{q} Method", *Proceedings of Power Electronics Specialist Conference, PESC 97*, pp. 1096–1101.

[7] Hyosung Kim, Frede Blaabjerg, Birgitte Bak-Jensen, and Jaeho Choi, "Instantaneous Power Compensation in Three-Phase Systems by Using p–q–r Theory", *IEEE Transactions on Power Electronics*, Vol. 17, No. 5, 2002, pp. 701–710.

[8] Fang Zheng Peng, and Jih-Sheng Lai, "Generalized Instantaneous Reactive Power Theory for Three-phase Power Systems", *IEEE Transactions on Instrumentation and Measurement*, Vol. 45, No. 1, 1996, pp. 293–297.

[9] Andrea Cavallini, and Gian Carlo Montanari, "Compensation Strategies for Shunt Active-Filter Control", *IEEE Transactons on Power Electronics*, Vol. 9, No. 6, 1994, pp. 587–593.

[10] Souvik Chattopadhyay, and V. Ramanarayanan, "Phase-Angle Balance Control for Harmonic Filtering of a Three-Phase Shunt Active Filter System", *IEEE Transactions on Industry Applications*, Vol. 39, No. 2, 2003, pp. 565–574.

[11] Ambrish Chandra, Bhim Singh, B. N. Singh, and Kamal Al-Haddad, "An Improved Control Algorithm of Shunt Active Filter for Voltage Regulation, Harmonic Elimination, Power-Factor Correction, and Balancing of Non-

linear Loads", *IEEE Transactions on Power Electronics*, Vol. 15, No. 3, 2000, pp. 495–507.

[12] Z. Li, H. Jin, and G. Joos, "Control of Active Filters Using Digital Signal Processors", *IECON'95*, pp. 651–655.

[13] S. Mohammad-Reza Rafiei, Hamid A. Toliyat, Reza Ghazi, and Tilak Gopalarathnam, "An optimal and Flexible Control Strategy for Active Filtering and Power Factor Correction Under Non-Sinusoidal Line Voltages", *IEEE Transactions on Power Delivery*, Vol. 16, No. 2, 2001, pp. 297–305.

[14] S. M. R. Rafiei, Reza Ghazi, and Hamid A. Toliyat, "IEEE-519-Based Real-Time and Optimal Control of Active Filters Under Nonsinusoidal Line Voltages Using Neural Networks", *IEEE Transactions on Power Delivery*, Vol. 17, No. 3, 2002, pp. 815–821.

[15] F. Z. Peng, H. Akagi, and A. Nabae, "A New Approach Harmonic Compensation in Power Systems: A Combined System of Shunt Passive and Series Active Filters," *IEEE Transactions on Industry Applications*, Vol. 26, No. 6, 1990, pp. 983–990.

[16] Sincy George, and Vivek Agarwal, "A Novel, DSP Based Algorithm for Optimizing the Harmonics and Reactive Power under Non-Sinusoidal Supply Voltage Conditions", *IEEE Transaction on Power Delivery*, Vol. 20, No. 4, 2005, pp. 2526–2534.

[17] Sincy George, and Vivek Agarwal, "A DSP Based Optimal Algorithm for Shunt Active Filter under Non-Sinusoidal Supply and Unbalanced Load Conditions", *IEEE Transactions on Power Electronics*, Vol. 22, No. 2, 2007, pp. 593–601.

[18] Sincy George, and Vivek Agarwal, "A DSP Based Control Algorithm for Series Active Filter for Optimized Compensation under Non-Sinusoidal and Unbalanced Voltage Conditions", *IEEE Transaction on Power Delivery*, Vol. 22, No. 1, 2007, pp. 302–310.

[19] Sincy George, and Vivek Agarwal "Optimum Control of Selective and Total Harmonic Distortion in Current and Voltage under Non-sinusoidal Conditions", *IEEE transaction on Power Delivery*, Vol. 23, No. 2, 2008, pp. 937–944.

[20] Jun-Zhe Yang, Member, IEEE, Chi-Shan Yu, and Chih-Wen Liu, Senior Member, IEEE, "A New Method for Power Signal Harmonic Analysis", *IEEE Transactions on Power Delivery*, Vol. 20, No. 2, 2005, pp. 1235–1239.

[21] José María Maza-Ortega, José Antonio Rosendo-Macías, Antonio Gómez-Expósito, Sergio Ceballos-Mannozzi, and Manuel Barragán-Villarejo, "Reference Current Computation for Active Power Filters by Running DFT Techniques", *IEEE Transactions on Power Delivery*, Vol. 25, No. 3, 2010, pp. 1986–1995.

[22] Julio Barros, and Ramón I. Diego, "Analysis of Harmonics in Power Systems using the Wavelet-Packet Transform", *IEEE Transactions on Instrumentation and Measurement*, Vol. 57, No. 1, 2008, pp. 63–69.

Chapter 8

Conclusion and current trends

Introduction of converter-based technologies significantly deteriorates the power quality that adversely affects the performance of equipment and systems. Development of converter-based compensator paved the way for an efficient solution.

This final chapter concludes the book, summarizing its main contributions and discussing the evolving trends in the field of power quality.

8.1 Introduction

As discussed in previous chapters of the book, our reliance on an uninterrupted power supply makes power quality a major concern in today's era. Traditionally electricity is generated, supplied and utilized either as DC or as AC with a sinusoidal waveform of frequency of 50 or 60 Hz world over. At the time of its inception, electrical power system hardly had any waveform distortion. However, today, in addition to the traditional methods, electricity is generated, transmitted and utilized in many different ways. Driven by the global warming concerns and decarbonization efforts in power sector, distributed renewable energy (RE) sources, such as solar photovoltaic generation (SPV) and wind energy turbines (WET), are being introduced into the power system in recent times. Distributed generation (DG), along with its associated systems, has paved the way for the formation of microgrids and smart grids. This has resulted in a variety of equipment and systems being integrated with the power system. On the generation side, the conventional generation using synchronous and induction generators is supplemented with electronically coupled SPV and wind turbine generators (WTG). On the transmission side, extra high voltage transformers, reactors, capacitors and new technologies like FACTS, STATCOM and HVDC etc. are being introduced. Similarly, on the utilization side, the traditional incandescent lamps, heaters, and motors are being replaced with LED devices, power electronic motor drives, electric locomotives, arc furnaces, etc.

Other than this, there is a large-scale induction of new loads such as EV battery charging stations, semiconductor industries, converter-based power supplies, UPSs, controllers for automation, communication, information technology equipment (ITE), data centres, Internet of things (IoT), SCADA, robotics, aerospace, health care, etc. These new devices depend on PLC, microprocessors, or DSP-based controllers, as well as intelligent electronic devices (IED). As a result of this, they are very sensitive to waveform distortions or other power quality issues. Thus, a major difference in power system has arisen with the increased presence of electronic equipment and sensitive controllers. Performance of both of them has an effect on power supply waveforms or power quality. Hence, the knowledge of causes, effects and mitigation of poor power quality are very important aspects in modern power system, which are summarized in the next sections.

8.2 Cause and effect of low power quality

The front end of much of the new equipment employs power-electronic-based AC-DC converters that produce a significant amount of harmonics in the power system. They also introduce notches, noise, and DC offset depending upon the type of converters, leading to distortion in supply waveforms. These distortions in the connected system adversely affect the performance of other equipment and systems. These effects can be broadly summarized as:

(a) Harmonics produce distortion volt-ampere, increase reactive power, and reduce power factor and thus the overall available capacity of the system.
(b) Harmonics increase losses, temperature rise and overstressing of components.
(c) Harmonics increase neutral current, necessitating a larger conductor cross-sectional area.
(d) Harmonics increase the probability of series and/or parallel resonance.
(e) Higher-order harmonics increase the possibility of interference on nearby metallic signal circuits, spurious operation of electronic components and systems, data loss or data corruption. This may cause operational safety concerns.
(f) Low power quality accelerates ageing of equipment, reduces its useful life and increases frequency of maintenance or predictive maintenance.
(g) The need for high power quality leads to the requirement of its continuous monitoring in the power supply system.
(h) It also calls for qualifying the equipment by its design, testing and validating it for the power supply quality.
(i) Low power quality reduces the overall efficiency and thus increases the operational cost.

Another major issue of poor power quality resulting power supply interruption reported world over is reactive power mismatch[1]. It is estimated that majority of electrical power is processed by power electronic converters and controllers that are sensitive to waveform or power quality. Thus, an interruption of power supply often results in huge economical loss[2] both to the utilities and the users.

[1]https://eepublicdownloads.entsoe.eu/clean-documents/pre2015/publications/entsoe/RG_SOC_CE/RG_CE_ENTSO-E_Defence_Plan_final_2011_public.pdf, https://www.cercind.gov.in/2012/orders/final_report_Grid_Disturbance.pdf

[2]https://leonardo-energy.pl/wp-content/uploads/2017/08/Poor-power-quality-costs-european-business-more-than-150-billion-a-year.pdf

8.3 Power quality mitigation strategies

Having realized the importance of power quality, industry stakeholders and researchers have jointly formulated comprehensive standards, limits and measuring techniques for power quality. Several countries impose economic penalties for violation of their regulations based on reliability indices[3]. Hence, it is essential to maintain stipulated sinusoidal waveforms in the system. Loads and generation in the system vary haphazardly, and accordingly, the conventional solution under multiple harmonic frequencies using tuned passive filters has only limited use. Also, with large-scale integration of electronically coupled renewable energy generators and variable frequency motor drives, the overall system inertia is drastically reduced[4].

Active compensation using voltage source converters is the solution to compensate for voltage or current harmonics, reactive power and imbalance in voltage or current in the system. Generally, current compensation is done using a shunt active compensator, and voltage compensation is done by a series active compensator, and both voltage and current can be controlled at the same time using shunt-series active compensator. The main components of these compensators are the voltage source DC to AC converters, which inject the required amount of voltage and/or current into the system to achieve the desired compensation. Under sinusoidal conditions, shunt compensators control reactive power to improve power factor and series compensators control reactive power consumed by the transmission line to control active power flow through the line. When voltage is sinusoidal and current is non-sinusoidal, shunt compensator eliminates distortion power in addition to compensating for the reactive power to achieve unity power factor and distortion-free current. However, when both voltage and current are distorted, efforts to make a unity power factor in the system using shunt active compensator produce current similar to voltage waveform with same phase. It may result in higher THD. On the other hand, efforts to make harmonic-free current using shunt compensator may result in a low power factor. A similar trade-off is observed in the case of a series compensator too. We have discussed an algorithm (Chapter 7) which optimizes this trade-off between the power factor and THD achieving the best compromise under non-sinusoidal voltage and current conditions. With this technique, an optimal operation of shunt, series and shunt-series compensators is possible. The algorithm is validated under various supply and load conditions as per the requirements of IEEE 519.

8.4 Summary of the book and future development including use of artificial intelligence

In summary, the following topics are covered in the textbook.

(a) An overview of electrical power generation, transmission, distribution and utilization. Presenting the past, present and challenges – power quality and cyber security.

[3]https://www.oeb.ca/oeb/_Documents/EB-2010-249/PEG_OEB_Service_Quality_Report.pdf System Reliability Regulation: A Jurisdictional Survey, May 2010, Pacific Economics Group Research

[4]https://www.ese.iitb.ac.in/~gil/reports/posoco.pdf "Report on Assessment of Inertia in Indian Power System", January 2022, Power system operation corporation Limited in association with IIT Bombay.

(b) Power quality issues in modern power supply systems: Transient voltage sag, undervoltage, voltage swell, overvoltage, voltage interruption, voltage fluctuation, frequency variation, unbalance, harmonics, inter-harmonics, notches, noise, DC offset, etc.
(c) Various sources of voltage and current harmonics in power generation, transmission and utilization and their adverse effects on various components of the power system.
(d) Power quality monitoring, indices and waveform analysis, power supply interruption and system reliability.
(e) Basic definition of power factor under linear loads; power factor = cos(ϕ) and its changing definition, power factor = distortion factor × displacement factor, under the voltage and current distortion due to nonlinear electronic loads.
(f) Tuned passive filter for compensation of harmonics, control of reactive power and voltage variation.
(g) Recent techniques of converter-based STATCOM solutions for dynamic control of voltage variation, reactive power, and active filters for mitigation of harmonics.

The book also covers recommendations corresponding to the various standards, viz., IEEE, IEC, etc., and numerical examples with simulations using MATLAB®.

Following suggestions are made for further investigations in this field [1,2].

(a) Monitoring of power quality data with the possible use of Internet of things (IoT). (Chapters 1 and 2)
(b) Assess power quality contribution from renewable energy generation, EV charging stations, energy storage system, etc. (Chapters 1–3)
(c) Carry out a power quality survey and, using the PQ data, assess the ageing of equipment and condition monitoring to avoid incipient failure and possible system interruption. (Chapters 2–4)
(d) Payback examination of power quality solutions.
(e) Study on system inertia and frequency response with large-scale penetration of electronically coupled renewable energy generation, motor drives, battery energy storage systems, synchronous condenser, etc., and replacement of conventional generators and directly connected large motor. (Chapters 1 and 3)
(f) Application of Artificial intelligence (AI) on monitoring of PQ (Chapters 1–3), system reliability (Chapter 4), reduction in power factor due to harmonics and system capacity (Chapter 5), improving response characteristic, efficiency, grid condition monitoring, power quality enhancement (Chapters 6 and 7), etc.

References

[1] Manual on Transmission Planning Criteria 2023, Central Electricity Authority, India https://cea.nic.in/
[2] National Electricity Plan (Draft) Volume II Transmission, December 2023, Central Electricity Authority, India https://cea.nic.in/

Index

www.ingramcontent.com/pod-product-compliance
Lightning Source LLC
LaVergne TN
LVHW011843200926
848484LV00002B/395